Birkhäuser

Uta C. Merzbach

Dirichlet

A Mathematical Biography

Birkhäuser

Uta C. Merzbach (Deceased)
Georgetown, TX, USA

ISBN 978-3-030-01071-3 ISBN 978-3-030-01073-7 (eBook)
https://doi.org/10.1007/978-3-030-01073-7

Library of Congress Control Number: 2018958343

Mathematics Subject Classification (2010): 01-XX, 31-XX, 30E25, 11-XX

This book is published under the imprint Birkhäuser, www.birkhauser-science.com by the registered company Springer Nature Switzerland AG
The registered company address is: Gewerbestrasse 11, 6330 Cham, Switzerland

Image made available by the University of Toronto—Gerstein Science Information Centre

Preface

Among the questions a potential reader frequently asks about a book, let me single out the following three: Why this subject? What brings the author to the subject? How is the subject presented?

Dirichlet was a mathematician who shared responsibility for significant transformations in mathematics during the nineteenth century in Western Europe, a notable period in the history of mathematics. Yet there is no book-length study of his life and work and only a small percentage of his publications tends to be cited.

I first saw his name in conjunction with Dedekind's Eleventh Supplement to Dirichlet's Lectures on Number Theory. As a graduate student in mathematics, interested in algebra, I was puzzled by what this Supplement might have to do with Dirichlet. Shortly thereafter, while assisting Garrett Birkhoff with the *Sourcebook of Classical Analysis*, I studied Dirichlet's own writings on convergence, functions, and potential theory. That did not answer my original question but led me to give a series of talks on Dirichlet in the 1970s, including an invited lecture at a meeting of the American Mathematical Society. This in turn led to an encouraging conversation with the late Saunders Mac Lane and numerous chats with the late Walter Kaufmann-Bühler of Springer-Verlag New York. It was he who urged me to expand my earlier research into a book. After some primary source studies in libraries and archives, I became enthusiastic about such a project, realizing that Dirichlet was a man who produced his conceptual contributions while living in an unusually fascinating intellectual environment.

Although I had to lay aside this intended, expanded study because of numerous other commitments, I recently was able to return to Dirichlet, to relearn many things I had forgotten, but to take the same pleasure in learning more about the man, his mathematics, and his environment that I had been granted before.

The presentation of a "mathematical biography" produces the same challenges as most biographies, especially with regard to the questions raised repeatedly among historians concerning the relationship of history to biography, of internal versus external studies, of psychohistory vis-a-vis a "factual" listing of events, or of the thematic content of a historical study. I have tried to avoid such methodological arguments. Rather, I have attempted to record as accurately as possible, using the tools in my possession, the life of a man who loved mathematics, who through accidents of time and place was able to pursue it as a career, who did not promote himself but, as noted in our concluding chapter, had a strong support system throughout most of his life. Many of those who attended his lectures in Berlin and Göttingen were proud to call themselves his students, no matter how widely their subsequent fields of specialization diverged.

As I realized the rare opportunity Dirichlet presents to convey to the contemporary reader a sense for certain nineteenth-century lives, my originally intended more mathematically oriented study turned into a fuller biography. I hope the result may interest not only mathematical specialists but readers with other backgrounds. I have chosen to alternate the chapters dealing with Dirichlet's publications with those discussing the corresponding biographical aspects of his life. This decision was based on an early conversation and the example of Walter Kaufmann-Bühler which persuaded me that this approach can facilitate the reading for both groups of readers.

In writing about the individual publications, I chose to condense the main threads of Dirichlet's mathematical discussions sufficiently to guide those who wish to work through his arguments and to suggest changes in his methodology vis-a-vis his main predecessors and his successors. Readers may find access to the full versions of his papers through the references in these chapters pointing to the bibliography at the end of the book. At the same time, I have described in greater detail his introductions to the mathematical memoirs, frequently non-technical and often including references, that reflect his own interest in history. These not only are largely understandable to the general reader but also shed considerable light on the development of the subject and on Dirichlet's own thought processes. Though at times repetitive, they are stylistically more accessible than the technical parts of his publications.

In contrast to the lectures published posthumously by some of his students, many of Dirichlet's memoirs have been ignored in the secondary literature. While covering his memoirs in the odd-numbered Chaps. 5 through 13, I have attempted to convey the purpose and major statements of each with only occasional details of his proofs sufficient to convey his methodology and to interest the curious general reader. More involved mathematical readers may attempt to fill in proof details or read up on them in the extensive multilingual bibliography at the end of this volume.

It should be noted that Dirichlet was not an emulable stylist. He was careful in detailing his arguments, but this very care, combined with his adhering to terminology quite different from the smoother, generalized one developed by later generations, frequently resulted in a certain awkwardness. This may have been reinforced by a lack of interest and consequent experience in writing. As will be noted in the following, this was more evident in his formal presentations than it was in his lectures and conversations.

Acknowledgements

Before acknowledging specific sources of support, I should like to call the reader's attention to some comments at the beginning of the bibliography, where I single out the meticulous publications of Kurt-R. Biermann and the more recent, concise, biographical sketch by J. Elstredt.

The original motivation for a detailed study of Dirichlet's mathematics arose from a set of lectures I gave at the invitation in the 1970s from Judith Victor Grabiner and the organizers of a Southern California lecture series.

A sabbatical leave from the Smithsonian Institution provided me with the opportunity to do research in the relevant archives and manuscript collections of Berlin, Göttingen, and Kassel named at the end of the bibliography. Special thanks are due to Dr. Haenel in Göttingen for many helpful services and to Rudolf Elvers in Berlin, who, among other courtesies, provided me with access to his typescript of Fanny Hensel's diaries before their publication in 2002.

Numerous conversations with Harold M. Edwards in Göttingen and Washington, DC, during the early phase of related studies furnished considerable food for thought.

This volume benefitted considerably from the advice of Jeanne LaDuke, who read over large portions of the draft and provided needed grammatical, orthographic, and stylistic advice. There would have been considerable delay in producing a readable copy without the assistance of Judy Green, who devoted many hours to translating my coded text to the appropriate LaTeX format.

I owe special thanks to the Birkhäuser editor, Sarah A. Goob, for her unfailing patience and courtesy. It pleased me greatly that, in a conversation at one of the Joint Mathematics meetings, Dr. Anna Mätzener had expressed immediate interest in adopting this project that had been orphaned after the death of Walter Kaufmann-Bühler, the former New York editor of Springer-Verlag, New York.

Georgetown, TX, USA — Uta C. Merzbach
June 2017

Publisher's Acknowledgements

This book would not have been possible without the dedication of Prof. Dr. Jeanne LaDuke and Prof. Dr. Judy Green. After Dr. Merzbach passed away in 2017, Dr. LaDuke and Dr. Green continued to proofread the manuscript and transfer the manuscript from Microsoft Word to LaTeX. They spent long hours reading through Dr. Merzbach's notes and filling in the gaps to finalize the book, all in honor of their colleague and dear friend. We are grateful to these two mathematicians for providing the mathematical community with a deeper understanding of Dirichlet all the while adding to Dr. Merzbach's legacy.

Contents

Abbreviations and Conventions

Institutions

Académie: Académie royale des Sciences de l'Institut de France
Akademie: Königlich Preussischen Akademie der Wissenschaften zu Berlin
EP: École Polytechnique

Publications

Bericht ...Kgl. Preuss. Akad. Wiss. Berlin.: *Bericht über die zur Bekanntmachung geeigneten Verhandlungen der Königlich Preussischen Akademie der Wissenschaften zu Berlin.*[1]
Crelle's *Journal*: *Journal für die reine und angewandte Mathematik*
C. R.: *Comptes rendus*
D.A.: *Disquisitiones Arithmeticae*
Liouville's *Journal*: *Journal des Mathématiques Pures et Appliqués*

Place Names

Breslau: now Wrocław, Poland
Königsberg: now Kaliningrad, Russia
Liegnitz: now Legnica, Poland

[1]This shortened form appears in the bibliography. In the narrative the expression the "Akademie's *Bericht*" is used to refer to individual issues of the journal and "Akademie's *Berichte*" (Akademie's Reports) is used to refer to the journal in general.

Chapter 1
Rhineland

He was born of contrast, in a time of conflict. The birth occurred half-way between Paris and Göttingen, in Düren, a community of nearly 4600 people west of the Rhine. The year was 1805.

1.1 Düren

Düren prides itself on a long existence, certified by records dating back to the days of Roman occupation, and a continuous history since the time of Charlemagne. In the eighteenth century, the town, part of the Duchy of Jülich, had belonged to the Electorate of the Palatinate until 1777, when it became part of the Electorate of Bavaria. In those years, the town partook of the changes that affected most of the Rhineland. In October 1794, after the area had been taken over by France, Düren came under French occupation. French became the official administrative language soon thereafter, and, in 1798, the Napoleonic code with civil registers and the Republican ("Revolutionary") calendar was introduced.[1]

Napoleon abolished the Republican calendar on January 1, 1806, less than a year after the name of the newborn had been recorded as Jean Pierre Gustave Lejeune de Richelet and the baptismal date as 24 pluviose an 13 [13 February 1805]. The event was entered in the baptismal register of the Church of St. Anna, the center of spiritual life of this predominantly Catholic town. For several hundred years, Düren's beacon of stability was the steeple of that church, reputed to hold a relic of Saint Anne, the grandmother of Jesus.

Parents and Grandparents

The lives of Dirichlet's parents and grandparents reflect the changing patterns of the eighteenth-century Rhineland. His paternal grandfather, Antoine Lejeune Dirichlet,

[1] Kessler 1968 provides a detailed history of Düren with an extensive bibliography. Müller-Westphal 1989 has considerable genealogical information on Düren families.

U. C. Merzbach, *Dirichlet*, https://doi.org/10.1007/978-3-030-01073-7_1

had been born thirty miles from Düren, in Verviers, in the bishopric of Liège, known for several centuries as a textile manufacturing center. Antoine was one of a number of cloth manufacturers from Verviers who established themselves in Düren. In 1749, he married the daughter of one of Düren's citizens and requested permission from the Prince Elector Carl Theodor to manufacture "Verviers cloth" in Düren. Five years previously, a Josef Coenen had been instrumental in establishing the industrial manufacture of cloth in Düren, over the opposition of the guild of cloth cutters. Although grandfather Lejeune Dirichlet's hometown of Verviers had been part of the Austrian Netherlands and his arrival in Düren coincided with the period of the Wars of Austrian Succession, he soon became well-established in his new community and its major church.[2]

The congregants of St. Anna's included not only many of the cloth manufacturers and merchants of Düren but also municipal and regional officeholders, including town councilors, tax assessors, and district judges. It was in that congregation that Antoine married Anna Margareta Koenen; it was in the same church that their children were baptized and married. These children included Dirichlet's father, Johann Arnold Remaklus Maria Lejeune Dirichlet, who would establish himself as a merchant as well as a town official, successively serving as junior councilor, municipal representative, postal supervisor, and postmaster. In 1788, he married Elisabeth Lindner, noted in the church register as "acatholica."

Dirichlet's mother came from a different religious, but similar commercial, background. Her parents, Carl Gottlieb Lindner and Maria Gertrud Lindner, née Hachtmann, had been among the numerous immigrants who had come to the Rhineland from Saxony in the early part of the eighteenth century. Many settled in Düren and established manufacturing concerns in textiles, paper, and iron. They were at the core of a growing Protestant enclave, not always accepted during the turmoil of wars surrounding the area.

Reformed Protestants on her father's side, Lutherans on her mother's, Dirichlet's mother was descended from several generations of cloth merchants, their products ranging from wool to linen and lace. Born in Vaals, itself known as a cloth manufacturing center in the Aachen district, her strongest social ties linked her to the Reformed Protestant communities in Aachen, Burtscheid, and Düren. Yet, in accordance with a prenuptial contract, her children, of whom Gustav was the seventh recorded in St. Anna's registry, were to be baptized in the Roman Catholic faith.

Dirichlet's father appears as amiable, upright, and moderate; when the town council of which he was a member was fired for expressing support for their Elector at the time of the French occupation, he was one of a few members not sent to prison. The entire council was reinstated subsequently. On the basis of extant materials, it appears that he produced the appropriate documents that formed part of his professional life but was not given to personal correspondence—a trait that would become a

[2] Antoine 's father, one Remacle Arnold from the small community of Richelette, had added Lejeune to his name so as to distinguish himself from his father who bore the same name. At the time of his baptism, our Gustav was still recorded as Lejeune de Richelet; in Düren, the family subsequently would often be referred to merely by the surname Lejeune.

well-known characteristic of his son Gustav. The mother was more outgoing, carried on the correspondence with Gustav when he was away from home, stayed abreast of financial, political, and personal affairs that would be of concern to him, and saw to it that he was supplied with the books as well as the clothes he needed.

The lives of Dirichlet's siblings reflect aspects of the area's demography. Two brothers became functionaries in nearby towns; his older sister's husband, J. C. A. Baerns, pursued a career in the postal service and, after a number of years in Iserlohn, became the postmaster of Aachen. His other sister married Carl Carstanjen, son of the founder of a well-known firm in Duisburg; its sales and products ranged from tobacco to sugar and fabrics. Their daughter Mathilde married a member of the successful paper manufacturing family Schoeller in Düren.

Education

The textile industry had prospered with the fusion in manufacturing of the largely woolen "Verviers cloth" and a variety of other fabrics throughout the area. Longer established iron mills had continued to thrive, and a paper industry had begun to emerge. It was the Rhineland where the tracks of modern nationalism would cross those of the industrial revolution so noticeably that it became a focus for Max Weber's 1905 study, later translated as *The Protestant Ethic and the Spirit of Capitalism.*[3] But education had suffered.

Throughout the eighteenth century, most of the Rhineland's intellectual progress had been the victim of conflict among and within its leading religions. Under Napoleonic rule, the weakened secondary educational institutions were disrupted further by enforced secularization. Dirichlet received his early schooling in Düren, where French had become mandatory in all elementary schools in 1810. By the time Dirichlet reached high school age, a lack of facilities for secondary education persisted.

With the Treaty of Paris in April 1815, Düren had become a part of Prussia. The Ministry of the Interior's Section of Culture and Education wasted no time in extending to this part of the Rhineland the educational reforms, largely due to its chief, Wilhelm von Humboldt, that were being instituted throughout Prussia. Particular attention was paid to two institutions within less than a thirty-mile radius from Düren: the gymnasium in Bonn (now known as the Beethoven Gymnasium) and the Marcellen Gymnasium in Cologne, the former Jesuit school once known as the "Tricoronatum." They had the reputation of being superior to the other high schools in the region, but they had ranked poorly even when compared to most of the suffering secondary schools in Prussia or surrounding lands. Dirichlet attended both.

---

[3] Weber, Max. 1930.

1.2 Bonn

In 1817, the twelve-year-old took up residence in Bonn. At the time, the town had a population of fewer than 10,000 inhabitants. Within a year, he would live in the same house as Peter Joseph Elvenich, a native of Nideggen in the Düren district, at that time a student of philosophy at the newly reestablished university in Bonn.

Elvenich

The older student looked after Dirichlet to the apparent satisfaction of the young teenager's parents, and the association between Elvenich and Dirichlet would grow into a lasting friendship. It is not known to what extent Elvenich's philosophical and theological leanings may have affected Dirichlet. Bonn's university, in contrast with its earlier predecessor, boasted two schools of theology–one Catholic and one Protestant. The Catholic Elvenich would become known as a leading follower of the iconoclastic theologian Georg Hermes, in the 1830s even unsuccessfully appealing in person to the Vatican to rescind a Papal bull issued against Hermes's writings.

The Gymnasium

Conforming to the new Prussian guidelines, the curriculum at the Gymnasium in Bonn consisted of thirty-two hours of classroom instruction per week. The bulk of this was devoted to classical languages: eight hours weekly of Latin, five of Greek. An hour a day was scheduled for mathematics and another one for geography and history. There were lessons in German three times a week, and minimal instruction (twice a week each) in religion, natural science, and drawing. In 1816, the year before Dirichlet entered, the school had boasted 126 pupils. The teaching staff was modest. History and geography were taught by the coach.

One of the strongest instructors was the mathematics teacher, Wilhelm Liessem, a bachelor in his forties with two decades of teaching experience, who had an assistant. Dirichlet's class in Bonn appears to have been introduced to plane geometry and to the principles of algebra through the solution of linear equations with several unknowns.

By 1819, Liessem, who would teach for another two decades and receive an honorary doctorate from the university in Bonn, was subject to serious disciplinary problems in his classes. Whether or not this was a factor in their decision, Dirichlet's parents that year chose to send him to the Marcellen Gymnasium in Cologne.

1.3 Cologne

The curriculum in the larger city of Cologne was similar to that outlined for Bonn, but the faculty was stronger. Highly regarded was the teacher for mathematics and physics, Georg Simon Ohm, who had been at the school since 1817. After receiving homeschooling in mathematics and science from his father, a self-taught locksmith, Ohm had studied mathematics, along with physics and philosophy, in Erlangen for three semesters, before being sent to Switzerland to teach because his father felt that he was not using his time properly at the university. He wished to return to formal studies by following his Erlangen mathematics instructor, Karl Christian Langsdorf, to Heidelberg, but Langsdorf advised him that he would be better off increasing his mathematical knowledge by reading the works of Euler, Laplace, and Lagrange. When he arrived in Cologne, he had more than ten years' teaching experience and was the author of a book on teaching geometry.[4]

G. S. Ohm's Educational Methodology

As Ohm explained in a lengthy preface to his book, published in 1817, his aim in writing it was to most effectively guide teachers to bring out the fundamental characteristics of geometry as a means of developing the students' thinking. He stressed that this cannot be done through "the yoke of memory" but that students should be led to explore those geometric properties that lead to theorems they can then prove on their own. He favored utilizing pictures of basic geometric objects that will be familiar to students from observation and can then lead them to the "higher" parts of the subject. Not surprisingly, Ohm mentions an eighteenth-century author's work titled *Socratic Conversations about Geometry* and thanks his father, the practical locksmith, for having guided his first studies in mathematics.[5]

It is doubtful that the young Dirichlet had a chance to see Ohm's book, but there is no doubt that the philosophy there expressed was reflected in Ohm's courses. Ohm had a reputation for devoting himself fully to his teaching, spending extra time with his stronger as well as his weaker students. There is no evidence that he was particularly impressed with the fourteen-year-old Dirichlet; we do know, however, that, at Ohm's request, Dirichlet asked his mother to procure a copy of Monge's *Application de l'Analyse à la Géométrie* for his teacher. In a letter to Dirichlet written years later by one of his classmates in Cologne, there is a reference to their Socrates, an indication that Ohm's model and methodology had not gone unnoticed.

It is unclear to what extent Ohm passed on to his better students the advice of reading the masters. Initially, he had used a textbook, but by the time Dirichlet arrived Ohm taught without a textbook. In this, he had the consent of the reform-minded Prussian Ministry for Education which had earlier stipulated that the French method requiring uniformity in the use of textbooks interfered with freedom in teaching and limited progress.

Ohm, who tried to keep abreast of the latest developments in mathematics and physics in his own studies, attempted to lead his students along the same paths to the extent that they were able to follow, but he preserved flexibility when it was clear that their limitations exceeded his goals. For example, in submitting his syllabus for the "Prima" class to the Ministry in 1823, Ohm stated that he wished to cover the doctrine of combinations, series, higher equations, continued fractions, and indeterminate equations in his two weekly sessions titled "Arithmetic"; analytic geometry, solid

[4]Ohm, G. S. 1817.

[5]Michelsen 1778 is the work cited by Ohm; it is one of several works by this author on "Socratic Conversations."

geometry, and spherical trigonometry in two other weekly sessions on “Geometry”; and “Excursions into the Past” once a week. He noted that in some of the more difficult subjects “the obstacles increase in inverse proportion to the means of removing them”; for that reason, the teacher requires flexibility and needs to utilize lectures in the history of the subject.

Physical Apparatus

In physics, Ohm had at his disposal the apparatus which was the pride and joy of local administrative officials. It seems that some of it had been salvaged from the days of the Jesuits, who were known for establishing centers of training in mathematical and astronomical measurements, consonant with their leadership in geographic-astronomical exploration in previous centuries. After their suppression in 1773, it was apparently the priest and science instructor Ferdinand Franz Wallraf, who, along with preserving a wide variety of objects of significance for the history of Cologne, continued his attempts to maintain the physical apparatus. This was subsequently pulled together and renovated, largely by Christian Kramp, who taught at the Marcellen Gymnasium (then the “Ecole Normale” of the Department of the Roer) until 1809. The apparatus included, among other pieces, three air pumps, a rather “complete” set of magnetic apparatus, and a large optical chamber (which presumably was the reason it had won notice from Goethe). The availability of this material may have drawn Ohm more and more to physics, the area for which his name is remembered to this day.

The Choice: Mathematics

Dirichlet, too, may have found the unusual exposure to these instruments useful in subsequent endeavors. Meanwhile, however, he became determined to study mathematics. According to his nephew Sebastian Hensel, his parents had hoped he would study business or law, as the sons of their established cohorts were doing. But the generally acquiescent young man resisted these options, finally telling them that, if they insisted, he would be a businessman by day, but study mathematics at night.[6] That apparently convinced them.

Once agreement had been reached as to his future course of study, there was little question where he should go. The German states had little to offer in the field of mathematics. The most renowned mathematician in those states was Carl Friedrich Gauss, professor of astronomy in Göttingen, busy with astronomical and geodesic measurements for the State of Hanover, who had little time or inclination for teaching mathematics to students with minimal background in the subject. Paris, on the other hand, had a reputation as a center of mathematical excellence, whose mathematicians had been respected for their teaching as well as their research. Moreover, numerous old acquaintances of the Dirichlets from the days of the French occupation, who could provide necessary contacts and guidance to the sixteen-year-old, lived in or frequently visited the French capital.

---

[6]Hensel, S. 1908, 1:416.

Dirichlet left the Cologne gymnasium in 1821. He received a certificate, signed by the recently appointed director, the classicist A. R. J. Heuser, on August 20 of that year, which stated that he was a member of the senior class; demonstrated great diligence in ancient languages, mathematics, history, physics, and German; had made remarkable progress in all his subjects; and behaved very commendably.[7]

[7]Kassel. Dirichlet Nachlass. Box 2:Ve.

Chapter 2
Paris

Dirichlet arrived in Paris on the last Sunday of May in 1822. The city that greeted him had a population of nearly three-quarter million inhabitants, almost fifteen times that of Cologne, the largest town he had known up to then. It shared the problems of other European cities of the period: poor sanitation (open gutters in the middle of the street), pollution, inadequate housing. While English visitors welcomed the relative lack of the coal pollution that hung heavily over London, and commented on the "crystalline green" of the Seine, Dirichlet may have found any contrast between the Seine and the Rhine, only in initial stages of surrounding industrializaton, less startling.

2.1 Early Reports Home

Whatever Dirichlet's initial impressions were, he did not dwell on them in the report home that announced his arrival. Like most letters to his parents, the epistles from Paris show him conscious of his obligation to make the most of his time and his parents' money.[1] He dutifully reported his expenses and projected needs. After only four days, he informed them that he would be following lectures at the Collège de France and the Faculté des Sciences; that he was engaged in finding suitable lodging; and that he had paid the appropriate visits to those he had been charged to call upon. These included numerous former Rhinelanders and members of the Napoleonic occupying forces.

[1] Kassel. Dirichlet Nachlass. Box 2:VIb and c.

U. C. Merzbach, *Dirichlet*, https://doi.org/10.1007/978-3-030-01073-7_2

2.2 Madame Lorge and the Deutgens

Foremost among his early Paris contacts was "Madame Lorge," the former Johanna Elisabeth Deutgen, who in 1794 had married General Jean Guillaume Thomas Lorge. The Deutgens were well-established manufacturers who formed part of the Reformed Protestant community in Düren. Three men of the family had been members of the Masonic Lodge in nearby Aachen. Johanna Deutgen Lorge's brother Eberhard Deutgen was counted among the top taxpayers in the area. Her niece Elvira had been a close friend of Dirichlet's since childhood. General Lorge had been a cavalry commander in the Napoleonic army, participating in some of the major battles fought, especially on the Iberian Peninsula. He was subsequently in charge of overseeing the return of French prisoners from Spain and Portugal. Now Madame Lorge, well acquainted with Dirichlet's mother, served as a conduit for transferring funds; her brother, on trips between Düren and Paris, not only aided in this but also brought books that Dirichlet had requested. Specifically, in the fall of 1822 Dirichlet asked his mother to have Mr. Deutgen bring him the two volumes of Bossut's *Histoire de mathématiques*, one volume of Kramp's *Arithmétique universelle*, and, in hindsight most significantly, Gauss's *Disquisitiones Arithmeticae*.

Others who made Dirichlet feel welcome in the strange city included a member of the Blankart family, long established in the Düren area, and François Larcher de Chamont, an acquaintance of the Lorges and of his parents from the days of the French occupation. Within two months after his arrival, Dirichlet could report home that he was seeing Mr. Blankart regularly, that hardly a day went by without their studying together for an hour, although it was nothing but repetition for Mr. B. As we shall note shortly, Larcher's acquaintance was to be of more lasting influence.

2.3 Professors

In Dirichlet's correspondence home, four professors are mentioned. They are Biot, Francoeur, Hachette, and Lacroix. We note that none were supporters of the Bourbon monarchy. Two had served in the post-revolutionary army, one (Biot) had taught in the artillery school at Besançon and been an Examiner for the artillery corps. Three had studied at the Ecole Polytechnique in its early days; Hachette was active in its founding.

Jean-Baptiste Biot in 1800 had become professor of mathematical physics at the Collège de France and had taught both physics and astronomy at the Faculté des Sciences of the University of Paris. (While still in Cologne, Dirichlet had also asked his mother to obtain Biot's book on experimental physics, possibly for Ohm.)

Louis Benjamin Francoeur, after his studies that had included work with Gaspard Monge, in 1798 had been appointed "Répétiteur" (teaching assistant) to Gaspard de Prony and to Lacroix, and subsequently became Examiner and Professor at the EP; he had been Professor at the Faculté des Sciences since 1808 and distinguished himself as author of a wide array of textbooks.

With the restoration of the Bourbon monarchy, Francoeur, Lacroix, and Hachette in 1816 had lost their positions at the Ecole Polytechnique, which, being designed as a military training establishment, eliminated faculty members with presumed revolutionary, or at least Napoleonic, sympathies. The EP also required special certifications for foreign students.

Both Hachette and Lacroix had worked closely with Gaspard Monge, assisting him with his descriptive geometry courses since the early days of the Ecole Polytechnique. Monge had been one of the cofounders of the EP and had taught descriptive geometry there since 1794. He had served as Director of the EP, his term of office being disrupted by Napoleon's request to have him join the Egyptian Expedition, where he was one of twelve members of the mathematical section of the Institut of Cairo, along with Fourier, Malus, and Napoleon himself. After their return from Egypt, Monge had continued his mathematical and educational efforts on behalf of the EP; he also filled a variety of political posts and received honors from Napoleon, to whom he remained loyal even past the Hundred Days. Monge had not been teaching at the EP after 1812 because of deteriorating health. Yet the Restoration government expelled him from the EP and the Institut de France. He died in 1818.

Hachette, who had published Monge's descriptive geometry lecture notes while Monge was in Egypt, continued to teach at the EP as professor of descriptive geometry until he, too, was dismissed in 1816. He had held the position at the Collège de France since 1810. There he continued to propagate and expand Mongean geometry, but he also provided valuable services in the study of machinery and was concerned with applying hydrodynamics to the study of water flow under a variety of conditions.

Lacroix had been professor at the EP from 1799 until his dismissal, but he also had taught at the Collège de France since 1812 and became chair of the mathematics department there in 1815. His textbooks, ranging from algebra and descriptive geometry to probability and the calculus, among others, were widely used and praised for his exemplary exposition, often cited as being more lucid than his lectures. Most famous was the one on the calculus which, reissued with numerous editions, was adopted for decades not only in France, but notably in the English-speaking world after members of the Cambridge Analytical Society translated this and a number of his other textbooks.

One of the distinctive features of Lacroix's calculus textbook was a list preceding the main text which contained the names of those who had contributed to the growth of the subject, with a brief reference to their contributions. More than twenty years later, in a letter on an unrelated subject, Dirichlet wrote to his "très vénéré Maître" what an impression this had left on him and on his own teaching style.[2]

Lacroix and Hachette would become especially aware of, and useful to, the young Dirichlet. Also, both included him on social occasions. A. A. Cournot, with whom Dirichlet established a friendly relationship, wrote in his *Souvenirs* that he and Dirichlet were the two Hachette students invited to attend a reception for Ørsted when the latter was being feted in Paris for his discovery of electromagnetism.[3] In addition,

---

[2]For the text of Dirichlet's letter to Lacroix see Taton 1954.

[3]Cournot 1913.

there is an extant note from Guillaume Libri, telling Dirichlet how to get to Lacroix's house, where apparently several of the younger men had been invited for a dinner.[4]

2.4 Smallpox

By fall 1822, Dirichlet, along with the Lorges and several thousand other Parisians, had fallen victim to smallpox. Free vaccinations had been available since 1817 and the fact that Dirichlet had been vaccinated probably made the attack less virulent than it would have been otherwise. More than a thousand Parisians were killed by the outbreak of 1822, the second of three major ones between 1819 and 1825. At first, in October 1822, Dirichlet was thought to have had German measles; but once it was recognized for the far more serious illness, he had to report home in early December that he still showed red spots and had incurred more expenses, forcing him to draw 400 francs on Mr. Deutgen.

The illness apparently did not set him back too seriously in his studies. Even before it struck, Dirichlet had wished to attend courses at the Ecole Polytechnique, but, since he was a foreigner, this required special permissions. When he spoke to Hachette about this, Hachette, the expelled founding member of the EP, advised him there would be minimal import in going there, as the same material would be covered at the Collège de France; but Hachette suggested that if he wished to do so he should turn to the Prussian envoy.[5] Because of the death of the predecessor in that position, there was only an interim official, which meant Dirichlet's application had to go to Freiherr von Stein zum Altenstein, the Prussian Minister of Cultural Affairs. Fortuitously, Dirichlet did not wish to bother. This brought him even closer to Hachette and Lacroix.

2.5 Water Flow

In 1815, Hachette had begun a series of experimental studies on water flow, supplementing his earlier theoretically oriented studies in hydrodynamics. The subject had been of more than theoretical interest in both France and Germany for some time. Regulating water flow of rivers, canals, harbors, as well as of the slowly emerging urban plumbing systems, was of increasing importance with growing industrialization.

Hachette had previously published summaries of his theoretical work and there had been commentaries by the mathematicians most closely involved with similar studies, notably Poisson and Cauchy. The chief French contributor to the subject

---

[4] Berlin. Staatsbibliothek. Handschriftenabteilung. Dirichlet Nachlass.

[5] Kassel. Dirichlet Nachlass. Box 2:VIc.

paving their research path had been Gaspard de Prony; in Germany it was the engineer J. A. Eytelwein.[6]

In 1814, a particularly relevant memoir by Eytelwein had been published in the *Mathematische Abhandlungen* of the Berlin Akademie. Lacroix suggested to Hachette that he have their young bilingual student Dirichlet provide a translation of this work. That was done.

In May 1823, Hachette presented excerpts from Dirichlet's translation with his own commentary to the Société Philomathique of Paris; these were printed in that Society's *Bulletin*, without mentioning Dirichlet. However, in 1825 the memoir was published in the *Annales des Mines*, and Dirichlet was given credit for the translation rendered in April 1823. In addition to carrying his name as the translator at the end of the lengthy title, a footnote attached to the title noted that

> Ce mémoire a été indiqué par M. Lacroix au traducteur, qui suivait alors les cours de la faculté des sciences, et la traduction a été faire sous les yeux de M. Hachette.[7]

2.6 First Employment

There was another reason why May 1823 proved to be of special significance for Dirichlet. François Larcher de Chamont, one of Napoleon's chief military engineers and head of the occupation of Jülich, who became known in Düren for the opulent restoration of its Burg Gladbach, had a friend, a M. Levelle, who was also a friend of General Maximilien Sebastian Foy, a distinguished veteran of the Napoleonic Wars and leader of the liberal opposition in the Chamber of Deputies. As Dirichlet would write his mother the following month, while visiting Larcher he had met Levelle who told him General Foy was looking for a young man who could teach his children German. If the man was not German himself he should have thorough knowledge of the language, should teach it as well as elements of Latin, French grammar, and arithmetic. He should be of good family. He would live at the Foys, have a regular salary, and would have time for his own studies, as the general did not want his children occupied all day. The oldest, a girl, was eleven. Larcher thought Dirichlet met all the requirements and asked Levelle to write the general. On May 23, Levelle wrote to Dirichlet informing him that the general wanted to hire him.[8] Dirichlet accepted.

[6]For an overview of the history of hydrodynamics "from the Bernoullis to Prandtl" see Darrigol 2005. This includes description of many nineteenth-century experimental studies. For the eighteenth century, Truesdell's introductions in Euler's *Opera omnia* (1954, 1955, 1960) remain standard references.

[7]Eytelwein 1825.

[8]Kassel. Dirichlet Nachlass. Box 2:VIc.

In the summer of 1823, Dirichlet moved from his modest lodging at the Clôitre St. Benoit in the Quartier St. Jacques to the Foy home in the Chaussée d'Antin to begin his job as tutor of the Foy children. The Chaussée d'Antin of the period has been described as "the fief of the big bankers where liberalism dominated."[9] The Foy home was a frequent meeting place of fellow ex-Napoleonic officers and members of the moderate "liberal" opposition, many of whom would shape the political future of France in the coming decades. Dirichlet was present at some of their discussions, although spending most of his time teaching his young charges and working on his research.

When Johanna Lorge learned that Dirichlet's mother was concerned about him, she responded by noting that the Foys were pleased with him and that, although she did not know her personally, she understood that Mrs. Foy was intelligent, attractive, thirtyish, and a bit of a coquette.[10] Years later, Foy's widow still remembered the lanky tutor sitting on a narrow stove, supervising the work of the children while pursuing his own studies.

2.7 Obligations at Home; Draft Call

Dirichlet's mother had a twofold reason for being worried about him. One was simply her undoubtedly being aware that, whereas he used to write home on a biweekly basis while in Bonn and Cologne, his letters now came monthly, a pattern that would persist whenever he was away from home.

His draft status was a second, more serious concern that she shared with Heinrich Syo, as town secretary of Düren one of her husband's former associates on the town council. At the time that Dirichlet had received his job offer from General Foy in May 1823, his mother wrote that she had had no response from him as to the beginning of his vacation. She had heard it might run from August to October and suggested he should spend it at home because he needed to report for the one-year draft call before the end of his nineteenth year. They would leave him in peace until he was twenty-three years old, but if he deferred reporting beyond his twentieth birthday they would get him for three years instead of one. She stated that it was probable he would be declared unfit because of his nearsightedness. Meanwhile he should prepare to leave his books but to bring his clothes and linens so that she could again undertake "a small reform" with them. Syo added a postscript suggesting he come in August.[11]

---

[9] Bertier de Sauvigny 1967:318.

[10] Kassel. Dirichlet Nachlass. Box 2:III.

[11] Kassel. Dirichlet Nachlass. Box 2:IV.

2.8 The Mysterious Research Project

A year later, on August 22, 1824, Dirichlet wrote that he would have to postpone his trip home. He explained that he was occupied with a work that had made considerable progress while he was in the country for two months. He continued by noting that "for reasons which you will find out about in time" he wished to complete this work before leaving for home; with greatest effort, he most likely would not be able to do so before mid-winter. For now, he could only say that he expected this to have some influence on his future existence, which influence he would like to know of before his trip so that they could take it into account during the determinations that should take place while he was at home.[12]

[12] Kassel. Dirichlet Nachlass. Box 2:VIc.

Chapter 3
First Success

The research project that occupied Dirichlet so thoroughly between 1822 and 1825, and because of which he delayed and heavily abbreviated his trip home, was his contribution to proving Fermat's so-called Last Theorem for the case $n = 5$.

3.1 Fermat's Claim

The interest of modern mathematicians in the theorem goes back to the seventeenth century when Pierre Fermat's son published his father's annotations to a recently produced edition of the *Arithmetic* of Diophantus. Fermat noted that he had "a truly marvelous proof " of the statement that for n an integer greater than 2, there are no positive integers x, y, and z such that $x^n + y^n = z^n$, but that the margins of the book were too small to contain his proof. This became one of the most frequently repeated stories in the history of mathematics. As Dirichlet noted years later, because Fermat had been so successful when stating apparently similar theorems, and had made major contributions in other mathematical areas, he felt that Fermat's claim had to be taken seriously. So did dozens of other mathematicians who attempted a general proof both before and after Dirichlet, until the matter was settled in the 1990s.

3.2 Lacroix and Legendre

Although we do not have the exact sequence of events that caused Dirichlet to undertake the challenging task of working on Fermat's "Last Theorem," the following occurrences are suggestive.

Study of Number Theory

As observed in Chap. 2, Dirichlet in 1822 had requested and received three volumes, one of which was Gauss's *Disquisitiones Arithmeticae*; his intensive study of Gauss's

U. C. Merzbach, *Dirichlet*, https://doi.org/10.1007/978-3-030-01073-7_3

number theory dates back to this acquisition. At some time prior to the spring of 1825, Lacroix asked Dirichlet to come see him about a paper Dirichlet had sent to Legendre.[1]

Legendre

By the time Dirichlet's work came to Legendre's attention, the seventy-year-old Legendre was known for contributions to a wide array of mathematical topics. His geometry textbook, first published in 1794, had already passed through numerous editions in several languages other than the original French. He had been the author of a three-volume work on the integral calculus that brought to the attention of mathematicians his early researches on elliptic functions. Forty years had passed since he had won the Berlin Akademie's prize for a study of the trajectories of projectiles in resisting mediums. He had written on astronomical orbits and the figure of planets, the attraction of spheroids, the calculus of variations, and other analytic topics. He had represented the Académie in the collaborative geodetic measurements involving the observatories of Greenwich and Paris. In addition, he was one of those supervising the production of revised mathematical tables after introduction of the decimal system and had remained an active member of the Institut throughout his later life.

The main reason Lacroix was interested in having Legendre take notice of Dirichlet's work, however, was Legendre's research on the theory of numbers and his related set of publications. The first of these was a memoir titled "Recherches d'Analyse indéterminé" appearing in the *Mémoires* of the Académie for 1785, published in 1788. Although its subtitle refers to Fermat's Last Theorem, it is remembered primarily for an initial presentation of the law of quadratic reciprocity. It was succeeded by two book-length volumes titled *Essai sur la théorie des nombres*. The first edition of these was published in the years 1798 (an 6) and 1799 (an 7). It was followed by a second edition in 1808, which was succeeded by two supplements, the first dated 1816. The Second Supplement, with limited circulation in 1825, was published in the Académie's volume 6, of memoirs for 1823, that only appeared in 1827. A footnote reference to Dirichlet's memoir and the Académie's approval of 1825 may have been added while Legendre's publication was in print.

By 1830, all these publications would culminate in Legendre's two-volume work titled *Théorie des nombres*, referred to as the third edition of his Theory of Numbers.

3.3 The Draft Board and the Institut of the Académie

In 1825, Dirichlet had finally presented himself to the draft board in Düren and, as his mother had predicted, was considered "invalid" because of nearsightedness. Dirichlet wrote his mother that when he went to the Académie shortly after his return to Paris,

[1] Berlin. Staatsbibliothek. Handschriftenabteilung. Dirichlet Nachlass.

he there learned from Lacroix that the Académie had assigned him and Legendre to investigate Dirichlet's work and that Legendre, who had known part of Dirichlet's paper before, had taken over the writing of their report and had already submitted it barely fourteen days after Dirichlet's departure. Lacroix had further told Dirichlet that this report had concluded very favorably and that he would have forwarded it to Dirichlet had he known his address.

It is not clear whether the reference to Legendre's already knowing part of Dirichlet's work has to do with an excerpt, a prior draft, or a conversation. As pointed out below, it is most plausible that it refers to the paper Dirichlet had sent to Legendre, according to Lacroix's earlier note.[2] The favorable report resulted in Dirichlet's being invited to present his memoir to the Académie's Institut at its meeting of July 11, 1825.

Prior to Dirichlet's arrival in Paris, the Académie had chosen the solution of Fermat's Last Theorem for its Grand Prize in Mathematics. Since none of the submissions were successful either of the two times the Prize was offered (in 1816 and 1818), it had been canceled. Nevertheless, it had served to renew interest in the long-standing puzzle, solution of which was said to have escaped the slim margins of Fermat's volume containing Diophantus's *Arithmetic*.

Dirichlet's memoir was titled "On the impossibility of some indeterminate equations of the fifth degree." It is doubtful that Dirichlet read his entire memoir to the assembled fifty-one members of the Institut. The minutes ("Procès-Verbaux") of the meeting for July 11, 1825, indicate that he "addressed" the members. As there were numerous other memoirs offered at the same meeting, it is most likely he merely provided the attendees with the explanatory introduction and the text of the three theorems his memoir contained. Some of the other memoirs at the meeting are described in the minutes as having been "presented," others as having manuscript pages "delivered," but only a few as having been "read."

It was determined that Dirichlet's memoir should be remanded to a review committee again consisting of Lacroix and Legendre. Their report appeared in the next issue, containing the minutes for the meeting of July 18, of the "Procès-Verbaux."

3.4 The Review Committee's Report

Cosigned by Lacroix and Legendre as "rapporteurs," but written by Legendre, the report began with a succinct summary of Dirichlet's memoir. Legendre began by noting that the author's first studies of the matter under consideration had for their object the proof of Fermat's theorem for the case $n = 5$. This statement did not appear in Dirichlet's published memoir and presumably is based on Legendre's previous awareness of Dirichlet's work on the topic. The report continued with the reminder to the reader that if equality held between a fifth power and the sum of two similar

[2]Berlin. Staatsbibliothek. Handschriftenabteilung. Dirichlet Nachlass.

fifth powers, one of the three indeterminates would have to be divisible by 5. Since one also has to be divisible by 2, the first case considered was that where the same indeterminate is divisible by 2 and by 5.

In treating this case, Legendre noted that Dirichlet here uses a method "analogous" to that Euler had used to prove Fermat's theorem for the case $n = 3$. But Legendre pointed out, as had Dirichlet, that the case for fifth powers offers some special difficulties, because the same number can appear in infinitely many ways in the form $t^2 - 5u^2$, whereas it can appear only once or a small number of times in the form $t^2 + 3u^2$. He continued by stating that "M. Lejeune has succeeded in vanquishing these difficulties, and has proved in a rigorous manner ... that the assumption of a solution of the equation leads to an absurd consequence."

Next, Legendre remarked that if a similar impossibility could be obtained for the case that the indeterminate divisible by 5 is odd, Fermat's theorem would be completely demonstrated for the case of fifth powers. But "M. Lejeune acknowledges that his efforts to demonstrate the impossibility of the equation in the second case have remained fruitless."

Dirichlet had concluded his memoir with an analysis of several other fifth degree equations, showing that his arguments could be extended to a large class of such equations. Legendre called Dirichlet's analysis "exact and founded on the true principles of the matter." His conclusion, on behalf of himself and Lacroix, asserted that

> we think that this Memoir, which contains some new results in material that is difficult and little cultivated until now, merits being approved by the Académie and being printed in the *Recueil des Savants Étrangers*.[3]

The Académie approved the conclusion of the report.

3.5 Legendre's Proof; Dirichlet's "Addition"

For the remainder of July and for August, Dirichlet went to the countryside. In September 1825, Legendre offered his own proof of Fermat's entire theorem for the case $n = 5$. This became known as the Second Supplement to the 1808 edition of his *Essai*. It was published in the sixth volume of the Académie's *Mémoires*. The title page of this volume reads "Mémoires de l'Académie royale des Sciences de l'Institut de France. Année 1823. Tome VI. Paris, Chez Firmin Didot, Père et fils, Librarires, Rue Jacob, No. 24. 1827." A footnote in this extensive publication, possibly added by either Legendre or Fourier while in press, reads

[3]*Mémoires de l'Académie des Sciences*, v. 8 (1829), *Procès-Verbaux, séance 18 juillet 1825*:241.

> Par une analyse semblable á celle dont nous venons de faire usage, on pourrait démonstrer l'impossibilité de l'équation $x^5 + y^5 = Az^5$, pour un assez grand nombre de valeurs de A; c'est ce qu'a fait M. Lejeune Dieterich [sic], dans un Mémoire présenté récemment á l'Académie, et qui a obtenu son approbation.[4]

In the "Procès-Verbaux" for November 14, 1825, we find that Dirichlet had sent an "Addition" to his earlier memoir, which was forwarded to the same review committee. In this "Addition," Dirichlet had noted that, since his own preceding memoir was presented to the Académie, Legendre had published a Second Supplement to his Theory of Numbers in which he proved the impossibility of the equation $x^5 + y^5 = z^5$. Dirichlet here had also informed the reader that the case where the indeterminate is divisible by 2 as well as 5 is treated in Legendre's work as in Dirichlet's own memoir and that Legendre then proceeded to prove the impossibility of the equation for the other case by means of a new analysis, although "of the same genre" as that used in the first case.

Dirichlet had continued by explaining that the object of his "Addition" was to establish two new theorems on indeterminate equations of the fifth degree which comprise the proof of Fermat's theorem for fifth powers as a special case. He proposed to do this by basing himself on the results obtained in his previous memoir and using an analysis similar to that which Legendre had utilized in his work, but which Dirichlet would now present in a manner that showed the great analogy it has with the method he himself had propounded in the preceding memoir.[5]

The November version of Dirichlet's memoir, meaning the paper presented in July with the supplementary "Addition," was printed independently, appearing as a separate pamphlet early in 1826, instead of being inserted in the Académie's publication, as recommended. This decision may have been prompted by the fact that the next volume of the *Recueil des savants étrangers* would not be ready for publication until 1827 and, as will be seen, it was in Dirichlet's interest to distribute his completed proof as soon as possible. Perhaps these delays in the Académie's publications may have been the reason as well that Legendre's Second Supplement was inserted as the first entry in volume 6 of the Académie's *Mémoires*, published in 1827.

So it came about that in 1827 readers of the Académie's publications had brought to their attention both Dirichlet's and Legendre's proofs of Fermat's Last Theorem for the case $n = 5$. What may be of equal importance is that a number of mathematicians and physicists playing a role in Dirichlet's subsequent relationship to the Académie attended both of the relevant July 1825 meetings—that of July 11, and that of July 18 with the laudatory comments by Legendre. These men included Arago, Fourier, Fresnel, Lalande, Laplace, Mathieu, Navier, Poinsot, Poisson, de Prony, and, of course, Lacroix and Legendre, among others.

It is of interest to note that the published versions of Dirichlet's July memoir contain neither of the references concerning his initial failed attempt to prove Fermat's theorem for fifth powers that Legendre reported.[6] This suggests that the report pub-

---

[4]Legendre 1827:35n.

[5]See the discussions of **1826** and **1828c** in Chap. 5 for further details.

[6]See Chap. 5 below.

lished in the "Procès-Verbaux" for July 18 was largely based on the version Dirichlet had sent to Legendre earlier. It would explain why it had taken Legendre less than a week to prepare this report, perhaps an expanded version of the earlier one that had led to Dirichlet's invitation to appear before the Institut.

André Weil would remark in 1983:

> This had perhaps been a modest Everest to climb, and Dirichlet had guided him almost to the top. But Legendre got there first.[7]

[7] Weil 1983:338.

Chapter 4
Return to Prussia

Ernst Eduard Kummer, while preparing Dirichlet's eulogy for the Berlin Akademie, consulted Peter Elvenich about their youthful days together. He was told that Dirichlet had combined his love for mathematics with a deep interest in history, particularly modern French history.[1]

Whereas it is doubtful that Dirichlet found the time to follow historical events too closely during his first months in Paris, once employed as a member of General Foy's household, he not only was witness to numerous political discussions, but, as previously noted, found himself surrounded by leaders of the political and philosophical movements that were shaping much of the future of France. It did not require any special interest in history to notice that these were interesting times.

4.1 Political Background

Early in 1821, the year before Dirichlet's arrival in Paris, the Restoration government had issued an ordinance stating that "the bases of a college education are religion, the monarchy, legitimacy, and the Charter." The following year the government passed laws furthering existing press restrictions, reduction of opposition newspapers, suppression of some courses, and expulsion of students. By June, the office of the Grand Master of the university was reopened with appointment of Denis Antoine Luc de Frayssinous. Already responsible for the new interpretation of the purpose of education, Frayssinous now determined personnel appointments and curricula. He had first come to public attention as a compelling orator, known for a series of lectures given at St. Sulpice from 1803 to 1809. Although respected by Napoleon, he had been silenced, returning to Paris and official favor, with growing power, only after the Bourbon Restoration.

[1] Kummer 1860; see Dirichlet *Werke* 2:313.

U. C. Merzbach, *Dirichlet*, https://doi.org/10.1007/978-3-030-01073-7_4

Throughout the year 1822, there were continual episodes of student unrest. The School of Medicine was suppressed from late November to the next February, and, despite gains of the opposition in legislative elections, the war with Spain served to provide renewed strength to the Restoration government the next year.

The Institut of the Académie, too, had been affected by political actions. For example, after Hachette had been elected to membership in the Section of Mechanics in 1823, Louis XVIII refused to accept the nomination and Hachette did not receive his membership until after the July 1830 revolution, in October 1831, under Louis Philippe. The seventy-two-year-old Legendre who voted against a candidate favored by the administration in 1824, soon thereafter had his pension canceled, though subsequently reinstated with reduced terms.

4.2 The Death of Foy

Two weeks after Dirichlet's "Addition" was officially received by the Académie, on November 28, 1825, General Foy succumbed to a heart ailment. On the day after his death, the *Constitutionel*, one of the few opposition newspapers that had not been shut down earlier, carried a black border. The funeral procession, said to constitute 100,000 followers, brought to public attention a memorably curious mix of hero worship and peaceful rallying around the liberal opposition. Foy's body was laid to rest in the Père Lachaise Cemetery. Casimir Perier, the future premier of France, who at the time was sitting with the moderate opposition in the Chamber of Deputies, was among the eulogists. Opposition papers took up subscriptions for the welfare and education of Foy's children. Foy's name is inscribed with that of other national heroes on the west side of the Arc de Triomphe, as is that of Foy's fellow veteran and national hero of the Napoleonic Wars, General Lorge, who died a year later.

It was time for Dirichlet to move on.

4.3 Fourier and Humboldt

Dirichlet's mathematical prowess, his courtesy, and the deference he showed toward his elders had earned him the goodwill of some of the leading mathematicians of Paris. In addition, personal association brought him to the special attention of those who had been socially and politically allied with his employer, General Foy.

Aside from the mathematical contacts, he had made at the Collège de France, with his emergence at the Institut Dirichlet had become close to Fourier and to Alexander von Humboldt. After Fourier's death in 1830, Victor Cousin mentioned Dirichlet as one of the young mathematicians whose presence Fourier enjoyed in his last, relatively isolated years. While Fourier reviewed Dirichlet's manuscripts with him and discussed mathematical topics, Alexander von Humboldt, who liked to stress that Fourier and Poisson were among his oldest friends in Paris, undertook to guide his young countryman in the social and strategic skills that would be a significant factor in shaping his future.

Both Fourier and Humboldt reacted to word that Dirichlet planned to return to Prussia. Although their advice had in common an unrealistically rosy outlook if he would follow the course each suggested, their suggestions were diametric opposites. Fourier hoped to attract promising mathematicians to Paris and to the Académie to regain the reputation it had once enjoyed worldwide. Humboldt, on the other hand, wished to build up Prussia's reputation in mathematics and the sciences.

Fourier

Fourier had received his early education in his birthplace, Auxerre, where he began his first mathematical studies at the Royal Military Academy. Educated by Benedictines, he contemplated joining the priesthood; instead, he continued to support the Revolution. During the Terror, he was briefly imprisoned by opposing factions. In 1795, he was appointed to the Ecole Normale Supérieure and had been teaching at the Ecole Polytechnique for three years when, in 1798, he joined Napoleon in the Egyptian Expedition. There, he became a member of the Cairo Institute's mathematics division, was elected its secretary, and impressed Napoleon with his administrative as well as his mathematical abilities before returning to France in 1801. He attempted to resume his teaching position at the EP, but Napoleon appointed him Prefect of the Isère. In this position, he supervised various civil engineering projects, including road constructions and swamp drainage.[2]

At the same time, Fourier began his work on heat conduction and by the end of 1807 read his first paper on the propagation of heat in solid bodies to the Institut of the Académie. This precipitated controversy. Lagrange and Laplace questioned Fourier's use of trigonometric series for the expansion of functions. Biot and Poisson took issue with his heat equations. Nevertheless, when in 1811 the Institut chose the question of heat propagation in solid bodies for its Grand Prize, Fourier submitted the same memoir with addenda, including additional proofs, and was awarded the prize.

For the next several years, Fourier was kept busy with his duties as Prefect of the Isère. After the defeat of 1812, Napoleon was about to return to Paris by way of Grenoble. Fourier advised against it on the grounds that it would be unsafe. Napoleon again wished to go through Grenoble after his escape from Elba and the first Bourbon Restoration. This time, Fourier, who had just posted an announcement to the citizens of the Isère asking them to be loyal to their Bourbon king (LouisXVIII) and then prepared a bed for Napoleon, simply left town. Napoleon now made him Prefect of the Rhone and, upon learning that Fourier had resigned, awarded him a pension. This was to go into effect July 1815, which was at the end of the Hundred Days, immediately after the battle of Waterloo. Fourier did not get his money.

Fourier's equivocal position after 1812 had the advantage that he did not suffer a large number of retaliations such as those to which Monge had been subjected. Fourier

[2]Herivel 1975.

was elected to the Académie in 1817, and, after the death of J. B. J. Delambre who had been the Perpetual Secretary for the Class of Mathematical Sciences, Fourier was elected to that position in 1823. This meant that he was in charge of reading and reporting on the memoirs presented to the Class just in time to become aware of young Dirichlet.

In the Académie's *Histoire* for 1825 (published in volume 8 of the *Mémoires* in 1829), Fourier noted Legendre's Second Supplement. Legendre there had mentioned two contemporary workers on the topic of Fermat's Theorem: One was Sophie Germain, the other Dirichlet. Both Legendre and Fourier recalled the contributions Germain had made to a variety of mathematical topics besides number theory, including her prize-winning memoir on elastic surfaces. Having reminded the reader of Legendre's previously having "enriched [number theory] with his discoveries and which he has treated in a work justly regarded as'classic'," Fourier then used the occasion to call attention to the rising star on the mathematical firmament:

> [Legendre] cites a very interesting Memoir of M. Dirichlet, who with remarkable success is occupied with difficult questions of indeterminate analysis, and who has succeeded in proving rigorously the impossibility of a large number of equations of the fifth degree.[3]

Humboldt

Shortly after their first meeting, Alexander von Humboldt (Wilhelm's brother) discovered to his apparent delight that Dirichlet intended to return to Prussia. Humboldt by this time was internationally known as a naturalist, exploring traveler, a supporter of science, and, being a second-generation Chamberlain to the King, someone close to the Prussian throne. Humboldt, who had spent nearly two decades in Paris working on the multi-volume account of his journeys in New Spain, was about to return to Berlin. He assured Dirichlet, and perhaps himself, that it would not be difficult to find, in one of the still weak Prussian universities, a position commensurate with Dirichlet's mathematical talent and future promise. The question was where this would be.

In Humboldt's eyes, Berlin was the prime target. But he was aware of potential difficulties to be overcome. Only shortly before, a tight purse string, procrastination, and misjudgment of Gauss's priorities had kept the Prussian administration from luring Gauss away from the Hanoverian Göttingen. It would be difficult to justify the kind of salary Humboldt wished to suggest for Dirichlet as appropriate, unless Dirichlet was willing to extend his service to the State and the king well beyond the strictly academic requirements of teaching and research at the university.

Humboldt knew also that he needed to convince the Prussian authorities not only of Dirichlet's mathematical ability, but of his potential for bringing international recognition of Prussia's reawakening intellectual strength. In addition, Humboldt was aware of the fact that Dirichlet had no Prussian academic credentials; he lacked a doctorate, and he had no teaching experience within Germany.

[3] Fourier 1829a:x.

4.4 Approaches to Prussia

In mid-May 1826, Dirichlet sent copies of the just-issued pamphlet (**1826**) that consisted of the November version of his 1825 memoir to the Königliche Akademie der Wissenschaften zu Berlin and to Freiherr von Stein zum Altenstein, Prussia's Minister for Cultural Affairs and Education.

In a letter to Altenstein, Dirichlet professed the honor of transmitting with his "innermost admiration" his first mathematical effort, which the Institut had honored with its approbation. He combined the transmission with the hope that Altenstein would accept his services for their fatherland, his offer of which he considered a "holy obligation." He mentioned the support of Humboldt and the interest Parisian scholars, notably Fourier, Lacroix, and Poisson, had taken in his scientific endeavors. He explained that, despite having spent several years abroad, his thoughts had remained focused on the Prussian state and his absence was prompted only by the desire to further his scientific training as much as possible so as to be that much better prepared to enter into the sphere of activity he hoped Altenstein would assign him upon his forthcoming return to the fatherland.[4] Humboldt added a handwritten note, "daring to commend to his Excellency's fatherly protection, ... this exceedingly gifted young man whose analytic works have drawn to him the attention of the Institut since his nineteenth year, and who recommends himself by his manners, his modesty and his neediness."[5]

4.5 Gauss

There was no immediate response from Altenstein. However, in May, Dirichlet also sent the pamphlet to Gauss, with an accompanying letter from Humboldt. Here, too, Dirichlet referred to his first mathematical attempt and mentioned Humboldt's recommendation to him to have Gauss read and evaluate it.

This letter to Gauss lacked the Humboldtian diplomatic phraseology used in the epistles to Berlin. Instead, Dirichlet here spoke in his own voice, expressing his preference for the study of indeterminate analysis, his hope that Gauss will give him attention, his learning of the difficulties in working on the higher arithmetic, but finding that upon daily occupation with the subject it has grown into such a passion that he could not easily decide to abandon it. Some of these statements are reminiscent of Gauss's own, especially those expressed in the introduction to the *D.A.*

Using Humboldt's frequently reiterated, demurring self-assessment, Dirichlet asked for Gauss's support which would be more meaningful than that of a non-mathematician like Humboldt. At the same time, he noted that he had learned many excellent mathematicians have less interest in indeterminate analysis than they have

[4] Biermann 1959a:13.

[5] Biermann 1959a:14.

in astronomy or integral calculus and therefore would be less likely to give him a very positive evaluation. He closed by offering whatever services he could render while still in Paris.[6]

Gauss, without immediately informing Dirichlet of his impression, acted. On July 9, he wrote to his one-time student, the astronomer Johann Franz Encke in Berlin, that he had received a small memoir on higher arithmetic from a young German named Dirichlet, currently in Paris, which gives a clue to an extraordinary talent. Gauss went on to state that the fewer individuals there are who become knowledgeable in this area—he knows of no one in Germany—the more convinced he is that this subject provides the best means to sharpen mathematical talent for other branches of mathematics; thus, he is all the more pleased by this phenomenon [Dirichlet] and thinks it would be all the more grievous for Prussia, his fatherland, to be outdone by France taking possession of this extraordinary talent. He continued by suggesting to Encke that steps be taken to have Dirichlet settle in Berlin.[7]

4.6 The Cultural Ministry

Encke was struck by the significance of this letter. He lost no time in sharing its content with Johannes Schulze, the relevant functionary in the Ministry for Education and Cultural Affairs. Encke added that in his own opinion this judgment by Gauss had such a high value because Gauss had always distinguished between results achieved by sheer effort and true genius. Encke observed that as long as he had known Gauss, despite Gauss's respectful appreciation of the work of others, he had never known him to express an opinion with such warmth. Encke concluded by noting that Dirichlet had sent the pamphlet to the Akademie in Berlin and by mentioning Humboldt's addendum with the remark of Fourier's and Poisson's appreciation of Dirichlet's talent. Encke noted as well that he had not been solicited in this matter.[8]

When in mid-August Dirichlet still had not heard from Altenstein, he wrote a follow-up letter explaining why his first letter might have been delayed or lost, giving various reasons that, as he made a point of mentioning, Humboldt had suggested to him. The substance of the letter was largely the same as his May letter, with two exceptions: He now included Legendre in the list of those with whom his work had found approbation, and he pointed out that because of the death of Foy he would have to find a more time-consuming but lucrative occupation; continuing as a preceptor in another home or two would allow no time for mathematical research. He reiterated his strong desire to return to his home country, his never having lost sight of his fatherland and the service he owed it, noting as an example his return to Düren the previous year to meet the demands concerning his military obligation.[9]

---

[6] *Werke* 2:373–74.

[7] Gauss *Werke* 12:70.

[8] Letter of July 17, 1826, quoted in Biermann 1959a:14–15.

[9] Biermann 1959a:15–16.

4.7 The Breslau Appointment

November 1826 marked the beginning of a convoluted sequence of events. During September and October, both Altenstein and Humboldt had been away from Berlin. But on November 5, Humboldt wrote a short note to Altenstein, reminding him of the young Dirichlet, "whose exceptional mathematical talent is so highly praised by Gauss, Eitelwein [sic] and Enke [sic]" and recommending him for Altenstein's help forthwith. Humboldt mentioned that Dirichlet had just left Paris for a few months and was spending the time with his father, the *Postkommissar* in Düren.

Altenstein, notorious for letting important decisions germinate on his desk, but aware not only of Humboldt's reputation as a scientist, but also of his proximity to the Crown, replied by return mail "to the Royal Chamberlain Mr. A. von Humboldt" enclosing the copy of a job offer for Breslau sent to Dirichlet. In addition, having obviously noted that Dirichlet did need help and that too much time had passed, he added that if the travel expenses from Düren to Breslau should cause Dirichlet discomfort, he would give the appropriate order for assistance. Altenstein concluded with the comment that the decision concerning Dirichlet had been delayed because of an oversight by his correspondence secretariat. Humboldt's response included a suggestion that funds be found for Dirichlet to be paid between 600 and 700 Thaler. In this, he was not successful.

Altenstein's communication to Dirichlet acknowledged the May and August letters and thanked Dirichlet for his reprint. Considering the unequivocal support from the "foremost living mathematicians," Altenstein wished to expedite Dirichlet's devoting his activities to the fatherland and acting as teacher of mathematics. He suggested a position as privatdozent with the philosophical faculty in Breslau. If Dirichlet agreed, he would assure him of an extraordinary remuneration of 400 Thaler annually, to be paid quarterly, and to continue until he could advance him with a fixed salary as extraordinary professor. He looked forward to Dirichlet's speedy response, noting that he wished Dirichlet could go to Breslau either this or the following month to start his activity as privatdozent in the philosophical faculty. If he did not yet have a doctorate of philosophy, he should seek to obtain that first. Altenstein declared he was certain that the university in Bonn would be pleased to respond to the application he would have to send for that purpose and that it would grant him "every easement consistent with existing rules" to obtain the degree.

Dirichlet responded on November 22, expressing his gratitude, explaining that he had already returned to Paris before Altenstein's offer reached him, making clear his interest in accepting the terms, but noting that, although he would turn around for Düren immediately to await the result of the application that he had just sent to Bonn, he feared he would not be able to do much good in Breslau before the end of the winter term. For that reason, he would like to spend a few months with his aging parents, from whom he had lived so far apart for over four years. If Altenstein still wished him to start work in Breslau during the winter, however, he would depart as soon as possible.

Meanwhile (on November 17), Humboldt had thanked Altenstein, with his usual diplomacy noting that it would be best to stay with Altenstein's salary decision for the time being, that everything else would depend on the talent and industriousness of Dirichlet himself. He commented that Dirichlet had taught Latin, although he did not know whether Dirichlet spoke it. He thought it was unlikely, as he had been busy with his contributions to mathematics. If the prognostications of Gauss, Legendre, Poisson, and Fourier were fulfilled, he was certain that Dirichlet would show himself worthy of the minister's goodwill and one day become a member of the Akademie der Wissenschaften.[10]

4.8 Bonn and the Doctorate

By January 28, 1827, Dirichlet reported a modification in the Bonn plan to Altenstein. When his request for the doctorate had been received in Bonn, the friend who was to transmit the application to the faculty decided not to do so, because a number of faculty members, including the professors of mathematics, felt that it would be appropriate and save time to nominate him for the title *doctor honoris causa*. Dirichlet was able to write to Altenstein that this honorific suggestion had been accepted by the faculty, although several more weeks might pass before his diploma was ready, because every faculty member had to subscribe to the reason for the award and that might take some time. He was awaiting arrival of the diploma and planned to start his journey immediately upon receipt, so that he could have his habilitation out of the way before the Easter recess and begin the sphere of activity assigned to him with the summer term.

In fact, matters had not progressed so smoothly in Bonn as Dirichlet's letter may suggest. The application had been transmitted by Peter Elvenich, Dirichlet's "Big Brother" from his earlier school days in Bonn, who had meanwhile joined the university faculty in Bonn and by this time was an extraordinary professor for philosophy and grammar. The suggestion that the degree be converted to an honorary one was initiated in early December by the two professors of mathematics, Karl Friedrich von Muenchow and Adolf Diesterweg, and by Johann Jacob Noeggerath, the professor of mineralogy, who was well known in governmental circles for his leadership in establishing Prussian mining regulations. They enclosed a reprint of Dirichlet's memoir and Altenstein's offer, furnishing a brief summary of Dirichlet's background, noting among other details that both of his parents were Germans from birth. Without mentioning a missing *Reifezeugnis*, they slid over his transition from leaving the Gymnasium in Cologne before the end of his last year to immediately going to Paris to continue his mathematical studies. They justified their request by

[10]The preceding exchanges with the ministry are found in Biermann 1959a:16–19; as noted there, these are based on manuscripts in the Merseburg Archive, as are those cited earlier (Biermann 1959a:13–15.

the extra time it would take to administer an oral exam, necessary for the regular degree but not the honorary one.

Their new application was passed on to the faculty by the dean with a positive recommendation and request for a vote. The first faculty member to receive the papers was the philologist C. H. Heinrich. In a detailed statement, he set forth reasons for questioning the suitability of the request: (1) the applicant was too young to have demonstrated the experience and contributions of others who had been approved almost automatically; (2) he had studied neither in Bonn nor another Prussian university; and (3) should the small memoir that had received such approbation in France not be evaluated in some detail by their own mathematical colleagues? He cast doubt on the precedent of having such a short paper considered such an extraordinary contribution as to be likened to a doctoral thesis. He suggested that if the faculty chose to allow the French memoir instead of a Latin doctoral dissertation and simply ordered a supplementary set of theses for the disputation, that would abbreviate the process of the faculty examination and could disallow the cost. This, he felt, would mean the faculty had done everything to conform to the minimum of the existing prescriptions and the candidate would be easily awarded the degree. Heinrich added that he would concur to the application only if these conditions were met.

The mathematicians quickly provided the requested commentary concerning the memoir. It was signed by von Muenchow and appears to have had some input from Dirichlet. They noted that Dirichlet would have been pleased to send in a Latin paper on a different subject had he thought this was permitted under the regulations. Without the disputation, there was no other way to help him out except by the *honoris causa* path.

Heinrich seemed satisfied with this statement from the mathematicians; he explained that he simply did not wish to go against the various legal prescriptions that were being violated, as an even higher authority might object to such an action later. After the rest of the faculty had exchanged its various assents and reservations, a unanimous consent was finally reached toward the end of January 1827. During February, a draft for the text of the diploma was circulated, completed, and printed, so that the final version reached members of the faculty by the twenty-fourth day of the month. Dirichlet was congratulated and sent a copy.[11]

4.9 Political Suspect

There had been another potential problem concerning Dirichlet's appointment, of which he may have been unaware. In January 1827, Johannes Schulze, on behalf of Altenstein's Cultural Ministry in Berlin, asked the Prussian envoy in Paris to make

[11] Schubring 1984 contains an extremely detailed account of the Bonn episode, based largely on the archive of Bonn's university, as well as on material pertaining to Dirichlet's earlier schooling. For the earlier period, it does not include some information used in our Chap. 1 on the basis of documents in the Kassel Dirichlet Nachlass and of supplementary material related to G. S. Ohm.

inquiries about Dirichlet's political connections, as it had been reported that he had been employed by General Foy. By mid-March, the necessary investigations had been completed. The envoy reported that there was nothing in the police records reflecting negatively on Dirichlet's orientation or lifestyle. For that reason, despite the fact that Dirichlet had lived in the house of one known to have been one of the most ardent opponents of the royal regime, the envoy believed it was safe to assume that he had lived only for his scholarship without having allowed himself to be dragged into political matters. He attached the result of the investigation which noted not only that Dirichlet had indeed been preceptor to the children in the Foy household, for which he had received favorable reports, but that he had left Paris in November to return to his natal country hoping to assume a position as preceptor in a house of distinction in Breslau![12]

4.10 The Visit with Gauss

In March 1827, Dirichlet left for Breslau. On the way, he stopped in Göttingen and called on Gauss.

The face-to-face meeting on March 18 appears to have been agreeable on both sides. Dirichlet reported to his mother that Gauss had received him with great kindness.[13] Gauss wrote to his fellow astronomer Olbers about the interesting visit he had had from young Dirichlet.[14] What seemed to please Gauss particularly was the fact that when they were chatting about various mathematical issues, and about recent publications by others such as James Ivory, he found out from Dirichlet that Fourier shared Gauss's opinions on the topics they discussed.

4.11 Breslau

Before reaching Breslau, after an additional stopover in Berlin, Dirichlet was informed that he was expected to present a sample lecture and a Latin *Habilitationsschrift* and also to defend the latter in an oral Latin discourse. He requested Altenstein to provide a dispensation from the Latin discourse, which was granted. His appointment as privatdozent, with an annual salary of 400 Thaler and the promised travel support of 75 Thaler, was effective as of April 1, 1827.

Franz Passow

The dispensation from the Latin defense of the habilitation raised questions among the faculty in Breslau. Not only was there objection to the dispensation, but certain

[12]Biermann 1959a:19–20.

[13]Kassel. Dirichlet Nachlass. Box 2:VIc.

[14]Gauss *Werke* 8:133.

members felt Dirichlet had not acted properly in requesting it, as he had given them to understand that he would provide the defense "within a few weeks."

The matter came to a head with a lengthy brief, signed November 12, 1827, and sent to the Ministry by Franz Passow, the distinguished classical philologist, best known for his Handbook of the Greek Language and respected among fellow scholars for numerous treatises on the Ancients. Passow, on the faculty since 1815, had become Acting Dean for the term 1827/1828. The statement he sent to the ministry partly resembled a modern legal document, occasionally a persuasive Ciceronian argument. He outlined the sequence of events as follows:

On August 1, "the undersigned faculty" had agreed in a meeting that they would transmit a history of the relationship between the privatdozent Dr. Dirichlet and the faculty. Since a quarter year had passed without any follow-up, the current dean [Passow himself] felt he must immediately act on the will of the faculty.

Passow began his documentation with the copy of a letter Dirichlet had sent to the faculty on April 22, in which he confirmed his wish to join the faculty as docent in mathematics, to hold his sample lecture, then to begin his lectures immediately, but to be permitted to hold the prescribed disputation "only a few weeks later." As Passow stressed, Dirichlet promised in this letter to use the allowed extra time with utmost conscientiousness to gain the needed fluency in the Latin language. Passow continued to observe that since the young man had come with the best recommendations, and there had been no reason to mistrust his assertions, the faculty had taken the unusual step to go along with the delay, for which Dr. Dirichlet thanked them on 30 April, renewing his promise. He then began his lectures before "some" listeners.

Passow continued by explaining that, although the Royal Ministry informed the faculty in a communication dated 28 April that it had granted a dispensation from the statutory disputation, since this dispensation included no prohibition keeping Dirichlet from fulfilling the promise made of his own free will, and since the faculty could not believe that he himself had applied for the dispensation, and since the Ministry seemed unaware of his communications with the faculty, and since Dr. Dirichlet did not even see fit to inform the faculty of a change of mind, they thought he would still produce the Latin defense. By the time most of the summer had passed, the dean had brought the matter before a faculty meeting on 12 July, and it was decided to ask Dirichlet in writing whether and when he planned to present his disputation according to his promise.

Passow also attached Dirichlet's response to the July inquiry. Dirichlet had begun by apologizing for not having explained his actions concerning the habilitation at an earlier date. First of all, he noted that he had thought he would give the defense shortly after having written out the *Habilitationsschrift*. But while preparing a clean manuscript he found new points of view, which simplified his presentation considerably, but necessitated a complete rewriting of the entire paper. This, he explained, frequently happens in a mathematical work. Yet, it delayed his passing in the *Habilitationsschrift* itself. He thought they would not mind the delay but would take it as a sign of his high regard for them that he had decided to give a careful treatment to his chosen subject. He had thought he might also be allowed the opinion that the faculty would have recognized the degree of scholarly attainment (*Bildung*) to be expected from an academic instructor because of an earlier work that the Academy of Sciences in Paris had honored by including it among its memoirs to be published, and which also had gained the approbation of "our Gauss who, in the branch of scholarship to which that memoir refers, is the most competent judge among all now living mathematicians."

The second issue to which Dirichlet had replied in this letter was that of the oral defense of the *Habilitationsschrift*. He had been told that some had called his handling of this matter duplicitous and he must reject this. He felt that he needed no excuses but should recall the events in question. When he first was offered the position in Breslau, he knew of no habilitation requirements except for those in Berlin and Bonn, neither of which included a disputation. He learned of the Breslau requirement only shortly, and somewhat vaguely, before he was leaving Berlin on his way to Breslau. This was confirmed by the [previous] dean after his arrival. Since the whole direction of his studies, and the fact that he had undertaken them in France, had held him far distant from speaking Latin, the wish arose that he be dispensed from this requirement. The dean assured him, however, that the faculty could not give this dispensation. That left him nothing but to address the ministry, and since he thought it might be quite possible that he would be turned down, he decided to ask the faculty for the extra time, in part to use it for the necessary linguistic practice and secondly not to let the semester pass without any academic activity.

He would have provided a Latin oral disputation, for better or worse, if the ministry had denied his request. In order to prepare it in accordance with proper philological standards, he would have needed at least half a year. He did not think he should spend that much time away from the studies in his own field only to gain competence in a subject that in itself has no scholarly significance and not the remotest usefulness for his discipline.

The conclusion of Dirichlet's letter, which must have raised the hackles of his opponent even further than this last statement, contained the hope that, after this explanation, the faculty would find it justified that he make use of the dispensation. He expressed the thought that he may assume this, particularly since he is convinced that an enlightened faculty refutes the assumption that a scholarly education not connected with a Latin conversational competence is invalid. This, he noted, has been shown by precedents when the faculty had no objection to a dispensation from the requirement of an oral Latin defense even for those whose fields lie closer to grammatical studies.

Having attached this letter of Dirichlet's as well, Passow stated that "we noted with regret that he thought himself free of all previously entered commitments toward the faculty, who had met him with open goodwill." Passow continued his exhortation with an urgent and respectful formal request that stated if in future similar appeals for dispensation of requirements set forth by statute were to be dared, the relevant faculty first be asked for a report concerning the conditions in question. Passow concluded with a further lengthy discourse, reiterating his defense of use of the Latin language, then commenting that if a young man comes from abroad and has not satisfied common academic achievements at any university, then one must assume that he is unable to meet the legal requirements. This led Passow to the conclusion that having such a person in their midst would have an ill effect on both students and faculty members.

Passow signed his letter “the philosophical faculty of the University of Breslau,” with his name and notation as Temporary Dean. This prompted a shorter but forceful response from Henrich Steffens.

Henrich Steffens

Steffens, Norwegian by birth, had spent his early studies and career in Denmark and Germany, but had been on the Breslau faculty since 1811. He had come to public attention that same year for leading his students en masse to join the Prussian troops in the Wars of Liberation. After this voluntary service in the military, for which he received the Iron Cross, he returned to full-time teaching duties in Breslau. Considered an anthropologist and philosopher, he was associated with the so-called Naturphilosophen, had been supported by Schleiermacher as far back as 1804, but had additional scientific training and interests.

Steffens apologized to the Ministry for daring to append a totally opposing view to the Dean’s communication. He proceeded by stating that he does not minimize the fervor of the dean to maintain the old forms but he feels that an instructor who had been appointed by the ministry with a fixed salary even though the Minister was not ready to give him the title of Professor, should be considered as belonging to the category of an extraordinary professor whose achievements the ministry would evaluate, and not as someone who comes merely “highly recommended.” Secondly, he emphasized that he subscribed to the principle the ministry should allow exceptions, as indeed it had done in the past.

Steffens remarked that in preparing the report an irregularity had taken place. The plurality of members of the faculty had thought the writing of it was too harsh. The Dean only modified one spot and informed “us—the disapproving plurality” that the unchanged version had been sent. Steffens noted that he was certain had the report been sent after due agreement among the faculty, its tone would have been quite different. He remarked also that he considered it superfluous to defend Dirichlet. He pointed out that Dirichlet’s factual explanation of the sequence of events showed that he had overlooked the proper form but had acted in good faith and that such an oversight of the proper form had its precedent which tended not to have been judged so harshly.

Steffens wished to reject the implication that Dirichlet would lose respect among students or faculty if he accepted the dispensation. Quite to the contrary, Steffens thanked the ministry on his own behalf and that of many of his colleagues for having awarded their university with the presence of a young man who had drawn the attention of the great masters and justified highest hopes.

Steffens concluded with two closing points. The first stressed something frequently repeated during Dirichlet’s lifetime: that he was not only respected for his thorough knowledge but popular because of his modesty. In this connection, Steffens observed that, whereas it was true that, like Gauss, Dirichlet attracted few students, those who occupy themselves seriously with mathematics know to treasure and use him. Secondly, Steffens, too, tackled the main issue:

> No teacher at a University, no philologist, will awaken so strange an image before students as when one may be allowed to evaluate a mathematician according to his competence in speaking Latin. And how can, at a University where a number of teachers—and by no means the least significant—have never disputed, the respect for the young man be diminished by this?
>
> I disputed, but I acknowledge that the serious preoccupation with my discipline has robbed me of my aptitude to speak a poor Latin. Over a thirty-year-long teaching career I have never remarked that I lost in respect or influence among my listeners because I do not dispute—yes, excellent philologists, some of whom have gained a high reputation, belonged to my most eager listeners, and have honored me with their lasting loyalty and friendship.[15]

Steffens ended by stating he thought it necessary to avoid a point of view that would harm the university more than the young man.

4.12 Confirmation and Recognition

Responding to Steffens' letter that had been sent with copies to the dean and the faculty, in December 1827 Passow reiterated his request to the ministry to disallow the dispensation. The ministry refused. Passow made another appeal, which was again rejected. Clearly, the ministry was not inclined to respond to continuing communications from Breslau that not only objected to its previous decisions but requested additional new policy guidelines, especially when such communications seemed to resurrect the embarrassing antagonisms of the "Breslauer Turnfehde" that had involved Passow and Steffens as opponents almost a decade before.[16]

On March 26, 1828, Dirichlet sent his *Habilitationsschrift* (**1828b**) to Altenstein, along with a related number-theoretic memoir (**1828a**) on biquadratic residues that had been published in the January issue of Crelle's *Journal*. His appointment as extraordinary professor became effective on April 1, and Humboldt informed Dirichlet on April 4 of that action.

The publication **1828a** in Crelle's *Journal* had been particularly well-timed. Upon reading it, Friedrich Bessel, Gauss's oldest student and Prussia's most renowned mathematical astronomer, active in Königsberg, wrote to Humboldt on April 14, 1828:

> Dirichlet's work in Crelle's *Journal* has pleased me very much; who would have thought that the genius would succeed in leading back to such simple considerations something that seems so difficult! The name Lagrange could appear over this memoir and no one would notice the error.[17]

---

[15]Biermann 1959a:30.

[16]The Breslau documents here cited are reproduced in Biermann 1959a; Biermann commented that the strident note of the Passow letters was more of a throwback to the "Turner" controversy than any predominant concern with Dirichlet. See Biermann 1959a:21–22. For a discussion of the "Breslauer Turnfehde" of 1817, see Schnabel 1964:193–196 or other histories of Germany for that period.

[17]Biermann 1959b:91–92.

Humboldt quickly made copies of this letter, sending one to Dirichlet, and one each to Altenstein and Major Radowitz.

4.13 Radowitz and the Kriegsschule

J. M. Radowitz, in Prussian service since 1823, at the time taught at the Kriegsschule but was also a member of the highest commission for military education and had an interest in mathematics and mathematical tables and formulas. A friend of the later King Friedrich Wilhelm IV, he would rise through the ranks, retiring as general, active in the 1848 Frankfurt Diet, but always involved in military education. As time went on he would represent an increasingly conservative point of view, especially on foreign policy.

Humboldt realized that acquainting Radowitz with Dirichlet and his work would be helpful in bringing Dirichlet to Berlin. His efforts in this direction were successful. For, on July 16, Dirichlet thanked Altenstein for his appointment as extraordinary professor and for allowing him to teach at the Kriegsschule in Berlin while on a leave of absence from Breslau.

At the same time, Dirichlet asked for permission also to give lectures at the university in Berlin. This was granted. By July 27, the ministry informed the philosophical faculty in Berlin that Dirichlet, the extraordinary professor at the university in Breslau, had a ten months leave from there, lasting from October 1, 1828, to July 31, 1829, to be active in teaching mathematics at the Kriegsschule and that, at his request, he was also given permission to lecture at the University of Berlin under the auspices of the philosophical faculty during this period.

4.14 Departure from Breslau

Dirichlet was ready to leave Breslau. He had made arrangements with Julius Scholtz, a fellow privatdozent of mathematics in Breslau, to have a book box shipped to Berlin. A letter from Scholtz of August 30, confirming that he had done so, attests to Dirichlet's having acquired good friends among members of the teaching staff in Breslau. Scholtz addressed Dirichlet as "Lieber theurer Freund." After providing details as to the departure, weight, and arrival of the book box, he let Dirichlet know that he and Heinrich Goeppert, then privatdozent for medicine and botany, would come to Berlin on September 9.

> Your departure from here has left a larger gap and emptiness for me than I had initially anticipated, and my stay in Berlin that will bring us together again, although only for a short time, provides no substitute for the cheerful and cozy life with one another which your

> transfer from here tore apart. Our friends who along with me feel the loss in the circle of their friends are all well and send greetings.[18]

The friends were not only young but largely scientists. Among these, Scholtz, after becoming an extraordinary professor in Breslau, assumed a full professorship in 1834. Goeppert, too, remained in Breslau, achieving a highly successful sixty-year-long career at the university where, in addition to his acclaimed research activities, he came to popular notice for his expansion of the Botanical Garden. They addressed one another with "*Du*," which in their case probably had less to do with being gymnasts or fraternity brothers—the two groups identified with use of the familial address form at the time—than it did with their common age, scientific interests, and comfortable friendships.

Steffens had been requesting a transfer to Berlin for some time. It was granted in 1832, and, after lecturing on philosophy and anthropology there for three semesters, he was appointed rector of the Friedrich Wilhelm University in Berlin. Passow died in Breslau in 1834.

[18] Berlin. Staatsbibliothek. Handschriftenabteilung. Nachlass Dirichlet.

Chapter 5
Early Publications

The pamphlet **1826** that constitutes Dirichlet's first publication in his own name is a small, twenty-page-long brochure containing his proof of Fermat's Last Theorem for $n = 5$. It had a very limited circulation, but, as we noted in the two preceding chapters, it reached the individuals and institutions considered important by Humboldt in finding a suitable position for Dirichlet on leaving Paris.[1]

In 1828, Dirichlet came to more wide-spread attention by the publication in Crelle's *Journal* of four memoirs. In addition, his Latin *Habilitationsschrift* **1828b** was sent to Altenstein in March of that year. The publications appearing in Crelle's *Journal* were written in French, as were all of Dirichlet's publications in that *Journal* until 1840; in that year, Crelle began to republish the reports that had appeared in the Akademie's *Berichte*. One of the memoirs, **1828a**, published in Crelle's *Journal* in 1828 was related to the *Habilitationsschrift* **1828b**. The memoir **1828c** was a more elaborate version of the pamphlet **1826**. The final two memoirs of the year, **1828d** and **1828e**, pertained to Wilson's theorem. Like **1828a**, they were explicitly designed to prove and expand results announced by Gauss.

5.1 Some Indeterminate Equations of Degree 5

1826

The 1826 pamphlet was called "Memoir on the impossibility of some indeterminate equations of the fifth degree." The title page indicated that it had been read ("lu")

[1] Aside from the copies already mentioned as having been sent to the Berlin Akademie, to Altenstein, to Gauss, and to the Breslau faculty, Kronecker, in the preface to Dirichlet's *Werke*, later related that he subsequently located copies in the Dirichlet *Nachlass*, the Berlin Staatsbibliothek, and in the Fonds Huzard of the Paris Institut's library (the last-named easily explained since the pamphlet was produced by the Imprimerie Huzard-Courcier).

U. C. Merzbach, *Dirichlet*, https://doi.org/10.1007/978-3-030-01073-7_5

at the Institute of France of the Royal Academy of Sciences on July 11, 1825.[2] An additional note informed the reader that, in accordance with the report of Mssrs. Lacroix and Legendre, the memoir had been approved and was to be printed in the *Recueil des Mémoires des Savans étrangers*.

As would be his custom in most later publications, Dirichlet preceded the chief content of his memoir with an introduction in which he provided some general background for the issue involved. He stated at the outset that the theory of equations of degree higher than two was still very little advanced. He observed that there is an infinity of equations of all degrees for which impossibility can be demonstrated by showing that whatever the values one attributes to these, the two sides of the equation cannot give the same value when one divides by the same number or modulus. But when this treatment does not work, it becomes very difficult to prove the impossibility, and so far this has only been possible in a very small number of cases. What these have in common is that one is led to one or more quadratic formulas which must be set equal to perfect powers. He briefly discussed the methodology Fermat and Euler had used in proving the impossibility of some of the more difficult cases of degree three and four. He outlined Fermat's method of least descent (without using that term) and the modified technique Euler had used for the solution of $x^3 + y^3 = z^3$. But, he added,

> one is stumped when trying to apply similar considerations to some equations of degree five and of a form analogous to those of the equations treated by Fermat and Euler. ... The quadratic formula at which one arrives and which needs to be set equal to a fifth power admits several different solutions, and among these, there is only a single one which leads to an equation similar to the proposed equation. In reflecting on this difficulty, I realized that it could be relieved simply by subjecting the determinate number which enters into the equation to some conditions. It follows from a theorem exposed in the preliminaries that when these conditions are fulfilled, the different solutions of which the quadratic formula is susceptible generally must be rejected, except for a single one which is precisely the one from which one derives numbers which satisfy an equation similar to the proposed equation. One thus arrives at establishing the impossibility of a quite extended class of indeterminate equations of the fifth degree. The left-hand side [first member] of these equations is the sum or difference of two fifth powers, and the second is the product of a fifth power and of a number subject to different conditions. On attributing to this number specific values compatible with these conditions one can obtain as many specific theorems as one wishes. This generality of our theorems is the more singular as the analogous equations of the third and the fourth degree whose impossibility has been demonstrated so far are only finite in number and very small.[3]

The first part of the 1826 pamphlet, containing the material Legendre had summarized in his July 18 report of 1825, is cast in the form of three theorems.

Theorem I states that if P and Q are two numbers relatively prime, one of which is even, the other one odd, and the odd one is divisible by 5, then in order for the binomial $P^2 - 5Q^2$ to be equal to a fifth power, it suffices to set

$$P + Q\sqrt{5} = (\phi + \psi\sqrt{5})^5,$$

[2] See, however, the fourth paragraph in Sect. 3.3 of Chap. 3 above.

[3] Werke 1:3–4.

where the indeterminates ϕ and ψ are relatively prime, one even, the other odd, and the first one not divisible by 5.

Theorem II states that if m and n are two positive numbers, the second one being different from 2, and if the number A is divisible neither by 2 nor by 5 nor by any prime number of any of the forms $10k \pm 1$, it will be impossible to find two numbers x and y relatively prime such that $x^5 \pm y^5 = 2^m 5^n Az^5$.

The third theorem, resulting from this, states that if the numbers m and A are subject to the same restrictions as in the statement of theorem II and if the number $2^m A$, being divisible by 25, gives one of the eight remainders 3, 4, 9, 12, 13, 16, 21, 22, it will be impossible to find two numbers x and y, relatively prime to one another, such that one has $x^5 \pm y^5 = s^m Az^5$.

From this, Dirichlet derived the impossibility of $x^5 \pm y^5 = z^5$ unless one of the indeterminates x, y, or z is not divisible by 5.

He concluded this portion of the memoir, the one reflecting the July 11 presentation, with the comment that the only thing still necessary would be to prove the case where the indeterminate divisible by 5 is odd: But "the method exposed in this Memoir seems insufficient for this case, and I do not see how one could complete the demonstration of the special case of Fermat's theorem in question."

The pamphlet next contained the "Addition," where the problem was resolved. The two additional theorems previously mentioned were proved and then stated as follows.

Theorem IV. Let the numbers P and Q be relatively prime and both odd, and let the latter be divisible by 5. I say that in order to let the binomial $P^2 - 5Q^2$ equal the quadruple of a fifth power with all suitable generality, it suffices to set

$$P + Q\sqrt{5} = \frac{(\phi + \psi\sqrt{5})^5}{2^4},$$

where the indeterminate numbers ϕ and ψ are relatively prime, both odd, and moreover, the first one is not divisible by 5.

Before proceeding to his main theorem, Dirichlet noted in a footnote that Theorem IV, like Theorem I, has analogues for many other primes. Theorem V. Let the letter n designate a positive number other than 0 or 2, and let the number A, not being divisible by either 2 or 5 or any of the prime numbers of one of the two forms $10k \pm 1$, it will be impossible to find two numbers x and y relatively prime and such that

$$x^5 \pm y^5 = 5^n Az^5.$$

Following the detailed proof of this theorem, Dirichlet pointed out that it includes Fermat's theorem for $n = 5$ as a special case.

Finally, he added that, just as Theorem III could be derived from Theorem II, so one can now state

Theorem VI. Let the number A be subject to the same restrictions as in the statement of Theorem V, and let this number have one of the following eight remainders,

3, 4, 9, 12, 13, 16, 21, 22, when it is divided by 25; then it will be impossible to find two numbers x and y among them such that one has $x^5 \pm y^5 = Az^5$.

1828c

The version **1828c** of **1826** that appeared in Crelle's *Journal* was more elaborate, presumably providing more explanatory details because intended for a wider readership. In fact, a footnote in **1826** had contained a remark that certain details were being omitted so as not to make the presentation too extensive. On the other hand, only the reference to Lagrange's "Additions" to the Algebra of Euler was retained, whereas previous brief references to Gauss's *D.A.*, Legendre's Theory of Numbers, and Euler's *circa divisores numerorum* were omitted; this, too, may have been because they were less widely available to much of the readership of Crelle's *Journal* than they would have been to those for whom the 1826 pamphlet was intended. However, it is most likely due to the fact that Dirichlet wished to call attention to his use of a different methodology from that used previously, which had depended more closely on the approaches by Euler, for the case $n = 3$, and by Gauss, while involving the more extensive manipulations of Legendre. Besides, Lagrange would have been considered an unimpeachable reference at a time when Legendre and Gauss were still edgy about sharing the limelight with one another.[4]

Comparison of 1826 and 1828c

The introductions to both of Dirichlet's memoirs are identical. Three preliminary theorems in **1828c** now lead up to a Theorem IV which corresponds to Theorem I of **1826**. Theorem V corresponds to the former Theorem II, Theorem VI corresponds to the former Theorem III, and the rest of the July 1825 presentation has been maintained except for omission of a footnote reference to Section 200 in Gauss's *D.A.*

The "Addition" contains a more detailed introductory explanation, followed by a Theorem VII which is identical to the former Theorem IV. Similarly, Theorems VIII and IX correspond to the former Theorems V and VI.

5.2 Biquadratic Residues

The memoir **1828a** that created the most interest and enthusiasm appeared in the first issue of Crelle's *Journal* for 1828. It was titled "Investigations concerning prime divisors of a class of formulas of the fourth degree." The most extensive of the 1828 publications, it dealt with biquadratic residues, specifically aimed at establishing the nature of prime number divisors of $x^4 - 2$.

---

[4]When Dirichlet mentions the Theory of Numbers (without naming the author) in the publications of 1826 and 1828 he is referring to Legendre 1808. Legendre's section numbers for his 1808 and 1830 publications correspond through the first part of his book V, and the content is almost the same except for additional examples in the 1830 edition.

Dirichlet noted at the outset that his work had been instigated by the extract Gauss had published without proof in April 1825 of an as yet unpublished memoir that dealt with the theory of biquadratic residues, in particular with the determination of the special characteristics of prime number divisors of the expression $x^4 - 2$.

The publication to which Dirichlet referred is Gauss's announcement (Gauss 1825) in the *Göttingische Gelehrten Anzeigen* of April 11, 1825, of the first part of his "Theory of Biquadratic Residues." Gauss had begun his announcement by stating that the theory of quadratic residues, well known for being one of the most interesting parts of higher arithmetic, could now be considered completed. He referred the reader to prior notices of 1808 and 1817 (Gauss 1811 and 1820), pointing out that he had there also provided preliminary words about investigations concerning the "equally fruitful and interesting but far more difficult theory of cubic and biquadratic residues."

Gauss had remarked that although he had already been in possession of the main points of these theories, he had been prevented by other work from publicizing these at an earlier time and had only lately been able to work out a part of these investigations. Now beginning with the theory of biquadratic residues, which is more closely related to the theory of quadratic residues than is that of cubic residues, Gauss cautioned the reader that this is not intended as an exhaustive treatment of the subject:

> The development of the *general* theory, which requires a special expansion of the field of higher arithmetic, must be saved for later continuation.[5]

At the end of his announcement, Gauss had further reinforced this remark by concluding with the advice that "friends of higher arithmetic" consider this topic for further research, since

> succcess therein will at the same time open a productive source for new expansions of this beautiful part of mathematics.[6]

It is easy to see why these introductory remarks would have ignited Dirichlet's curiosity. He observed that Gauss's extract contained two theorems Dirichlet considered "extremely elegant." They were to decide whether a prime number divisor of $x^2 - 2$ does or does not divide $x^4 - 2$. As Dirichlet explained, when he first read Gauss's announcement which did not contain Gauss's proofs, he decided to try establishing Gauss's theorems on his own. He succeeded with a proof he considered quite simple and probably quite different from Gauss's, whose own demonstration seemed to require a good number of preliminary, "very delicate and quite extended" investigations. Dirichlet reported next applying analogous considerations to other questions. Specifically, he was interested in exploring properties that distinguish the prime divisors of the expression $\alpha x^4 + \beta x^2 + \gamma$ and told the reader that through this he arrived at a large number of interesting theorems.

Dirichlet remarked that he had had knowledge of Gauss's announcement "in the course of the year that was just completed." Whether or not Dirichlet had read Gauss's

---

[5]Gauss 1825; *see* Gauss *Werke* 2:166.

[6]Gauss 1825; *see* Gauss *Werke* 2:168.

1825 announcement itself at that time, he would have been intrigued by learning of specifics it contained from a letter Gauss sent him in September 1826.[7] In this, Gauss referred to the theory of biquadratic residues as "a new subject"—as contrasted with those portions he had had in mind in 1801 when considering a continuation of the *D.A.*—and his thought to publish his findings in a series of three memoirs.

Dirichlet divided his rather lengthy memoir into five sections and an addition. The first section begins with the definition of a biquadratic residue: "If one can attribute to the indeterminate x a value such that $x^4 - A$ becomes divisible by B, A is said to be a biquadratic residue with respect to B."[8] It continues with several simple properties of biquadratic residues.

The second section is devoted to examining the characteristics that distinguish the prime divisors of $x^4 - 2$. Since the prime divisors of $x^2 - 2$ are either of the form $8n + 1$ or $8n + 7$ and it has been shown that the second form can be disregarded, it suffices to deal with primes of the form $8n + 1$. Utilizing Legendre's notation and the Law of Quadratic Reciprocity (Gauss's Fundamental Theorem), he arrived at the theorem

> Let p designate a prime number $8n + 1$; if one sets $p = t^2 + 2u^2$, I say that ± 2 will or will not be a biquadratic residue with respect to p according as to whether t is either one of the forms $8n + 1$, $8n + 7$ or of one of the forms $8n + 3$, or $8n + 5$.[9]

This is the first of Gauss's two theorems Dirichlet mentioned in his introduction. To derive the second theorem from this, Dirichlet assumed that p is the sum of two squares. This allowed him to derive the second theorem in a fairly straightforward fashion, arriving at the statement:

> Let p designate a prime $8n + 1$; having set $p = \phi^2 + \psi^2$ (where ψ is assumed to be divisible by 4), ± 2 will or will not be a biquadratic residue with respect to p according as to whether ψ is of the form $8n$ or of $8n + 4$.[10]

Dirichlet next added a third theorem, which he did not bother to prove, instead commenting that it could be easily derived "in a direct manner, and by considerations analogous to those on which the demonstration of the first of the two preceding theorems is based." He added that it can also be deduced from each of the preceding ones, in nearly the same way as one has passed from the first to the second. The theorem states:

> Having arbitrarily set $p = t^2 - 2u^2$, ± 2 will or will not be a biquadratic residue with respect to p according as to whether t is one of the forms $8n + 1$, $8n + 3$, or of one of these: $8n + 5$, $8n + 7$.[11]

With Section 3, Dirichlet proceeded to some generalizations for which he depended heavily on Legendre and manipulations of the Legendre symbol. His chief result was a theorem he stated as follows:

---

[7] *Werke* 2:375.

[8] This is equivalent to Gauss's definition published in Gauss 1801, Section 4 and Gauss 1825.

[9] *Werke* 1:69.

[10] *Werke* 1:70.

[11] *Werke* 1:71.

> Designate by b a prime number $4n+3$, and by p a prime number $4n+1$, susceptible of being put in the form $t^2 - bu^2$. Having set $p = t^2 - bu^2$ (where t is assumed to be odd), I say that $-b$ will or will not be a biquadratic residue with respect to p according as to whether t is or is not a quadratic residue with respect to b.[12]

In Section 4, Dirichlet utilized the preceding theorem to produce one which, he noted, would make it possible to determine more easily whether or not $-b$ is or is not a biquadratic residue with respect to p. The derivation is more direct than the preceding one; he goes back to the original subdivision of cases, still utilizes Legendre symbols, and explicitly refers to the law of quadratic reciprocity with a reference to the *Theory of Numbers* of Legendre. The new theorem, which he singled out as Theorem I, states the following:

> Let b designate a prime number of the form $4n+3$, and p a prime number $4n+1$ such that $\left(\frac{b}{p}\right) = 1$; if one sets $p = \phi^2 + \psi^2$ (where ψ is assumed to be even) one will have the following rule for deciding whether $-b$ is or is not a biquadratic residue with respect to p: If ϕ is divisible by b, $-b$ will or will not be a biquadratic residue depending on whether b is of the form $8n+7$ or of this: $8n+3$. If ϕ is not divisible by b, one will seek a number χ such that one has $\chi^2 \equiv p \bmod b$. Given this, $-b$ will or will not be a biquadratic residue with respect to p, depending on whether one has $\left(\frac{\chi(\chi+\psi)}{b}\right) = 1$ or $\left(\frac{\chi(\chi+\psi)}{b}\right) = -1$.[13]

Dirichlet then provided examples to show that by setting values such as $b = 3$, $b = 7$, etc., one can obtain specific theorems analogous to Gauss's which, by the preceding, can be judged rigorously proved.

In Section 5, Dirichlet took on the somewhat more difficult task of establishing a statement that he designated his Theorem II:

> Let a denote a prime number $4n+1$, and p another prime number $4n+1$ such that $\left(\frac{a}{p}\right) = 1$; if one sets $p = \phi^2 + \psi^2$ (where ψ is assumed to be even) one can decide in the following manner whether a is or is not a biquadratic residue with respect to p. If ϕ is divisible by a, a will or will not be a biquadratic residue with respect to p depending on whether a is of the form $8n+1$ or of this: $8n+5$. If ϕ is not divisible by a, one will seek a number χ such that one has $\chi^2 \equiv p \bmod a$. Given this, a will or will not be a biquadratic residue with respect to p, depending on whether one has $\left(\frac{\chi(\chi+\psi)}{a}\right) = 1$ or $\left(\frac{\chi(\chi+\psi)}{a}\right) = -1$.[14]

Again, Dirichlet gave examples to show that by successively setting values such as $a = 5$, $a = 13$, one will have specific theorems that can be regarded as proved rigorously by the preceding, where, for these values of a, the expression $t^2 - au^2$ will have only a quadratic divisor of the form $t^2 - au^2$ or, in other words, that every divisor of $t^2 - au^2$ is itself of the form $t^2 - au^2$. He spelled it out by noting that "if p denotes a prime number of one of the forms $20n+1$ or $20n+9$, if one sets $p = \phi^2 + \psi^2$ (where ψ is assumed to be even), one easily is assured that one of

[12] *Werke* 1:74.

[13] *Werke* 1:78.

[14] *Werke* 1:84.

the numbers ϕ or ψ is a multiple of 5. Given this, I say that 5 will or will not be a biquadratic residue with respect to p depending on whether ψ or ϕ is divisible by 5."[15]

In the course of proving his two theorems Dirichlet had indicated that he had found it necessary to rely on a theorem that it seemed particularly difficult to prove rigorously. He now commented that he found this difficulty could be circumvented by considering the more general equations $t^2 - bu^2 = ps^2$ and $t^2 - au^2 = ps^2$, or even more simply $t^2 + au^2 = ps$ for the latter, that can always be satisfied when the conditions $\left(\frac{b}{p}\right) = 1$ and $\left(\frac{a}{p}\right) = 1$ hold. How do we know this?

"It results from the beautiful theorem that Mr. Legendre has given for judging the possibility or impossibility of the equation $\alpha x^2 + \beta y^2 = \gamma z^2$ (*Théorie des Nombres*, no. 27)."[16]

The memoir concluded with the remark that this modification of the proof, aside from being entirely rigorous, also has the advantage of greater simplicity, "as every reader who has absorbed the spirit of the preceding considerations could judge while developing this demonstration according to the indication that has been given."

Having reviewed the entire memoir—and possibly having received some comments from those who felt something was still missing in their absorption—Dirichlet supplied an "Addition to the preceding memoir," of about half the length of that entire memoir, in which he not only provided more detailed proofs of the preceding but additional specific examples whereby the reader could more easily be satisfied with the established theorems.

It is easy to see why this memoir was highly praised by Bessel. Two factors stand out: First of all, Dirichlet had succeeded in making considerable progress in dealing with an issue that Gauss had described as being particularly difficult. Secondly, it was unusual for a contemporary not only to provide lip service to achieving rigor, but to demonstrate it, as Dirichlet did explicitly at the end of his publication.

5.3 The *Habilitationsschrift*

In the *Habilitationsschrift* for Breslau (**1828b**) that Dirichlet sent to Altenstein in March 1828, Dirichlet observed that it follows from the theory of quadratic residues or of divisors of second-degree forms that these divisors are characterized by certain linear forms. He noted that when dealing with a degree higher than 2, this holds only for some specific forms such as $x^n \pm 1$ that Euler had examined. Dirichlet stated that while studying the relevant works by Euler he thought of a new sort of higher order forms having properties similar to those Euler had treated. Rather than discussing them in a formal introduction, Dirichlet wove in pertinent references to Euler as well as Lagrange, Legendre, and Gauss, along with Gauss's "opus egregius" [the *D.A.*].

[15] *Werke* 1:85.

[16] *Werke* 1:85.

Although closely related to the more restricted but more extended **1828a**, the *Habilitationsschrift* did not receive the same attention as had **1828a**, to which it refers. Aside from enabling Dirichlet to meet the requirement for his habilitation in Breslau, it had no historical significance in the sense of successors building on it with further developments. It was the basis, however, of a much later memoir by Kronecker, read to the Akademie and published in 1888.

While editing the first volume of Dirichlet's mathematical works, Kronecker studied the *Habilitationsschrift* and decided to try resolving the problem Dirichlet addressed by using his own system of modular forms. In this he succeeded. He briefly summarized the point of Dirichlet's work in two paragraphs:

The higher order forms Dirichlet used are the forms U and V, which come into being when one converts the expression $(x + \sqrt{b})^n$ to the form $U + V\sqrt{b}$, where x is a variable, n an arbitrary positive integer, and b a whole number that can be positive or negative but must not be a square.

Dirichlet considered the prime divisors of V and determined them under the condition that n is a prime number, and those of U for the case n is a power of 2. He remarked that his method can be applied to any other integral value of n, but is restricting himself to the stated case to save space.

Kronecker justified his short publication with two reasons: Every topic that Dirichlet treated draws interest, and this one provides new support for the applicability of modular systems.[17]

5.4 Wilson's and Related Theorems

In a memoir titled "New Demonstrations of Some Theorems Pertaining to Numbers," (**1828d**), Dirichlet introduced his topic by calling attention to Gauss's proof of Wilson's theorem found in Section 77 of the *D.A.* He observed that this proof by Gauss is by far the simplest of numerous proofs that had been given, but noted that, by a slight generalization of Euler's and Gauss's definition of *numeri socii* and then following steps analogous to those Gauss had taken in using these corresponding ("associated") numbers, he could prove not only Wilson's theorem but two additional ones which are of great use in number-theoretic studies.

Dirichlet proceeded by recalling Euler's definition of corresponding numbers: Let p be a prime number. Let m and n be two numbers less than p such that $\frac{mn}{p}$ leaves a remainder of 1. Then m and n are said to be associated or corresponding numbers. Dirichlet's generalization had m and n be numbers less than p whose product mn leaves the same remainder as a fixed number a which we assume not to be divisible by p. Dirichlet next considered the sequence

$$1, 2, 3, \ldots, p - 1.$$

[17]Kronecker 1888 provides the details of Krcnecker's modular forms treatment of Dirichlet's problem.

Letting m be any member of this sequence, he observed that it will have one and only one associated number n in the sequence. This is because the congruence $my \equiv a \bmod p$, in which neither mn nor a are divisible by p, always has one and only one root y less than p.

Essentially using a sequence of steps analogous to Gauss's in Section 3 of the *D.A.*, Dirichlet used congruences and his expanded definitions to prove Wilson's theorem as well as Fermat's and the more general one where $a^{\frac{1}{2}(p-1)} \equiv \pm 1 \bmod p$, which also includes that of Euler's which Dirichlet mentions as having special importance in the theory of residues.

This memoir was followed by a challenge problem.

5.5 A Challenge

The challenge **1828e** was titled "Question of Indeterminate Analysis." Dirichlet outlined two corollaries of Wilson's theorem that Lagrange had derived in 1771, in the memoir that contained Lagrange's proof of the theorem.[18] If Wilson's theorem is written as "$(p-1)! + 1$ is divisible by p" the corollaries state that

1. If p is of the form $4n + 1$, then $[\frac{1}{2}(p-1)^2]! + 1$ is equal to a multiple of p;
2. If p is of the form $4n + 3$, then $[\frac{1}{2}(p-1)^2]! - 1$ is equal to a multiple of p.

Dirichlet's challenge lay in providing a general rule for determining whether or not $\frac{1}{2}(p-1)!$ is or is not a quadratic residue of p. This and related issues would be discussed by numerous authors in the decades to follow.[19] Among publications of Dirichlet's associates, both Jacobi 1832 and Kronecker 1858 provided answers to the challenge. Kronecker did so, staying closest to Dirichlet's own work, by applying Dirichlet's later (1840) class number determinations in his 1857 note on complex multiplication of elliptic functions.[20]

[18]Lagrange 1771; *see* Lagrange *Oeuvres* 3:425–38.

[19]For further references *see* Dickson 1919–23 (2005) 1:275–76.

[20]Kronecker 1858; *see* Kronecker *Werke* 4:182.

Chapter 6
Berlin

Dirichlet had arrived in Berlin just in time to establish himself in his quarters before becoming involved in the preliminaries to an occasion that journalists of a later time might have described as "chic reform." The occasion was the seventh annual gathering of the Society of German Scientists and Physicians, the *Gesellschaft deutscher Naturforscher und Aerzte*.

Since the first meeting of the group in Leipzig in 1822, it had been the custom to meet in a different German city each year. Other previous host cities had been Halle, Würzburg, Frankfurt, Dresden, and Munich. Because a primary goal of the society was the promotion of personal acquaintance to facilitate exchange and furtherance of research results, the attendees were used to supplementing the scholarly lectures, heavily dominated by the founder Lorenz Oken and his *Naturphilosophen*, with suitable social activities. Yet none of the previous meetings had prepared them for the "happening" that was to take place in Berlin.

6.1 The 1828 Convention

The event was planned, staged, and executed by Alexander von Humboldt. He starred in his own production while sharing duties as local host with the zoologist Hinrich Lichtenstein. Special invitations had been extended to a number of distinguished foreign luminaries. Among these were the chemist J. J. Berzelius from Stockholm, the physicist Hans Ørsted from Copenhagen, and, because he unexpectedly had come to drop in on Humboldt, the mathematician Charles Babbage from England. The meeting was scheduled to run from 18 to 27 September. In fact, the foreign guests and particularly distinguished visitors such as Gauss had been invited to come a week early, to facilitate social and intellectual exchanges in a smaller setting.

On the evening of September 18, Humboldt gave a gala in the concert hall of the Schauspielhaus (the Royal Theater) for which he had the young Felix Mendelssohn Bartholdy compose a special cantata. The affair lasted from 6:00 to 9:00 p.m. and was attended by King Friedrich Wilhelm III and the Crown Prince, as well as Prince

U. C. Merzbach, *Dirichlet*, https://doi.org/10.1007/978-3-030-01073-7_6

Albrecht, the king's youngest son. Local gossip had it that there were 700 guests. The registered participants added up to 458, of whom 195 were from Berlin, and they included deputations from the schools as well as the university. Except for the opening feast and a few other social activities to which wives and daughters from abroad were admitted, ladies were excluded—a fact that drew wry comments from the musician Fanny Mendelssohn Bartholdy, who remarked on the scientists' all-male Mohammedan paradise, and her younger sister Rebecca, who alluded to the monks' lives led by the attendees.

This was only a foretaste of the social whirl in which the delegates were caught. The following day there was a trip to the Botanical Garden; another one on Sunday took them to the Kreuzberg, the popular hill that had assumed new significance with completion in 1826 of Schinkel's cast iron National Memorial to the victims of the Napoleonic Wars. Local reports had it that twenty-seven coaches sped the delegates to a tour of the Pfaueninsel (the peacock isle), the scenic Havel River spot in the Wannsee that, among other improvements to his father's romantic getaway, the king had stocked with a menagerie ranging from bears and monkeys to kangaroos and, of course, peacocks. The final evening, on September 27, was spent at the casino in Potsdam. In the eighteenth century, Frederick the Great had turned Potsdam from the garrison town of his father into a center of scenic and cultural attractiveness, with a focus on Sanssouci, the small palace built to his specifications as his own "get-away." There he had conversed with Voltaire, cultivated his gardens, and planned successful strategies to wage war against Maria Theresa's Austria and to induce Lagrange to Berlin.

There were six general sessions and seven special section meetings. It was the first year that this society held subject-oriented section meetings; but for the most part the section meetings were far less memorable than the social affairs. Not only did they lack an equivalent general appeal, but most lacked in scientific substance as well. There were exceptions, however. Heinrich Wilhelm Dove, a graduate of the university in Berlin, recently appointed to the faculty in Breslau, presented a solid meteorological study on winds, which undoubtedly did not hurt his chances for a subsequent transfer to the Berlin faculty; and Wilhelm Weber, a young extraordinary professor at Halle, presented a paper on acoustics that seemed to surpass the other, rather diverse, contributions in physics, and found favor with Gauss as well as Humboldt and Dirichlet. Three years later, aged twenty-seven, Weber joined Gauss on the Göttingen faculty as professor of physics.

There were drawbacks to the meeting in the Prussian capital. Accommodations were sparse, the streets were unpaved and alternately dusty or swampy, and gutters were filled with sewage and refuse. Unlike London, Berlin could not yet boast of gas lighting except in isolated houses. Rental carriages were few and difficult to obtain. But if the guests came from towns that presented a more pleasant environment, perhaps they were kept too busy and tired to reflect on such matters until they had returned home.

Humboldt achieved several results with this gathering. From the public relations point of view, the event was a spectacular success. Mutual exposure served to flatter king, court, and scientists. If dutiful members of the Ministry of Culture chose to interpret the king's patronage as a sign of support for science, it could only aid Humboldt's long-range plans. In the meantime, Humboldt's cleverly engineered bringing together of select younger and older scientists was to prove beneficial for several of his protégés. Similarly, his steering the more promising among the younger men toward one another was to have lasting effects in a number of cases. His carefully selected special guests balanced the dominance of the *Naturphilosophen* with a dose of younger men oriented toward the mathematical sciences. Finally, the meeting provided the final impetus he required for his well-developed plan for an international research project of his own. Dirichlet figured in and was affected by several parts of Humboldt's scheme.

6.2 Meeting Scientists

Dirichlet was one of the few mathematicians present at the meeting, and Humboldt used the opportunity to introduce his youngest protégé in the appropriate circles. This was facilitated by the unanticipated presence of Charles Babbage, who had missed Humboldt in Austria while both were traveling on the Continent earlier in the year, but had followed him to Berlin, leaving his calling card at Humboldt's home. Early the next morning, Humboldt had invited him to breakfast, where, as Babbage would write later in his autobiography:

> Humboldt himself expressed great pleasure that I should have visited Berlin to attend the great meeting of German philosophers, who in a few weeks were going to assemble in that capital. I assured him that I was quite unaware of the intended meeting, and had directed my steps to Berlin merely to enjoy the pleasure of his society. I soon perceived that this meeting of philosophers on a very large scale, supported by the King and by all the science of Germany, might itself have a powerful influence upon the future progress of human knowledge. Amongst my companions at the breakfast-table were Derichlet [sic] and Magnus. In the course of the morning Humboldt mentioned to me that his own duties required his attendance on the King every day at three o'clock, and having also in his hands the organization of the great meeting of philosophers, it would not be in his power to accompany me as much as he wished in seeing the various institutions in Berlin. He said that, under these circumstances, he had asked his two young friends, Derichlet [sic] and Magnus, to supply his place. During many weeks of my residence in Berlin, I felt the daily advantage of this thoughtful kindness of Humboldt. Accompanied by one or the other, and frequently by both, of my young friends, I saw everything to the best advantage, and derived an amount of information and instruction which under less favorable circumstances it would have been impossible to have obtained.[1]

Dirichlet, too, found that many of the men he met before and during the meeting would leave a lasting impression, and several had a noteworthy influence on his later career. In particular, these included those he saw at some of the meals Humboldt had pre-arranged. Dirichlet was a guest of Humboldt's at a small gathering along with Babbage, Gauss, the astronomer Encke, and Crelle, founder of the three-year-old *Journal für die reine und angewandte Mathematik*. He had breakfast with Humboldt

[1] Babbage 1864:200.

and Babbage and accompanied Humboldt in taking Babbage to a local tavern for the afternoon meal. Another time, he dined with Babbage, Encke, and Lichtenstein at Humboldt's; yet again, he reported for breakfast with Gauss, Radowitz, and General Müffling, who five years previously had hoped to gain Gauss for Berlin and to establish a polytechnic institute in Berlin. These meals provided Dirichlet with an opportunity to chat privately with Gauss and to exercise his proven social skills in the presence of Radowitz, the "key man" used by Humboldt for arranging Dirichlet's appointment to the Kriegsschule.

In addition, Humboldt most likely was aware of Fourier's hope that teaching of heat theory would be developed in coming years. When he suggested to Wilhelm Weber, who was planning to spend the following year in Berlin, that he would benefit from attending lectures by Dirichlet on heat theory, he paved the way for a relationship that was to deepen into a lifelong friendship between the two younger men and played a significant role in Dirichlet's eventual move to Göttingen.

6.3 Geomagnetism

Between 1800 and 1807, Humboldt had conducted a series of geomagnetic observations that resulted in records of declination figures and epochs of unusual phenomena. His long-held dream was the establishment of recording stations covering the globe that, once the results of such observations were collated, would provide substantive insight into the nature of the magnetic variation as well as other geomagnetic phenomena. To this end, he persuaded numerous rulers and scientific institutions to establish such stations. His most far-reaching success would come in 1829 after the Imperial Saint Petersburg Academy of Sciences agreed to the establishment of magnetic and meteorological stations throughout the various European and Asiatic climatic zones of Russia, along with establishment of a central Physical Observatory in Moscow. A. T. Kupffer, who had been trained in Dorpat, Berlin, Paris, and Göttingen, would lead the Moscow Physical Observatory. As Humboldt reported in the fourth volume of his *Kosmos*, where he summarized many of these international efforts, conversion of the frequently isolated sites into scientific observation points called attention to interesting histories, such as that involving the establishment of a station in Peking in a building that had housed Greek monks since the days of Peter the Great.

Meanwhile, Humboldt wished to establish a station in Berlin. One of the requirements for the observations was an iron-free structure sufficiently far removed from carriage traffic and other vibrations to provide accurate magnetic measurements. Humboldt decided that the ideal place for such a building would be the spacious, park-like garden attached to the rear of the house of town councilor Abraham Mendelssohn Bartholdy. He was an old family friend. As young men, Abraham and one of his brothers had shared the same tutor with the Humboldt brothers. Abraham's oldest brother, the banker Joseph Mendelssohn, had provided Humboldt with credit for his scientific journeys to old and New Spain after another bank had denied his request.

This was the beginning not only of an enduring friendship between these two men, but also of a constant source of support for Humboldt whose scientific enterprises were costly. Now Abraham, who also had become a successful banker, consented to have a "magnetic house" erected at the rear of his home at Leipzigerstraße 3.

The small structure was built to specifications by Schinkel, Berlin's leading architect of the era, who ensured that it would be iron-free, that, instead of iron, pink copper would be used, even for small components such as keys, locks, door jambs, and the like. Humboldt's little copper house in their back garden provided the Mendelssohn family with numerous stories reflecting a mixture of pride and amusement at finding themselves at the center of a serious scientific enterprise. Initially, curiosity and excitement at seeing Humboldt and his younger scientific associates in close proximity, outside the conventional social settings, predominated. Daughter Rebecca wrote with some anticipation of waiting for Humboldt to establish himself in their blue room, where he would spend a few nights while beginning the nightly observations in their garden; she teased that she would be able to report what he looked like in the morning and what he ate for breakfast. Since he had the good sense not to attempt the enterprise until his "400 friends" of the scientists' gathering had departed from Berlin, her patience was tested for several weeks.

The preparations involved committing several potential observers to the project. Aside from Humboldt himself and Encke, Berlin's chief astronomer, these included younger men such as Dove, the physicist Peter Riess, the sixteen-year-old Paul Mendelssohn Bartholdy, and Dirichlet. Regular observations only began in the year 1829. By this time, Dove, Riess, and Dirichlet had class schedules to observe, but young Paul took over most of the day-time observations; Humboldt and Encke took turns at night, Dirichlet assisting Humboldt. Dove and Riess guided the entire enterprise when Humboldt left town again later in the year.

In February 1829, writing to the family friend Klingemann, Rebecca confirmed that Humboldt was letting brother Paul handle a major part of the observations, although Humboldt had just spent several days and one night at their home himself. What the family found more startling than Humboldt's breakfast menu, earlier anticipated by Rebecca, was his traversing the garden fourteen times between 3:00 p.m. and 8:00 a.m. The observations ended in November 1830.

Among the numerous anecdotes connected with this unusual enterprise in the family household, best known because published by Sebastian Hensel in his family history,[2] is the story, related by Fanny, of a night when, after her wedding to Wilhelm Hensel, Fanny's sister-in-law, the poet Luise Hensel, was visiting.[3] Awakened by Luise shyly wandering through their bedroom in nightcap and sleeping jacket, the Hensels learned from her that there were thieves in the house. Wilhelm Hensel wrapped himself in a red bedcover and from the bedroom wall took a weapon, a souvenir from his glory days as officer in the Jäger Corps during the Wars of

[2]Hensel, S. 1908, 1:238.

[3]She would later stay for more extended "visits," the longest lasting for most of the period 1833–1837.

Liberation.[4] Guided by the lantern his sister carried for him, he made his way to the room where the suspect spotted the weapon and attempted to escape through the yard. Cornered in the gardener's house, the intruder turned out to be Encke, whose turn it was to carry out the observations for the night. Had the family known then that a century later Luise was to remain best known for her song "Müde bin ich, geh zur Ruh" ("I'm tired, going to rest"), it could only have added to the hilarity when they teased her about her watchful night.

6.4 Leipzigerstraße 3

The three-months-long period of preparation before the start of the magnetic observations had provided Humboldt with a convenient opportunity to introduce Dirichlet to the milieu of Leipzigerstraße 3.

Abraham Mendelssohn Bartholdy had bought the stately property in 1825. At the time Dirichlet first entered it, the central, multi-story portion of the house was occupied by Abraham and his wife Lea, as well as three of their four children, Fanny, Rebecca, and Paul. Felix, the second oldest, could claim it as home, but tended to be away much of the time, especially after a long visit to England in 1829.

The structure of the house formed a rectangle. Opposite the high front was a "garden house" with a large parlor that lent itself to the various sizeable social events which were a part of the family's weekly activities. Two one-story side wings connecting these two main sides contained smaller units used as apartments for tenants and guests. Humboldt's magnetic house occupied a small space outside this rectangle, in the park-like garden (an extension of the Tiergarten) that bounded the rear of the "garden house." All that constituted Leipzigerstraße 3.[5]

The family entertained regularly. Dirichlet had become used in the Foy household to being part of highly civilized social gatherings. There, the conversations and discussions had involved largely political and, to a lesser extent, scientific topics. Here, there was still a strong strain of interest in affairs of state, with the orientation largely echoing that of the Foy circle; but music, art, history, and philosophy were represented conspicuously, with a noticeable emphasis on rationalism and dominance of the Hegelian school. In both households, the conversations were frequently punctuated by lively, though courteous, exchanges of opinion. Leipzigerstraße 3 lent itself to welcoming groups of visitors who came not only to talk, eat, and drink, but also to attend musical and theatrical performances. Smaller groups, usually composed of family or the "inner circle" of friends, met in the large room that served as Lea's drawing room when there were no theatrical performances. Large gatherings attended concerts in the huge parlor located in the middle of the garden house, with sliding glass doors that opened toward the garden in the rear.[6]

---

[4]Fontane 1880 (1967, 2:860).

[5]Cullen 1982.

[6]Hensel, S. 1908, 1:167–68.

Among the early tenants was a group from the Danish delegation in Berlin; this included the musically talented Karl Klingemann, a native of Hanover, who, after his move to London, became a regular correspondent of the younger generation and a special friend of Felix's. Another one-time tenant, who was a close friend of Felix's until the latter's death, was Eduard Devrient who had lived in one of the apartments with his wife, Therese, until they needed a place sufficiently large and airy for a growing family. Eduard Devrient worked with Felix on the famous 1829 performance of Bach's St. Matthew Passion, in which he also sang the role of Jesus. In later years, after his voice had failed, he turned primarily to managing theater productions. Both he and his wife, a well-known actress, remained friends of the Mendelssohn Bartholdy siblings over the years. The widely respected and influential philologist August Boeckh, whose company Dirichlet is said to have enjoyed especially, lived in Leipzigerstraße 3 from 1840 to 1846; and there were numerous other occupants, largely single men and women, who rented the smaller units of the house for varying periods of time. By 1846, most of these were replaced by the sizeable entourage of Count Pourtales, Master of Ceremonies at the Court, who had moved in during 1843 and remained, with a growing number of attendants, until the house was sold in 1850.

Dirichlet joined the social circle at the Mendelssohn Bartholdys soon after the scientific meetings had adjourned on September 27, 1828. We learn from Fanny Mendelssohn Bartholdy's diary that on the evening of October 1, the day he assumed his duties at the Kriegsschule, Dirichlet attended a party at the house in the company of Gauss and Humboldt, among others. Christmas, which traditionally gave rise to a large, festive occasion, presented an opportunity for the younger people to come to the fore. Rebecca reported to Klingemann about the youngsters' growing up. For example, her brother Paul, decked out in suit and tie, "crawls up on long Dirichlet" to copy his flirting so he can apply his new-found knowledge. She added that she herself, as the oldest of the younger generation (she was eighteen, just one year older than Paul), curbs some of these youthful excesses by still wearing her hair in a plain part and living significantly solidly. She noted, however, that since Klingemann left she has grown; and what is she to say to the fact that everyone around her is getting married?

"Everyone" may have been an exaggeration. But romance was in the air in Leipzigerstraße 3.

6.5 Fanny and Wilhelm Hensel

Rebecca's sister Fanny had met the painter and portraitist Wilhelm Hensel in 1821 in conjunction with the "Lalla Rukh" festivities, a spectacle produced to coincide with the visit of members of the Russian court, including the future Emperor Nicolas I. Hensel had been instrumental in the success, particularly of the "Living Pictures" portion of the exhibition, that had been a highlight of the preceding Berlin art scene and involved members of both the Russian and the Prussian courts.[7] In recognition of his artistry and his organizational ability, Hensel had been granted a Royal stipend for studies in Italy. Now, in October 1828, Wilhelm Hensel had returned to Berlin.

By the end of 1823, their contemporaries had claimed that Fanny and Wilhelm were secretly "engaged." At that point Wilhelm had left for Italy, financed by the Royal stipend which committed him to stay for five years. Before leaving, in 1822 he had been invited for Christmas to the Mendelssohn Bartholdys, but the parents felt he needed to prove himself in Italy before the relationship became too close. Direct correspondence between the two was not permitted. Instead, the letters he wrote had to be sent to Lea Mendelssohn Bartholdy, Fanny's mother, who passed on to Fanny what she deemed proper. For example, when Wilhelm enclosed a small self-portrait with some lines of poetry attached, Lea returned it with a kind note explaining that it was too intimate a present but that he should resend the letter and the book that he had enclosed with someone else's poems. Even before Hensel's return, it was clear that he had not been diverted by Italian attractions and that his and Fanny's feelings toward one another remained unchanged.

The problem the Mendelssohn Bartholdys had was not uncommon for families in their situation, even though their particular situation may have been considered unusual. It involved money, religious affiliation, and the cultivated social class that they represented. The unusual part lay in the fact that Abraham's father was the renowned Jewish philosopher Moses Mendelssohn. The family was proud of that heritage even though its meaning may have been subject to varying interpretations.

At the height of the Napoleonic Wars in March 1812 the Prussian government, in an effort at unifying its population, had lifted many existing restrictions by granting large classes of Jews citizenship. But with the fall of Napoleon, a series of anti-Semitic actions in parts of Germany, serious acts of violence in Italy identified with Catholic extremists, and reinforced prohibitions in Prussia that denied adherents of the Jewish faith entry to governmental service, including professorships at universities, the picture had changed once again. Abraham and his wife had had their four children baptized in a quiet Protestant ceremony in 1816, before they themselves converted a few years later, in 1822. Following the example of Lea's brother, the diplomat and art supporter, Jakob Salomon, who had taken the name Bartholdy when he himself became a Protestant, Abraham added the name Bartholdy as well. As the result, his family came to be known by the surname Mendelssohn Bartholdy, in distinction to his father, Moses Mendelssohn and to the other two sons of Moses, Joseph and Nathan.

The projected wedding of Wilhelm and Fanny complicated the financial and religious aspects even further. Abraham and Lea were sufficiently wealthy so that their daughters were considered desirable even without other known virtues. When Hensel had left for Italy, he had little money to his name. The Hensel siblings were children of a Protestant minister. Wilhelm's sister Luise had converted to Catholicism. She was a fervent convert, close to the well-known author Clemens Brentano, who had

[7] A list of the aristocratic participants in the Living Pictures portion is provided in Fontane 1880 (1967, 2:858–59).

returned to Catholicism in 1817. There was a twofold concern that Wilhelm, too, might change religion and that he might not have a regular income. This turned out not to be the case, however, and after renewed discussions and a certain reassurance obtained by the fact that Wilhelm Hensel had not only returned from Italy with two large paintings that pleased the king but continued to receive royal commissions, the official engagement was announced in January 1829. Hensel was appointed Royal Court Portraitist and elected to the Academy of Arts in February. The wedding followed in October.

6.6 Kriegsschule

Dirichlet meanwhile had begun to make his mark in the Kriegsschule. This "war school," designed to train a select group of candidates as officers for the Prussian army, in 1810 had been recast from an earlier institution. This meant it not only benefitted from the general educational reforms advanced by Wilhelm von Humboldt, at that time Head of Ecclesiastical Affairs and Public Education within the Ministry of the Interior, but, under the direct leadership of the distinguished General Scharnhorst, was intended to place major emphasis on the study of mathematics.

The curriculum was based on a three-year course of study. The months October through June were dedicated to teaching. For July through September, the emphasis lay on weaponry and its practical ramifications. This included acquaintance with each of the branches of the military so that infantrists during this time would become knowledgeable with principles of artillery and cavalry; therefore, horsemanship was an important factor. Learning was not to be restricted to listening to lectures but was to contain an applicable (*applikatorisches*) element. This did not refer to applications of the subject but to the student's applying the content as one would in a discussion group or seminar.

The mathematical curriculum called for algebra up to equations of the fourth degree, elementary theory of series, beginnings of stereometry and elements of descriptive geometry in the first year. The second year included trigonometry, conics, treated analytically and synthetically, and the introductory elements of three-dimensional analytical geometry; finally, the third year was to include mechanics of solids and fluids as well as mathematical geography. Another teacher also provided some elementary geodesy. Infinitesimal calculus was prohibited as it was considered "higher math," not suitable for the intended purpose of the school.

Among Dirichlet's fellow faculty members at the Kriegsschule, Friedrich Theodor Poselger, a largely self-taught mathematician, had been awarded an honorary doctorate in 1822; he was appointed codirector of the Kriegsschule and as such became a member of the Akademie. Then there was Martin Ohm, the brother of Dirichlet's former teacher in Cologne. In fact, G. S. Ohm had become dissatisfied with his prior teaching duties and now served as assistant to his brother at the Kriegsschule. Martin Ohm was the author of a logically structured pedagogical system of teaching mathematics. To successfully apply his system he tried repeatedly, but unsuccessfully, to obtain permission to retain the same students from the first through the third year. There is no known evidence that Dirichlet established any special personal relationship with either brother, however.

The *Applikation* was divided evenly between the regular instructor and the *Repetent*, each taking half of the load. Dirichlet was not hired as an independent teacher but only as a *Repetent* (teaching assistant) to Poselger, the instructor for the third-year students. There was a potential problem with this assignment. It was not just his lack of teaching experience, but the fact that he was a civilian.

The students had been picked in a class-conscious environment in which it was understood that it would be beneath the dignity of a future officer to have to be quizzed or to give direct responses to questions or orders from an instructor not of military rank. Under these circumstances, it was not only difficult for civilian teachers to maintain discipline, but on top of the poor mathematical preparation that most of the students brought when they entered the Kriegsschule many had little to show when they left it. In hindsight, it is clear that, since he could not ask the students for responses, a civilian *Repetent* had particularly little chance of success in the *Applikation* sessions.

In the view of his students, Dirichlet possessed some unexpected redeeming qualities, however. The young officer candidates knew enough of his background to have learned that he had spent time with some of the most distinguished French military officers from the previous wars. In addition, he spoke fluent French and had the social graces they had been raised to expect from equals. These were factors they admired and that gave him a certain *éclat* which none of the other civilian teachers could draw on. Given these conditions, even his age, which had been considered a questionable factor as he was barely older than his students, worked to his advantage.

6.7 Steps to a University Appointment

In 1828, when the philosophical faculty in Berlin was first informed by the ministry of Dirichlet's having permission to hold lectures in their midst while on leave from Breslau, not all of its members were pleased. Since he had neither a *venia legendi* nor a professorial appointment issued by them, they felt his lectures could not be recognized as being part of theirs and that the appointment went against the statutes. Everyone, including Dirichlet, agreed that they must observe the statutes.

After discussions in which Humboldt had some part, and a question whether to make use of the laudatory letter Bessel had written about the biquadratic residues memoir, it was agreed that although the ministry had the right to give Dirichlet a dispensation from the habilitation requirements, it would let the faculty determine what those should be. Accordingly, in October 1828 Dirichlet wrote a letter to the faculty requesting his appointment. The next day, the dean circulated a memo including the Bessel letter and suggesting that in view of Dirichlet's recognized achievements his appointment be confirmed with "designate" status and that he was to provide a Latin program and to present a Latin discourse in the Auditorium Maximus. Members of the faculty signed their agreement on the circular and Dirichlet lectured in Berlin as privatdozent while on leave from his position as Extraordinary professor in Breslau.

In May 1829, Dirichlet wrote to Altenstein requesting that he either have his leave from Breslau extended or be given a transfer to the university in Berlin. His leave was renewed in June. When this renewal was coming to an end in 1830, the officer in charge of the Kriegsschule appointments requested that Dirichlet be given a permanent appointment at the Friedrich-Wilhelm University of Berlin in order to enable him to continue providing his services to the Kriegsschule. It had been determined that he was an excellent teacher, had proved truly useful to the school, and had managed to circumvent the hazards known to greet a new civilian *Repetent* in the *Applikation*. These, and his lack of teaching experience, had been the reason his initial appointment had been considered probationary.

The request to transfer Dirichlet to Berlin on a permanent basis was given strong support by Crelle in several communications sent to the ministry. In a letter to Altenstein of June 1, 1830, he based his position on multiple grounds. He thought Dirichlet was one of those rare talents who know not only to be useful to the youth they teach but to advance science. His publications, which, Crelle pointed out, were well known to him since most of them had been appearing in his *Journal*, so obviously

> bear the mark of the most penetrating sagacity, and even gift of genius, and proceed with such firm steps on the path of the masters into the Unknown that they deserve to be set alongside the most superior recent productions in the wide field of the mathematical sciences. Mister Dirichlet usually chooses one of the most difficult parts of mathematics as subject of his research, namely indefinite analysis and theory of numbers. As is well known, this subject least lends itself to be treated by determined algorithms, and the strength of *method*, emerging from deep, penetrating ideas, here is less helpful than in other parts of mathematics. So the subject is one of those that simultaneously require the most intensity of the power of thought and the freest mobility in the capacity for deduction and abstraction. The excellent successes which Mr. Dirichlet has already had in this part of mathematics (for his memoirs on biquadratic residues and the numerical equations of the fifth degree contain recognized significant advances of these theories) are additional proofs of his extraordinary talent for which in my mathematical exchanges with him I have seen manifold surprising samples, combined with the sign of deeply comprehended acquaintance with mathematics in its various branches, especially extraordinary for one of his years.[8]

Crelle continued by remarking that such a decided talent would be of use to the Royal Akademie, which would be curtailed if Dirichlet were to be assigned to another town where he would not have the scientific exchanges possible in Berlin. Additionally, this would deprive him of the animation and courage for further efforts. Crelle concluded by pointing out that in teaching, too, it would be important to keep Dirichlet in the capital rather than in another university of the monarchy as significant progress emanates from there, so that it would be beneficial for the state as well as for science to firm up a position for Dirichlet in this location.[9]

On October 3, the Cultural Ministry responded that the transfer could not yet take place because of lack of funds. Dirichlet's leave from Breslau was extended to the end of June 1831, however.

---

[8]Biermann 1959:45–46.

[9]Biermann 1959:46.

6.8 The University

The level of mathematics at the university in Berlin when Dirichlet began to teach there had been mediocre at best.[10] When Dirichlet first lectured at the university in October of 1828, E. H. Dirksen was the only full professor for mathematics, a position he was to retain until his death in 1850. Jabbo Oltmanns held a professorship for applied mathematics until he died in 1833. Dirksen had studied in Göttingen, received his habilitation in Berlin in 1820, and was appointed extraordinary professor that same year. Oltmanns had worked for the astronomer Bode at the Berlin Observatory at the time Alexander von Humboldt returned from his travels in the New World. Impressed by Oltmanns's calculating talents, Humboldt had him do the computations for the astronomical portion of the first publication dealing with his American discoveries. This led to Oltmanns being appointed to both the Akademie and the university in Berlin. He had left to do miscellaneous geographic, geodetic, and astronomical calculations but in the 1820s asked Humboldt to help him return to Berlin. The result was the creation of a professorship in applied mathematics which died with him.

It seems clear that, aside from Oltmanns turning into a tippler, neither man's mathematical inadequacy had so much to do with personal weakness on his part as did a misunderstanding by Altenstein, who wished to build up a competitive mathematics department. He was misled by the aura of Gauss and Göttingen in the case of Dirksen, and by Humboldt's high recommendations in the case of Oltmanns. It was not recognized that both men's training and experience had been primarily oriented toward astronomy and numerical competence rather than to the abstract discipline making such strides in France.

There had also been a full professor for astronomy, mathematical geography, and chronology since 1821. That was Ludwig Ideler, who had been a member of the Akademie since 1813, had been an extraordinary professor at the university since 1817, but had given mathematical lectures for twenty years, from 1813 to 1833. Ideler has been considered the most scholarly of this group of men; however, his chief interest lay in chronology and in the history of mathematics and related sciences, especially for ancient and early medieval times.

E. G. Fischer had an extraordinary professorship in physics and mathematics from 1810 until his death in 1831, but did not teach mathematics. Most active in lecturing on elementary mathematical subjects, primarily intended for teachers of mathematics, was J. P. Gruson who had held the same position as extraordinary professor since 1816 and would continue for more than another twenty years. Martin Ohm had been the other extraordinary professor since 1824.

It was only in 1830 that a privatdozent joined this group who would help Dirichlet raise the quality of teaching and research at the university. That was Ferdinand Minding of whom we shall take more notice in Chap. 8.

---

[10]Biermann 1988:21–31.

6.9 Rebecca Mendelssohn Bartholdy

When it had become clear that Fanny and Wilhelm Hensel would marry, the attention of the young men in the social circle of Leipzigerstraße 3 focused more strongly on Rebecca. She was younger than her sister and said not to have the musical talent of her two older siblings, Felix and Fanny. But she was quick, low-keyed, well-read, had a faculty for objective criticism, often softened by her spontaneous witticisms, and was attractive. When she wrote to Klingemann that she had grown, she was not referring merely to her height. Even the ironic Heinrich Heine, not always given to complimenting members of the female sex (is not Lorelei's danger for the fisherman her mysterious beauty?), felt comfortable in Rebecca's presence on his occasional visits to the house. In an exuberant greeting to Berlin he wrote in 1829, after sending greetings to the Victoria statue on the Brandenburg Gate, "respectfully" to the "Stadträthin" Lea, "somewhat less respectfully" to Fanny's beautiful eyes, "among the most beautiful eyes" he had ever seen, then: "The round Rebecca, do greet that round person, the dear child, so pretty, so good, every pound an angel."[11]

As several potential suitors dropped out, it seemed clear that the chief remaining contenders for Rebecca's hand were Eduard Gans and Dirichlet. Had the elderly aunts and the young artists and academics who were part of the social circle been placing wagers, the odds would have favored Gans.

Born in Berlin and having studied in Berlin, Göttingen, and Heidelberg, where he became a follower of Hegel, Gans in 1819 had been one of the founders of the Society for Jewish Culture and Learning (*Wissenschaft*); but in 1824 he was baptized, and in 1825 had become an extraordinary professor of law at the university in Berlin. He had just recently been appointed ordinary professor in the law school. He was voluble, articulate, never failing to impress, a lecturer who drew large crowds, and was clearly considered the intended successor to Hegel.

Dirichlet was liked but had no firm appointment in Berlin. His lectures drew fewer than a dozen listeners. He did not display his knowledge. He came from modest financial circumstances. And he was a Catholic.

There were two women who disregarded such characterizations. They were both artists, one an actress, the other a musician. Both were keen observers of their surroundings and of human nature. They were Therese Devrient and Fanny Mendelssohn Bartholdy. Years later, Therese Devrient would recall having watched Rebecca and Dirichlet as the latter accompanied Humboldt to one of the nightly observations in Humboldt's magnetic house. She did not know that the nightly observations had to do with observing the effect of magnetism on a needle and were not devoted to star-watching; but she recognized the human element:

> Late in the evening, ... the old master came with his disciples into the garden house and spent a little time with us in most amiable small talk. As soon as he gave the signal, young Dr. Dirichlet, Humboldt's main support with these observations, with Rebecca's help, lit a small, dark lantern. The gentlemen excused themselves, and we saw the small, wandering light in the dark garden until it disappeared behind the bushes.

[11] Droysen 1929, 1:9; cited in Weisweiler, ed. 1985:64.

> When I once said to Rebecca that the young astronomer had again made totally new discoveries in the stars of the heaven he found in Rebecca's eyes, she pretended to be quite angry but did not dislike hearing it, and the subsequent engagement announcement of Rebecca Mendelssohn and Dr. Lejeune Dirichlet showed that I had prophesied correctly.[12]

Fanny, Rebecca's older sister and confidante, had observed the contenders with interest, remarked on Dirichlet and Gans at times tussling physically like schoolboys, and noticed the frequency with which the newcomer to the circle managed to attend family affairs. Already in January 1829, Fanny made a note in her diary that among many visitors on New Year's Day was Dirichlet "who arranged it so that mother invited him for dinner. He was very nice and amusing."[13]

Fanny had the benefit not only of knowing Rebecca well, of noticing the interactions of Dirichlet and Rebecca, but also of Rebecca confiding her feelings and related events to her. It is from Fanny's diary that we learn Dirichlet proposed to Rebecca at one of her great-aunt Levy's teas shortly before Christmas 1830, but that Rebecca had turned him down. The next day, she shared with Fanny her regret and concern that she had been overly hasty. Most likely, a major factor governing a following, lengthy period of inner anguish for Rebecca was the fear that Dirichlet might not understand the reason for her refusal. They did not see one another for several weeks until they met again at a ball given by one of the Mendelssohn cousins. At that point, there was no question in either one's mind about the other's feelings.[14]

6.10 Family Concerns

In hindsight, it seems clear that it was a second fear that had governed Rebecca's stress, just as it would later keep her brother Paul from revealing his secret engagement, and Felix being rather circumspect in having his family meet that of his bride. This was the anticipation of their parents' reaction. The parents liked Dirichlet, just as they had favored Hensel, on a personal basis. Their concern, always present in this closely knit family, was for the welfare of their children.

On February 13, 1831, Rebecca informed her parents of her feelings and intentions and immediately told her sister she had done so. The sisters thought all would be well because there had been no reaction so strong as the one they had anticipated. By evening, both discovered the truth.

Rebecca's mother spoke to Rebecca and, as Fanny put it, the storm broke. Lea painted the consequences for all concerned of Rebecca's marriage to Dirichlet in the darkest possible colors. Abraham did what, according to Fanny, was his habit. At first appearing to agree, he subsequently "lost courage" and wished to take no action. Wilhelm Hensel spoke to Dirichlet, then to Abraham, confirming Dirichlet's

---

[12]Devrient, T. 1905:350.

[13]Hensel, F. 2002:1.

[14]Hensel, F. 2002:31.

good intentions, but to no avail. Now Abraham even began to show displeasure with Rebecca, which made her unhappier, and the matter dragged on for months.

In retrospect, one gains the impression that Lea, despite initial remonstrations, by a natural instinct for order and management, as well as empathy for the human condition, found it easier to see the way to a rational conclusion in these matters than did her husband who had to fight internal conflicts concerning heritage, adaptation, and his own identity. But above all, he wanted his children to be safe. He never forgot that Felix during the previous decade had been harassed by anti-Semitic hooligans, once in company of his sister Fanny whom young Felix had defended physically. It is undoubtedly the reason why Abraham wrote Felix, on whom he doted, a stern letter to England when he found that Felix had dropped the "Bartholdy" in some of his concert programs there.

6.11 New Security

By mid-1831, Dirichlet's position in Berlin seemed somewhat more certain. In July, he had received his transfer as Extraordinary Professor Designate to the university in Berlin with the same salary as before (400 Thaler per year), but with renewal of his position at the Kriegsschule, now as independent teacher, received an additional salary of 600 Thaler. In this respect, he had reached a position similar to that of many of his fellow faculty members who often took on teaching jobs at secondary schools to supplement the meager salary of the university. He had been admitted to membership in the Hegelian *Societät für wissenschaftliche Kritik* in 1827 while still in Breslau, and from 1831 to 1843 his name appeared on the board of editors of its annual publication. He could now follow a balanced routine of teaching, research, social activities, and travel; in fact, after the summer term of 1831 Dirichlet took the first of many vacations to Switzerland.

In the intervening time, Rebecca's parents presumably had become aware of the fact that Dirichlet's religious orientation, informed by his mother's close bonds to Reformed Protestantism, had more in common with their observances than with Luise Hensel's devout Catholicism. What seems clear is that the basic beliefs of both sisters, their parents, and Dirichlet were surprisingly similar. Given the choice of form versus substance, they chose substance. It was what caused the sisters to be critical of over-dressed women, romanticism, and flashy music; it was what caused Dirichlet to turn from complicated techniques of mathematical manipulation to concepts that revealed "an inner substance." This, and a shared aversion to absolutism, was what had prompted Lea Mendelssohn Bartholdy to read the daily papers to her daughters and to Wilhelm Hensel during the days of the July Revolution in France in 1830, and what caused baby Sebastian's pillow to display a knitted French Revolutionary tricolor in 1832, to the chagrin of his nationalistic father, Wilhelm.

The widely read Abraham realized that his family's basic beliefs were still true to the message of reason, learning, and tolerance they had been left by his own father, the noted philosopher Moses Mendelssohn. But, though guided by the concern for his children, and aware that his household had met the expectations of *res publica eruditorum*, he could not escape an inner conflict—often coming to the surface with statements opposed to "Jewishness"—when it came to interpreting the tradition of his people, that small band at the Sinai of whom Moses Mendelssohn had written in *Jerusalem*.[15]

On November 5, 1831, Rebecca and Dirichlet announced their engagement. On February 13, 1832, Dirichlet was elected to membership in the Royal Academy of Sciences in Berlin, the Akademie. On March 14, he requested consent from the Cultural Ministry to marry the daughter of town councilor Abraham Mendelssohn Bartholdy. On May 22, 1832, they were married.

[15]Mendelssohn. M. 1983 and Baeck 1958:23.

Chapter 7
Publications: 1829–1830

In the years 1829 and 1830, Crelle published three more memoirs written by Dirichlet. Appearing in volumes 4 and 5 of his *Journal für die reine und angewandte Mathematik*, they served two related purposes: They were most likely prompted by an attempt to clarify and make more rigorous the concepts to which Dirichlet had had to introduce the students in his lectures. At the same time, they brought to the attention of a wider audience Fourier's work on analysis and heat theory, as well as the competing approach that Poisson utilized.

7.1 Definite Integrals

The first of the three memoirs, a note on definite integrals (**1829a**), is an extension of the work of Poisson and Poisson's teachers Lagrange and Laplace. As would be his custom when presenting a concept or technique presumed to be unfamiliar to his readers, Dirichlet introduced his subject by recalling preceding work that had been done. He observed that definite integrals were to be found among those that Poisson, Cauchy, and others had recently evaluated; but he justified his contribution by the extreme simplicity of his procedure, which he felt mathematicians ("géomètres") would appreciate. An extension of the method both Laplace and Poisson had used, it was based on the property double integrals have of being independent of the order of integration. He added that, while the men he had named were responsible for causing this method to be so popular because of the ingenious applications they had made of it, he felt justice demands that Euler be credited with the first idea of using this property of double integrals for evaluating simple definite integrals.[1]

[1]Dirichlet here provided the reference to Euler's paper in the *Novi Comment. acad Petrop., tom. XVI.*

U. C. Merzbach, *Dirichlet*, https://doi.org/10.1007/978-3-030-01073-7_7

By a sequence of intriguing but straightforward manipulations, Dirichlet arrived at the following formula

$$\int_{-\infty}^{+\infty} \frac{e^{-cx\sqrt{-1}}}{b^2+x^2} \cdot \frac{1}{(k+x\sqrt{-1})^p} \cdot \frac{1}{(k'+x\sqrt{-1})^{p'}} \cdot \frac{1}{(k''+x\sqrt{-1})^{p''}} \cdots dx$$
$$= \frac{\pi e^{-bc}}{b} \cdot \frac{1}{(b+k)^p} \cdot \frac{1}{(b+k')^{p'}} \cdot \frac{1}{(b+k'')^{p''}} \cdots$$

where c as well as $b, k, p, b', k', p', b'', k'', p'', \cdots$, or at least their real parts, are positive as well.

Giving specific values to the constants in his formula allowed Dirichlet to obtain in three pages results that Poisson had discussed in a lengthy memoir on definite integrals and complex numbers published in the *Journal* of the Ecole Polytechnique in 1820; to conclude this note, Dirichlet provided an example of "the numerous consequences" his formula could provide.

Poisson apparently appreciated Dirichlet's achievement. It may have contributed, however, to Sturm's feeling free, in May 1837, to comment in a letter to Dirichlet that Poisson writes "large volumes that one hardly reads and that sometimes leave the real difficulties aside."[2] By then, Poisson indeed had issued book-length publications on capillary action, heat theory, and probability, in addition to the substantial number of his previous memoirs.

7.2 Convergence of Fourier Series

The note on definite integrals was followed by the memoir **1829b**, also appearing in Crelle's fourth volume, that would attract particular attention over the years. Dated January 1829, it was titled "On the convergence of trigonometric series which serve to represent an arbitrary function between given limits." It is a rigorous formulation of the convergence of such trigonometric series. Dirichlet began with a forceful statement:

> The series of sines and of cosines, by means of which one can represent an arbitrary function in a given interval, among other remarkable properties enjoy that of being convergent. This property had not escaped the illustrious geometer who opened a new course to applications of analysis by introducing this way of expressing such arbitrary functions. It is found stated in the memoir that contains his first researches on heat.[3]

---

[2]Quoted in Grattan-Guinness 1990, 2:1283, based on the letter in Berlin. Staatsbibliothek. Dirichlet Nachlass.

[3]This refers to Fourier's work of 1807. [P1808] (by Poisson) provided the first published summary; a more complete account by Fourier followed in 1811 but was only published fully in the 1820s. It was one of Fourier's works that had been criticized by Poisson among others. For fuller details of these criticisms see Grattan-Guinness 1970, Chapter 5.

As the only previous attempt at a convergence proof for these series, Dirichlet called attention to Cauchy's work published in volume 6 of the *Mémoires* of the Paris Académie for 1823 (which had appeared in 1827), noting that Cauchy himself acknowledged that his proof fails for certain functions whose convergence is unquestionable nevertheless. Dirichlet went further in telling us that Cauchy's proof will not even hold for the cases Cauchy claimed.[4] In outlining the areas where Cauchy failed, Dirichlet first called attention to a number of instances where Cauchy replaced the variable x, in the function $\phi(x)$ to be expanded, by one of the form $u + v\sqrt{-1}$, without noting that arguments valid for the real case cannot necessarily be carried over for complex numbers, as Cauchy himself had pointed out in several places.

Dirichlet's strongest case was made with a counterexample. Dirichlet noted that, while considering imaginary quantities, Cauchy was led to a result on decreasing the terms of a series which is, however, far removed from showing that this sequence converges. Considering that the interval in question ranges from 0 to 2π, Dirichlet cited Cauchy's result as follows: "The ratio of the term of rank n to the quantity $A\frac{\sin nx}{n}$ (where A designates a determined constant, dependent on the extreme value of the function) differs from unity taken positively from a quantity which diminishes indefinitely proportionately as n becomes larger." Dirichlet now pointed out that from this result and from the one that the series which has $A\frac{\sin nx}{n}$ as general term is convergent, Cauchy concluded that the general trigonometric series is convergent as well.

> But this conclusion is not permitted, for it is easy to assure ourselves that of two series (at least when, as happens here, the terms do not all have the same sign) one can be convergent, the other divergent, although the ratio of two terms of the same rank differs as little as desired from unity taken positively when the terms are of very advanced rank.[5]

Dirichlet proceeded with the example of considering two series, one having as general term $\frac{(-1)^n}{\sqrt{n}}$, and the other, $\frac{(-1)^n}{\sqrt{n}}(1 + \frac{(-1)^n}{\sqrt{n}})$. The first of these series is convergent; the second, on the contrary, is divergent. This can be shown by subtracting it from the first which results in the divergent series

$$-1 - \frac{1}{2} - \frac{1}{3} - \frac{1}{4} - \frac{1}{5} - etc.$$

Yet the ratio of two corresponding terms, which is $1 \pm \frac{1}{\sqrt{n}}$ converges toward unity in direct proportion as n increases.

Dirichlet proposed next to enter into the matter at hand by beginning with an examination of the simplest cases, to which all the others can be reduced.

In a carefully sequenced number of steps, patterned after the more general argument Fourier had presented in the *Analytical Theory of Heat*—Dirichlet explicitly refers to the portion following Fourier's Section 232—Dirichlet developed an argument which resulted in the proof of a theorem more restricted than his opening statement may lead the casual reader to believe.

---

[4] *Werke* 1:119.

[5] *Werke* 1:120.

Dirichlet based his argument on consideration of the function $\phi(x)$, assumed to be bounded and single-valued ("finite and determined") for each value of x between $-\pi$ and π. Following Fourier, he expanded this function into a trigonometric series

$$\frac{1}{2\pi}\int \phi(a)da + \frac{1}{\pi}\Big(\cos x\int \phi(a)\cos a\, da + \cos 2x\int \phi(a)\cos 2a\, da + \dots$$
$$+ \sin x\int \phi(a)\sin a\, da + \sin 2x\int \phi(a)\sin 2a\, da + \dots\Big)$$

By summing the terms, taking the limit as their number increases, and performing changes of variables, he demonstrated the convergence of his series, and also established that for each value of x between $-\pi$ and π it is equal to

$$\frac{1}{2}[\phi(x+\epsilon) + \phi(x-\epsilon)]$$

while at the points $x = \pi$ and $x = -\pi$, it is equal to

$$\frac{1}{2}[\phi(\pi-\epsilon) + \phi(-\pi+\epsilon)].$$

What he concluded was that

> the preceding considerations prove rigorously that if the function $\phi(x)$, which is assumed to be bounded and single-valued ["finite and determined"] has only a finite number of discontinuities ["solutions of continuity"] between the limits $-\pi$ and π, and if besides it has only a determined number of maxima and minima between these same limits, then the series
>
> $$\frac{1}{2\pi}\int \phi(a)da + \frac{1}{\pi}\Big[\cos x\int \phi(a)\cos a\, da + \cos 2x\int \phi(a)\cos 2a\, da + \dots$$
> $$\sin x\int \phi(a)\sin a\, da + \sin 2x\int \phi(a)\sin 2a\, da + \dots\Big]$$
>
> whose coefficients are definite integrals depending on the function $\phi(x)$, is convergent and has a value generally expressed by $\frac{1}{2}[\phi(x+\epsilon) + \phi(x-\epsilon)]$, where ϵ denotes an infinitely small number.[6]

Dirichlet's statement of convergence was accepted as valid, and the memoir would remain a standard reference in discussions concerning Fourier's work.

Perhaps carried away by Gauss's similar comments when he predicted future, expanded results (which Gauss tended to achieve), Dirichlet did not stop with the preceding, carefully formulated, statement but concluded with a conjecture concerning cases where the assumptions concerning the number of discontinuities and of maximum and minimum values no longer hold: "These exceptions can be reduced to those which we have just considered."

---

[6] *Werke* 1:131.

This powerful statement tended to overshadow the immediately following qualifications, at times overlooked by subsequent readers. For his expanded convergence conditions to hold, Dirichlet put certain restrictions on the function $\phi(x)$. First he stated that

> $\phi(x)$ must be such that if one denotes by a and b any two quantities between $-\pi$ and π one can always place other quantities r and s between a and b that are sufficiently near to one another so that the function remains continuous in the interval from r to s.

He based the necessity of this restriction on the terms of the series being definite integrals. He pointed out that this makes no sense if, for example, the function $\phi(x)$ is defined so as to have a fixed value c for all rational values of x, another fixed value d for all irrational values. He concluded, however, that, as long as the function does not become infinite, his restricted conditions, which he indicated would be spelled out in greater detail in another note dealing with additional remarkable properties of the series, would be sufficient.

Cauchy had expressed doubt about the validity of his own convergence proof. Within two years after Cauchy's publication, Dirichlet had provided the counterexample found in this memoir. Dirichlet, almost never given to publishing conjectures, here expressed with certainty his own supposition concerning cases where his considerations concerning discontinuities and maximum and minimum values can be disregarded. As a result, the weakening of his original sufficiency conditions for convergence of Fourier series became the subject of numerous nineteenth-century efforts. In later years, he himself continued to provide examples where his restrictions could be lightened. Notable among these are the addition to **1837a** and an argument given in a letter to Gauss in 1852, made as the result of his interpretation of a remark Gauss had made in a preceding exchange.[7]

Perhaps the clearest nineteenth-century analysis of Dirichlet's conditions is that found in [Lipschitz 1864]. This and similar efforts were soon overshadowed by a critique Karl Weierstrass published in 1870, however, followed by Du Bois Reymond's example of a continuous function whose Fourier series has points of discontinuity. An increasing number of examples and counterexamples concerning Dirichlet's conditions, at times confusing his proof and his conjecture, followed. Many of these resulted in modifications and refinements of the concepts of continuity and of convergence. By 1904, Fejér, introducing his own theorem leading to later summability notions, could remark that "the question of sufficient conditions for the convergence of the Fourier series...has been treated almost too extensively in the literature."[8]

---

[7] *Werke* 2:356–57.

[8] Fejér 1904, reproduced in Birkhoff 1973:150–56. Nevertheless, the topic had led to significant contemplations of these issues, among which note Georg Cantor's early interest in trigonometric series, encouraged by Eduard Heine; see Dauben 1973, esp. Chapters 1 and 2.

7.3 A Problem from Heat Theory

Dirichlet's memoir **1830** is his one publication dealing exclusively with heat theory. Dirichlet in **1829a** did not mention heat theory at all and there is only a passing reference in **1829b**, in the context of Fourier having used the convergence of trigonometric series in his memoirs and book on heat theory. In the memoir **1830**, titled "Solution of a question relative to the mathematical theory of heat," however, Dirichlet not only dealt specifically with a problem from heat theory but with a pertinent memoir by Fourier that had been recently published in volume 8 of the Académie's Memoirs for 1825 [Fourier 1829b]. The question is that of dealing with a heated bar whose extremities are maintained at temperatures given as a function of time. While Dirichlet praised the method that Fourier used, "a singular kind of passing from the finite to the infinite," which, he stated, "offers a new example of the richness of this analytic procedure which led the author to so many remarkable results in his great work on the theory of heat," nevertheless Dirichlet thought that his treatment of the problem by a very different analysis gives rise to application of some "artifices of calculation" that appear to be useful in other researches. He proceeded by establishing limiting conditions for the function in question, making implicit use not only of convergence but of the function approaching definite values to prove his solution, and referring to the relevant sections of Fourier's *Théorie analytique de la chaleur* [Fourier 1822] at each step of the way.

This small memoir is of interest because it is a relatively early example of an approach Dirichlet used whenever he found a proof or discussion incomplete or cumbersome. He sought to find a technique to the given problem that would be either more rigorous or more straightforward in arriving at the desired outcome. He repeatedly referred to looking for a concept that would come closer to revealing "the true substance" of the issue at hand or simply, as in this case, reduce unduly complex manipulations by following a different course. Dirichlet appeared to favor spending the extra effort especially when it was a case of improving upon the work of someone whom he regarded highly, such as Gauss or Fourier. His repeated use of limiting conditions in analysis (already noted in the application to number theory, in Chap. 3, above) may have been encouraged by their occasional use of these as well.

7.4 Summary

The publications of those two years 1829–1830 dealt with issues of analysis rather than number theory. Aside from making more rigorous certain arguments of Cauchy and Fourier, Dirichlet at this time made significant contributions to the theory of convergence, both in dealing with Fourier series and in bringing out the—as yet unnamed—distinction between absolute and conditional convergence.

It will be seen in Chap. 9 that these studies would be of particular significance once Dirichlet built on Euler's and Gauss's applications of analysis to number theory in a cohesive fashion.

Chapter 8
Maturation

After the announcement of Dirichlet's engagement "to the rich Mendelssohn," the recently married mathematician Carl Gustav Jacob Jacobi wrote to his parents one must wish that, given a happy domestic condition just as his own, Dirichlet could now settle down to concentrate more fully, as he had been letting his talents lie unused up to then.[1] It was the kind of mixed message that was not so surprising emanating from Jacobi as it might have been coming from another source, since it was not uncommon for Jacobi to voice definite opinions or criticisms which he frequently modified later. The topics usually involved uncertain areas.

It appeared doubtful that Dirichlet would find more time for research than he had had before. Once married, outside the family his days were divided among obligations to educational commissions, to the Kriegsschule, to the university, to the Akademie, to new and old friends, and to acquaintances who welcomed visiting and corresponding with him as his future became more secure. It would turn out, however, that by the end of the 1830s Dirichlet had produced a remarkable body of research results, noteworthy not only because of quantity and quality but because of a cohesiveness that included the previously published, seemingly unrelated memoirs as part of a structure that clarified basic concepts of analysis and would result in a new approach to number theory.

8.1 Educational Commissions

On June 6, 1832, less than a month after his wedding, Dirichlet wrote to his mother that he was particularly busy because he had been appointed a member of the *Wissenschaftliche Prüfungskommission*. This testing commission had been established as part of the reforms instituted by Wilhelm von Humboldt in 1810. Its purpose was to examine those wishing to teach at secondary school levels, and Dirichlet's appointment to the commission may have been a natural outgrowth of his prior func-

---

[1] Ahrens, W. ed. 1907:10.

U. C. Merzbach, *Dirichlet*, https://doi.org/10.1007/978-3-030-01073-7_8

tion as sole academic representative on the preceding Commission of a Mathematical Teaching Plan, the *Lehrplankommission*. In writing to his mother, Dirichlet expressed his regret at being asked to take part in activities that he found unpleasant (*unangenehm*) and that frequently took up a whole day.

To understand his discontent, it helps to review the previous history of these educational commissions and their membership. When Wilhelm von Humboldt had set forth his educational reforms for Prussia in 1810, they included a mandate that future teachers were to be examined in three major fields: History, Mathematics, and Philology. The newcomer in this triumvirate was mathematics. By the 1820s, it became clear that those who had been traditionally in charge of the school curricula did not always know how to handle the appearance of that newcomer, not to mention the inferred threat to their traditional dominance, felt by classical philologists in particular. The Cultural Ministry, formed in 1817 from the subdivision on Ecclesiastical Affairs and Public Education of the Ministry of the Interior, received an increasing number of complaints about the inadequate mathematical preparation elementary and secondary school students were given.

Crelle and the *Lehrplankommission*

A. L. Crelle, whose recently founded mathematical journal was receiving laudatory international and domestic recognition, requested a transfer to the Cultural Ministry from the Ministry of the Interior, where his achievements during long service in overseeing building and transportation developments had been noted. The timing appeared perfect. The relevant ministers appealed to the king to grant the transfer, noting Crelle's awareness of the state of affairs with regard to mathematics as well as his competence in the subject. On November 8, 1828, the Royal Cabinet of Public Education approved the transfer and Crelle's new mandate.

In particular, Crelle had been charged with reviewing mathematical teaching in the secondary schools (the *Gymnasien*) and with developing guidelines for methods and syllabi. Crelle immediately went to work. By early August 1829, he had prepared a detailed report, which included the recommendation for a commission to consider a mathematical syllabus for teachers. This *Lehrplankommission* was appointed within a month. It consisted of two members of the school administration, two high school principals, and one university mathematician—Dirichlet. It met from October 1829 to February 1830; one of its major recommendations proposed the production of a textbook of arithmetic and mathematics which was to include theory of imaginaries, expansion of series for complex exponential and circular functions, solution of equations by approximation, and the principle of virtual velocities.

The *Prüfungskommission*

While these noble efforts were underway, the ministry continued to receive complaints about the criteria by which teachers were prepared and tested. In February 1830, just before the *Lehrplankommission* held its last meeting, the ministry sent a directive informing schools that the results of the examinations given to future teachers were being evaluated unfairly. Those who did not pass the mathematical requirements could nevertheless be accepted, whereas the expectations for passing the philological standards were higher and met more successfully.

Specifically, it was noted that of fourteen individuals tested in Bonn, twelve lacked even the most elementary mathematical knowledge. For that reason, those philologists with the best mathematical knowledge should be given preference. It appears that the extreme disparity of performance in Bonn could be attributed to the one-sided outlook of the director of the philological seminar in Bonn and his associate, C. H. Heinrich.[2] The situation in Bonn was not totally out of the ordinary, however.

Dirichlet was appointed to replace a member on the testing commission whose term had come to an end, and he would serve in 1832 and again for two years in the early 1840s. In 1843, while applying for a leave of absence for a forthcoming trip to Italy, he also reminded the ministry that a substitute needed to be found for his place on the commission. Actually, by the 1840s the difficult phase of the mathematical teacher preparation during the 1820s had been overcome, so that the ministry was freed from the overwhelming number of earlier complaints.

8.2 The Kriegsschule

In contrast to the charged atmosphere in the teacher training commissions at the beginning of Dirichlet's career, the discipline and mathematical preparation of the officers' candidates at the Kriegsschule may have provided a more relaxed environment for Dirichlet in his early years there. He got along well with the young officer candidates; as mentioned previously, they respected his acquaintance with French military officers as well as his ease in the environment that characterized the social class from which they were drawn. In addition, they appreciated the effort he made to bring his mathematically underprepared students up to par and to provide incentives to the more promising ones.

Enhancing the Mathematical Curriculum

By 1835 Dirichlet's quarterly report revealed that, as a result of requests from students, he was providing them with a summary exposition of the main aspects of higher analysis in the *Applikation*. Radowitz made a marginal notation on the report that this was to be permitted for especially talented students in the *Applikation* but not in the lectures.

The outcome seems to have been satisfactory, for Dirichlet's next report mentioned that students were receiving training in higher analysis. This appears to have been his contribution to a gradual change in the curriculum because it marks the first introduction of some calculus and still later led to a revamping of the mathematics curriculum that would remain in place until the end of the century. It is described as

[2]The latter was the same person who had led the objections to Dirichlet's being given the doctorate in Bonn only three years before; as it turns out, Heinrich would no longer serve on the testing commission (the *Wissenschaftliche Prüfungskommission*).

requiring algebra, theory of series, stereometry, spherical trigonometry, and analytic geometry in the first year; differential and integral calculus in the second year; and analytical mechanics in the third.[3]

8.3 The University

The mathematical environment at the Friedrich Wilhelm University of Berlin became more interesting during the period under consideration as well. Initially, most of the students were still ill-prepared, largely continuing to reflect the mathematical weakness of their prior teachers. But it soon emerged that, as the preparation of the secondary school teachers improved, Dirichlet, by judicious choice of the lectures he offered and the time he took with his students, was able to considerably raise the standard of mathematics offered at the university.

Students

During the 1830s, fourteen students were awarded their doctorates in mathematics at the university. Of these, several specifically mentioned, in the vitae forming part of their dissertations, having taken Dirichlet's courses on infinite series and partial differential equations; some also noted his beneficial advice. They included Adolph Goepel (Dr. Phil. 1835), Johann Foelsing (1836), Elias Mueller (1836), Carl Gerhardt (1837), Gustav Michaelis (1837), and Johann Boymann (1839). Most became secondary school teachers; Carl Gerhardt is remembered as editor of the mathematical works of Leibniz.

A number of chemistry and physics students also made a point of mentioning Dirichlet in their doctoral dissertations. They were students of the physicist Gustav Magnus and of the chemist Eilhard Mitscherlich. These students included Friedrich Wilhelm Barentin, later remembered primarily for his authorship, with Dove and Bischof, of a handbook of chemistry and physics, and Gustav Wiedemann, who took his degree in chemistry, thinking it a necessary prerequisite for doing physics. Aside from other research activities, Wiedemann would become best known as Poggendorff's successor in editing the *Annalen der Physik und Chemie* and as author of a fundamental work on electricity.

Mathematical Faculty

Among the faculty, Dirksen had been the only ordinary (full) professor in mathematics since 1824, and for that reason his name appears as adviser for all who took the doctorate until the fall of 1845. For a short period (1824–1833), there had been a professorship in applied mathematics as well, but that was abandoned after the death of its incumbent, Jabbo Oltmanns.

Dirichlet was joined as extraordinary professor by Martin Ohm who read plane and spherical trigonometry as well as algebra and analysis, both presented on the

[3] Poten 1900 and Lampe 1906.

basis of his sequentially structured textbooks. As previously noted, he, too, was on the faculty of the Kriegsschule.

It still was not uncommon for members of the university staff—especially the privatdozenten and extraordinary professors—to have a second position in one of the high schools so as to subsist more easily on the low payments they received. Although there were variations in different universities and subjects, privatdozenten usually had no fixed salaries paid by the institution but were paid by students directly. The salaries of extraordinary professors varied similarly, and there were additional exceptions. Notably, when Dirichlet was appointed extraordinary professor, his salary was fixed by the ministry, as it had been even when he became a privatdozent.

During 1832 and 1833, Julius Plücker, later known for his work on analytic geometry, also served as extraordinary professor, in addition to teaching at the Friedrich Wilhelm Gymnasium. He had previously received his habilitation in Bonn, where he had served as privatdozent and extraordinary professor. Had he stayed in Berlin, he and Dirichlet might have developed a closer relationship, as Plücker, fluent in French, earlier had overlapped with Dirichlet in Paris. There he, too, had heard Biot and Lacroix, and, at the Ecole Polytechnique, also Cauchy and Poisson.[4] Like Dirichlet, Plücker had shown an early interest in applications to physics; in fact, after returning to Bonn as professor of mathematics, he subsequently became professor of physics there.

Plücker left Berlin in 1833, returning to Bonn in 1834 after a year in Halle. In 1834, he was replaced as extraordinary professor in Berlin by the synthetic geometer Jakob Steiner. It has been noted that had both Steiner and Plücker been on the faculty at the same time, the atmosphere would have been less than congenial.[5] Despite the difference in their mathematical orientation, in terms of pedagogical methodology Dirichlet was closer to Steiner; both were influenced by their respective teachers—G. S. Ohm and Pestalozzi, who shared a strong visual, if not Socratic, bond.

Steiner, Swiss by birth, had received his early education from Pestalozzi and continued to be a loyal disciple of Pestalozzi and his methods. After spending less than three years in Heidelberg, where he attended regular mathematics courses at the university, he had come to Berlin in 1821. He had maintained himself primarily by private tutoring and teaching at various secondary schools for short periods. His Pestalozzian teaching style led to repeated disagreements with several principals, notably at the Werdersche Gymnasium. There he came into serious conflict with the principal for not using a textbook—particularly that man's textbook. But at the same time he became friendly with Jacobi, who was still studying at the university. As the years went on, it became clear that they shared not only their love for mathematics but also a decidedly undiplomatic form of discourse. By 1833, thanks largely to Jacobi (in Königsberg since 1826), Steiner received an honorary doctorate from Königsberg. Throughout the difficult decade of the 1820s, he had produced some of his

[4]Some recent publications refer to Dirichlet as a student of Poisson; this misunderstanding may be based on Humboldt's repeatedly stressing his own long-standing friendship with Poisson and noting Poisson's positive reaction to Dirichlet's early work.

[5]Biermann 1988:68.

best-known work in synthetic geometry, so that by 1834, when he joined Dirichlet at the university, he also was elected to membership in the Akademie. Continuing to be narrowly focused on synthetic geometry, he remained in the position of extraordinary professor until his death in 1863.

In addition to these men, the university counted three privatdozenten for mathematics in the 1830s. They included Samuel Ferdinand Lubbe, who had submitted the first doctoral dissertation at Berlin in 1818, was the author of some elementary textbooks, and as privatdozent taught elementary mathematical topics from 1819 until the year of his death in 1846. He had been a teacher at the Friedrich Wilhelm Gymnasium. Then there was Ferdinand von Sommer who was in and out of Berlin—he liked to travel—until 1843, but did little to contribute to the good of mathematics.[6] The most productive of the three, and the one who was closest to Dirichlet among colleagues of the decade, was Ferdinand Minding, who came to attention through his publications as well as his teaching.

Minding's book on introduction to higher arithmetic, published in 1832, was considered as satisfying some of the desiderata put forth by the earlier *Lehrplankommission*. It was followed in 1836 by a two-volume handbook on the calculus, with applications to geometry and mechanics. Minding observed that initially he had not intended to do much with definite integrals in this work as they lay outside the realm of what he wanted to accomplish. But the simplicity and rigor of the method that Dirichlet, who had advised him in other areas as well, had proposed to him convinced Minding that he should also include the chief properties of the Gamma function.[7]

Minding had a second job at the Allgemeine Bauschule where he taught theory of curves as well as dynamics and miscellaneous topics in analysis. His work there also led to some of his most significant publications in Crelle's *Journal*, contributing considerably to growing awareness of Gauss's work on differential geometry. On a more mundane level, he published a table of integrals while at the Bauschule, which would prove to be useful in many areas outside that school. In 1843, he accepted an appointment as full professor for applied mathematics in the German language Russian university in Dorpat where he taught until shortly before his death in 1885. Although he would become a member of the Saint Petersburg Academy in later life, Dirichlet's earlier attempt to have him elected to the Berlin Akademie was unsuccessful.[8]

---

[6]Biermann 1988:69–72 cites evidence that Sommer misrepresented his credentials, including his doctorate and a non-existent relationship to Gauss, and that he generally appears as a swindler.

[7]Minding 1836, 1:preface.

[8]For further details and references concerning Minding, see Biermann 1988:52–55. Biermann attributed Minding's inability to gain further career opportunities in Berlin to his not sharing in the growing popularity of special analytic functions and series. It is possible that in the 1830s and early 1840s this may not have been so significant a factor as Dirichlet is still lacking voting rights at the university.

8.4 The Akademie and the Académie

Dirichlet had been nominated for membership in the Mathematical Class of the Royal Academy of Sciences of Berlin (the *Königliche Akademie der Wissenschaften zu Berlin*) in November 1831; the nomination was signed by Poselger and Crelle, with countersignatures from Encke and the physicist Paul Erman. The election followed on his birthday, February 13, 1832.

He took his membership seriously. Beginning in the following year, he read a paper to the plenum each year until his trip abroad in the 1840s, and upon his return read to both the plenum and the mathematical class. In addition, he served on committees, such as one on finances; participated on proposals for membership; and generally made the wider world of scholarship aware that the Akademie had a resident mathematician in its midst of a caliber not noticed since the days of Lagrange.

In Paris, he was elected corresponding member of the Académie on May 6, 1833. In recent years, the Académie had received offprints of several of his publications; Lacroix had been asked to provide an oral report on the divisor memoir in 1828. Nominated in the section on geometry at the last April meeting in 1833, Dirichlet received 37 out of 43 votes the following week.

8.5 The *Repertorium*

There was a new commitment to which Dirichlet agreed in the mid-1830s. Gustav Fechner, the professor for physics in Leipzig for many years before becoming known for his interest in psychophysics, had produced standard contributions to physics and edited a survey journal for experimental physics. This was appropriately titled *Repertorium der Experimental Physik*. In 1836, Fechner decided to give up this editorship.

Heinrich Dove was approached to take over the editorial duties for the *Repertorium*. Initially, Dove resisted the invitation. Although in better health than Fechner, he too felt overloaded with lectures, laboratory work, and his basic research. However, after much prodding from colleagues in the field, the experimental physicist Ludwig Moser agreeing to serve as coeditor, and promises of collaboration from others, he agreed.

It was decided that the journal would not simply be a continuation of Fechner's. In 1837, it appeared with a slightly modified title and a new subtitle as *Repertorium der Physik: enthaltend eine vollständige Zusammenstellung der neuern Fortschritte dieser Wissenschaft unter Mitwirkung der Herren Lejeune Dirichlet, Jacobi, Neumann, Riess, Strehlke, herausgegeben von Heinrich Wilhelm Dove und Ludwig Moser*. It also started with a new numbering of volumes, beginning with volume 1, to reinforce its changed identity. It became clear with this first volume, dominated by representatives of the universities of Berlin and Königsberg, that the emphasis would shift from the purely experimental to include the more mathematical aspects of physics.

The initial volume edited by Dove listed Jacobi as taking over the responsibility for mechanics, Dirichlet for mathematical physics. Dirichlet was to do a review of Poisson's treatise on heat theory. As it turns out, neither man's contribution appeared. It is likely that both Dirichlet and Jacobi felt the readership would not be prepared for the technical details involved in either anticipated memoir. In addition, Dirichlet, always hesitant to enter a predictable controversy, likely was averse to commenting on Poisson's heat theory. The issue had less to do with mathematics than with physical interpretations of the movement of heat. Dirichlet was respectful of Poisson's mathematics, but he favored Fourier's approach to heat theory. Poisson had been repeatedly critical of Fourier's work on the subject but had commented favorably on Dirichlet's earlier memoirs, especially the one on definite integrals.

What Dirichlet contributed to the *Repertorium* instead was a paper on the representation of arbitrary functions by trigonometric series that was published in 1837 (**1837c**). It is ironic that this is often cited with reference to the history of the function concept, describing it as a new rigorous formulation of that concept. It is true that it was a clearer presentation than had been found in the prior literature. But it was not supposed to be something new. Recalling that the *Repertorium* was intended as a review of the then current state of knowledge, designed especially for those studying physics, the reason for Dirichlet's clear exposition lies in the fact that he wished to explain fundamentals to the readers in unambiguous, rigorous terms, rather than his having in mind the presentation of a new concept. At the same time, it implicitly supported Fourier's use of trigonometric series for the benefit of those who had not studied Dirichlet's 1829 memoir on the subject.

8.6 Gaussian Interactions

Among the individuals whose relationship to Dirichlet was strengthened during this period, we note Gauss first of all. This occurred primarily through Dirichlet's publications, most of which were based solidly on Gauss's *Disquisitiones Arithmeticae* and more recently inspired by Gauss's publications on biquadratic residues with his related expansion of number theory to complex numbers.

In addition, there were three men who communicated steadily with both Gauss and Dirichlet. They were the astronomer Encke, the physicist Wilhelm Weber, and the mathematician Carl Gustav Jacobi. While several of Dirichlet's friendships had their beginning in this period, none would be closer or more enduring than those with Wilhelm Weber and Jacobi.

Encke

Encke had been an astronomy student of Gauss between 1811 and 1816. His studies were interrupted numerous times by his volunteering for the Wars of Independence from Napoleon. Although this meant that he left Göttingen without a doctoral degree, his unquestioned ability as an astronomer led to his becoming head of the Seeberg Observatory in 1822; his investigations on comets had led to support from Gauss, and soon thereafter caused him to be put in charge of establishing a new observatory in Berlin, which would be completed by 1835. We have already noted (Chap. 4) a significant contribution to Dirichlet's appointment to Prussian service that Encke made through his letter to Johannes Schulze in 1826.

Dirichlet's arrival in Berlin gave Encke a welcomed opportunity to act as a self-appointed go-between, keeping Gauss abreast of affairs in Berlin, and providing him with miscellaneous bits of gossip. Once Dirichlet became established in Leipzigerstraße 3, Encke lost no time in calling on him and then reporting to Gauss about the home, Dirichlet's participation in the *Jahrbuch für wissenschaftliche Kritik*, his knowledge of least squares (helping Encke with some related procedures), as well as various events in Berlin, including university affairs, some colored by Encke's dislike of Dirksen.[9]

Encke suggested to Dirichlet a number of topics, of use in astronomy and other observational sciences, that Dirichlet would add to his course offerings at times. Notably, these included work on probability, with emphasis on optimal methods of observation, and aspects of celestial mechanics.

Wilhelm Weber

Dirichlet had encountered Wilhelm Weber at the 1828 conference, and, as previously noted, Weber had sat in on Dirichlet's lectures the following year. Weber would come to general notice during the decade of the 1830s for his collaboration with Gauss in constructing a "telegraph" from the Göttingen Observatory to the library. It was a sensational success, although there had been intermittent difficulties with the connecting line. As it turns out, among various changes that were made, some problems were overcome by the use of a copper wire that Weber had chosen upon advice from Dirichlet. Dirichlet's participation in the work within Humboldt's "magnetic house" had been well-timed to bring him back to thinking about physical problems and remembering the precautions needed even for elementary physical measurements, that he had probably learned as a pupil of G. S. Ohm. Therefore, it is not surprising that we find him making suggestions to Wilhem Weber.

Weber's other major achievement of the decade was the beginning of his establishment of standardized units of measure. In addition, what was of particular importance in his collaboration with Gauss was the series of reports on recent results in their study of geomagnetism. The observations in Göttingen were conducted somewhat more professionally and with an increasingly more sophisticated set of equipment than that in the use of which Dirichlet had participated earlier, in Humboldt's little magnetic house. However, just as Gauss and Weber had produced and begun to publish their first impressive results and inferences, in 1837 the peaceful activities in Göttingen were disrupted by the firing of the "Göttingen 7."

In 1837, Ernest Augustus, the uncle of Queen Victoria, who took over Hanover upon the death of his brother William IV, rescinded the constitutional guarantees recently established by William and demanded a loyalty oath from those in his domain, including all the professors in Göttingen. They had just a few years previ-

[9]Göttingen. NSUB. Gauss Nachlass. Correspondence.

ously sworn an oath to a newly established constitutional system. Seven professors explained that they could not go back on their previous oath. These included Weber; the brothers Grimm; the historian Dahlmann; Gervinus, professor of history and literature; Ewald, the orientalist and son-in-law of Gauss; and the jurist W. E. Albrecht. All were fired; some were expelled from Göttingen. Weber was allowed to stay in town and carried on his research with Gauss for another two years; but he had no income and was no longer in charge of the physics laboratory.

Gauss had depended on Weber especially for the experimental results that had emerged from the laboratory, but, despite protests by representatives of the university, the Societät, and other Göttingen institutions, Weber, like the other six, was out of a job. By 1839, Weber had to look for some income. For a decade, he found a place to live, eat, and work in Leipzig by occupying a professorship at the university there, where his two brothers Ernst Heinrich and Eduard were active, Ernst Heinrich as professor of human anatomy and Eduard as physician and anatomist.

Jacobi

Dirichlet had first met Jacobi prior to 1830. According to Kummer, probably in 1829, Dirichlet and Jacobi had taken a trip, first to Halle, where they joined Weber for a more extended visit to Thüringen. From then on, Dirichlet was in continuing direct and indirect communication with both men. He saw Jacobi, who regularly visited his parents' home in Potsdam, more frequently.

In Königsberg, Jacobi, having decided a few years earlier to give up ancient philology for mathematics, was busily churning out results in several different mathematical areas, although he did not find time to publish them all at that point. Some of his major number-theoretic results became known through his incorporating them in his lectures. Urged by his friend and colleague the astronomer Bessel, he sent in a report "On the Theory of the Calculus of Variations and Differential Equations" to the Berlin Akademie that let his contemporaries in on some of the outstanding results he had achieved. In addition, word began to trickle out about the correspondence between him and Legendre during the years 1827 and 1832, which showed Legendre's respect for the work Jacobi had done in regard to elliptic functions.

8.7 Family: 1833–1835

There were important family events in Dirichlet's life during the 1830s, with alternating joyful and tragic occurrences.

On July 2, 1833, the Dirichlets' first child, Walter, was born. He developed an early interest in, and talent for, drawing; it appeared for the first two decades of his life that he might follow a career as an artist. Aside from his aunt Fanny's husband, Wilhelm Hensel, there were two other prominent painters in the family, Philipp and Johannes (previously Jonas) Veit, the sons of his great-aunt Dorothea Schlegel, and there was hope that Walter might achieve similar success. As he was growing up, the Hensel household provided him not only with a firsthand view of an artist at work—his uncle Wilhelm Hensel—but also a playmate, the two-year-older cousin, Sebastian.

The first two years after Walter's birth appeared relatively tranquil for the Dirichlets. He obtained leave during the fall vacations to go to Aachen, where his mother became acquainted with little Walter. Since his father's retirement, Dirichlet's parents had lived, or at least spent much time, in Aachen where his sister Caroline had married J. C. A. Baerns, by this time the *Postkommissar* in Aachen.

In the spring of 1835, Dirichlet obtained leave for a trip to go bathing at the seashore. This vacation was not to be tranquil. While Dirichlet was finishing his classes for the term, Rebecca and Walter, along with her parents and the Hensels, headed for the Rhineland. Her brother Paul had recently married Albertine Heine, the sister of the mathematician Eduard Heine, and they remained in Berlin. The rest of the family had taken this rather strenuous journey to see Felix, who had been appointed civic music director in Düsseldorf in 1833, conduct a successful music festival in Cologne. He would lead a larger one on Whitsunday weekend in Düsseldorf, before leaving for his permanent appointment as head of the Gewandhaus Concerts in Leipzig the following month.

By the time Dirichlet arrived, the night after the last concert, the Hensels had left for France; Rebecca's mother, who had been ill for some time, was being cared for by Rebecca and improving. Just as the Dirichlets were beginning to enjoy their stay, during which they spent relaxing times with the family of Otto von Woringen, who had been hosting Felix since his arrival in Düsseldorf, Dirichlet received the news that his father had a serious case of a stomach illness that had been making the rounds and appeared to be very weak. Dirichlet rushed to Aachen by Rapid Post. During the following week Rebecca's mother had recovered sufficiently for her parents to be able to return to Berlin, and she joined Dirichlet in Aachen. Leaving Walter under the care of his grandmother there, they finally left for Ostende to begin bathing in the sea.

Dirichlet experienced a sore throat, and an eye, ear, and nose infection after his first bath. Rebecca decided to walk along the sea instead of going in—meanwhile observing with astonishment that men and women were bathing together. After about a week, Dirichlet had recovered sufficiently for them to enjoy their open-air breakfasts and for Rebecca to report that she was getting fat.

During this time, the Hensels had left Paris, where Wilhelm Hensel had had a multitude of society requests for small portrait sketches, and headed for Boulogne-sur-Mer for their own bathing *Kur*. This appears to have been no more "curing" than that of the Dirichlets in Ostende; Fanny received an eye infection, Wilhelm the stomach flu, and the weather turned bad.

The Dirichlets headed homeward, the initial part of the trip facilitated by rail travel. They saw Gent, Antwerp, spent several days in Brussels, home of the Quetelets, and finally arrived back in Aachen for a happy reunion with Walter and Dirichlet's parents. Dirichlet got into some nettles while walking, but, undeterred, after a short stopover in Düren, managed to hike to Bonn, where he attended the annual September meeting of the Society for German Scientists and Physicians. After everyone had been reunited in Bonn, the Hensels left for Leipzig before returning to Berlin. The Dirichlets, too, returned home via Leipzig, where they were joined by Felix and his visiting friend, the composer Ignaz Moscheles, who traveled with them to Berlin. During this time, they finally enjoyed some cheerful days.

Cheerfulness was short-lived. In November, Dirichlet's father-in-law, Abraham Mendelssohn Barthody, who had been plagued by a worsening cough, died. He had in recent years suffered from failing eyesight but retained his intellectual interests and mental acuity. He had relied on his two daughters to read to him in those years; Rebecca read him passages from Rousseau's *Emile* the night before his death.

The evening after Abraham Mendelssohn Bartholdy's burial, Felix, the Hensels, and Rebecca had discussed what they could do for their mother, and decided it would be best if the Dirichlets offered to move over to the front part of the house to live with her. They did. Felix served as persuasive intermediary, Lea Mendelssohn Bartholdy accepted, and everyone seemed pleased with the arrangement which went into effect January 1836. Fanny was relieved that Felix, who had been particularly hard hit by his father's death and concern for his mother, seemed more relaxed. For the next several years, the Hensels would spend evenings alternately in their upstairs apartment and downstairs with the Dirichlets and mother Lea. Meanwhile, Wilhelm Hensel's two sisters, who had been living with their mother, also recently deceased, moved into the rooms vacated by the Dirichlets.

8.8 Family: 1836–1838

Just as family affairs promised to calm down, in the spring of 1836 Rebecca had a miscarriage. The physician ordered a rest cure in the famous spa of Franzensbad (currently better known as Frantiskovy Lazne, the largest spa in the Czech Republic), where she went that summer, accompanied by little Walter and her sister. While there, she was not only bored, and irritated by the ostentatiously dressed ladies from Central Europe, but suffered from the painful facial neuralgia that would recur over the years. Once Dirichlet joined her, she began to enjoy herself and improved sufficiently so that she and Dirichlet could undertake a vacation trip to Munich and Salzburg. But now there was more bad news. In Munich, Dirichlet learned of the death of his last surviving sister, Caroline Baerns, and hurried to be with his mother, who soon would come to live with them as well.

While all of this was happening, in September 1836 Felix had become engaged and in March 1837 married Cécile Jeanrenaud, a member of a distinguished Frankfurt family. Her father, August Jeanrenaud, had been a French Reformed Protestant minister and her mother, Elisabeth, was a Souchay. Cécile's maternal great-grandfather, Jean Daniel Souchay, had preceded her father as minister of the same church in which Felix and she were married. When the Dirichlets met Cécile's maternal grandmother, Helene Souchay, née Schunk, she was a formidable presence. She was known as "Madame Souchay," had been married to Carl Cornelius Souchay, a wealthy wholesale textile merchant, maintained a salon in Frankfurt, and honored her husband's French Huguenot heritage.

Dorothea Schlegel now also lived in Frankfurt, with her son Philipp Veit, and corresponded with Rebecca fairly regularly. She represented the Mendelssohn family at Felix's wedding.

Within less than two years after Abraham's death, Dirichlet's own father died in Aachen. Dirichlet's father had shown signs of diminishing mental capacity for some time before his death in 1837. He had been forced to retire in 1829 and Alexander von Humboldt had sought to gain special consideration for his pension from Karl Friedrich Nageler, the supervising official in charge of Prussian Postal Services, pointing out that the sixty-eight-year-old Postcommissarius Lejeune Dirichlet had been in state service for forty-five years; that he was the father of a talented, humble, and amiable young man, the friendship with whom was the motivation for this most urgent request that His Excellency favor the father in the matter of his pension.

Before the end of 1837, both Rebecca and Cécile had given birth. Fanny had had another miscarriage and would have no other children besides Sebastian. She had in fact been fortunate when Sebastian was born in 1830; both Rebecca and Paul had come down with measles that year, but she and Wilhelm Hensel had escaped the problem, possibly because of a strict quarantine the physicians had imposed on the affected members of the family.

As will be evident in the next chapter, despite all these eventful family affairs, Dirichlet had been singularly productive in 1837.

The year 1838 began with the pleasant news that Dirichlet had been elected corresponding member of the Saint Petersburg Academy in Russia. Most of the rest of that year proved to be increasingly troublesome, however. The weather had been bad since the previous summer. Walter was bitten by one of the tenants' dogs, Dirichlet suffered from serious dysentery, and the rest of the family had the measles—again. By October 1838, Rebecca had had one of the more serious recurrences of her facial neuralgia, which, in hindsight, we may attribute to the presence of a shingles virus, or a similar virus located in the trigeminal nerve. This could account for the facial pain that continued to recur with increasing frequency and intensity as the years went on, usually after particularly stressful periods or events.[10]

In November 1838 came the worst blow the Dirichlets were to sustain: the death of their one-year-old, beautiful son, little Felix. Fanny in her diary would later describe the inordinate pain and suffering Rebecca underwent. She had barely recovered by the following May.

8.9 The Death of Gans

According to Fanny, May 1 was the first beautiful spring day of the year 1839. Eduard Gans stopped by at the Dirichlets was persuaded to join Rebecca and her mother for lunch; Fanny dropped in, joked with Gans as had been their habit over the years, and

[10]Rebecca had been the first one to be hit by the 1830 measles outbreak in the family; it seems far more plausible that in 1838 she was having another viral nerve attack rather than a second case of the measles.

left, so the others could enjoy their meal. His unannounced visits at the Dirichlets were not unusual. Both sisters had been conscious of his various deficiencies in social graces—one of the things that had been a sharp contrast to Dirichlet's impeccable manners—but they had been unsuccessful in their efforts to smooth the rough edges of one of their favorites.

After the first course, Gans, who, they felt, had aged considerably and lost strength over the winter, complained that his arm was falling asleep. Soon his right side became paralyzed and he collapsed. After he regained consciousness, Rebecca asked whether she should send for his mother; he answered, "no, wait till I'm dead." Wilhelm Hensel came and gave initial assistance; then, he, two servants, and a young doctor took Gans home to his recently occupied apartment in the Behrendstraße.

Gans had apparently suffered two strokes. He lasted four more days. During this time, he was visited by two more senior physicians and by Varnhagen von Ense, who noted the details in his diary. Other visitors were a group of Hegelians, including the art historian Heinrich Gustav Hotho and the theologian and orientalist Franz Benary; Alexander Mendelssohn; Madame Amalie Beer, a Berlin socialite in her own right, aside from being the mother of Giacomo Meyerbeer and of Heinrich Beer, one of Rebecca's cousins by marriage; as well as Alexander von Humboldt; and others. Despite all this well-intentioned activity, Gans managed to live until the morning of the following Sunday.

The burial took place on Wednesday, May 8. It was marked by a procession said to be a half-mile long with marshals of the student body taking the lead, followed by "professors, students, civil servants, merchants, artists and literati," as well as ninety-five wagons; two Ministers, von Altenstein and von Grolmann, were present. Philip Konrad Marheineke, the senior member of the theological faculty, gave the funeral address, later attacked in some circles because of perceived political overtones. Varnhagen observed that Berlin had not seen a turnout such as this since the death of Schleiermacher five years before. Gans was buried in the Dorotheenstaedtische Cemetery, near Fichte and Hegel.[11]

A few days later, Varnhagen stopped by at Lutter and Wegener's, populated by the usual mixture of citizens.[12] Someone came in and reported that one of the young princes who had been seriously ill seemed to be out of danger. An elderly Berliner eyed the newcomer and explained the surprising silence that had ensued: "You can always get another prince; but Gans is gone, there is no other."

The university honored a colleague and teacher. The followers of Hegel remembered their spokesman. Many Berliners grieved for Gans as one of their own. Along with their youth, the Dirichlets had lost a friend.

[11]Reissner 1965.

[12]Lutter and Wegener, the historic wine restaurant, is best known to later generations as the site of E.T.A. Hoffmann's *Tales*.

Chapter 9
Publications: Autumn 1832–Spring 1839

In 1832, Dirichlet completed two studies published in Crelle's *Journal*. The first was prompted by Gauss's expansion of the study of residues to complex numbers [Gauss 1832], the second was the proof of Fermat's Last Theorem for the case $n = 14$. These were followed by a total of seventeen publications appearing prior to the summer of 1839, of which twelve related to presentations given to the Akademie.

After Dirichlet's election to the Berlin Akademie in February 1832, he gave one address before the plenum (the *Gesammtsitzung*) of the Akademie each year from 1833 to 1855, excluding only the years 1843–45 of his trip abroad. Between 1833 and 1839, seven Akademie presentations would result in five full memoirs, and, beginning in 1834, four summary reports, as well as three translations into French.

The five full-length publications of the Akademie presentations given between 1833 and 1839 were published in the *Mathematische Abhandlungen* of the Akademie, their dates of publication (**1835**, **1836a**, **1837a**, **1839a**, and **1841a**) reflecting the Akademie's lag time of two years between the date of presentation and the resulting publication.

Of the presentations summarized in the Akademie's Reports (the *Berichte*), two were not included in the *Mathematische Abhandlungen*: **1836b**, a commentary on the method of least squares, presented in July 1836, and **1838a**, a discourse on the determination of asymptotic laws in number theory, read in February 1838. The other two dealt with the important memoirs on arithmetic progressions (**1837c**) and on the introduction of a discontinuity factor (**1839c**) for evaluating certain multiple integrals. Since these reports were published in the year of the presentation, they made known the substance of a presentation approximately two years before the publication of the corresponding full-length memoirs.

Of the translations ensuing from the Akademie presentations of the years 1833 to 1839, one (**1837d**) was a French version for Crelle's *Journal* of **1837a**, "On the application of definite integrals to the summation of finite or infinite series." The other two were abbreviated versions of Dirichlet's seventh Akademie presentation of the period, made in February 1839, on his "New Method," the use of a discontinuity factor.

U. C. Merzbach, *Dirichlet*, https://doi.org/10.1007/978-3-030-01073-7_9

All of Dirichlet's full-length Akademie publications prior to 1839 dealt with number theory. His memoir on arithmetic progressions (reported as **1837c** and fully published as **1839a**), that introduced Dirichlet series, initiated a decisive new approach to this branch of mathematics. It is at times referred to as constituting the beginning of analytic number theory, and it marks a new direction in his studies of number theory that affected his subsequent treatment of mathematics generally.

The final memoir of this group, presented to the Akademie in February of 1839, although not published until 1841, was his "New Method," an approach to integration theory in which he introduced the use of a discontinuity factor in the evaluation of certain multiple integrals with variable bounds.

In addition to the Akademie presentations, one memoir, **1837d**, was prepared for Dove's *Repertorium der Physik*, and there were four more memoirs that first appeared in Crelle's *Journal*. His contribution to Dove's *Repertorium* was an introductory account, intended primarily for physicists, on the representation of functions by trigonometric series.

The four memoirs in Crelle's *Journal* interspersed between 1836 and 1838 contained known results. They were largely intended to show how Dirichlet's recent methodology would render these results in a simpler and more straightforward manner than that used in the prior proofs indicated by Legendre and Gauss. Aside from providing examples of the advantage and wide usefulness of his *L*-series, Dirichlet there also showed how other series derived from special functions could be used to represent arbitrary functions between given limits. Additionally, these short memoirs clarified older concepts and served as a more detailed explication of results introduced in Dirichlet's lectures, where their usefulness in problem solving at times overshadowed his references to the limiting conditions that guaranteed the validity of the methods used.

Except for the memoirs and reports published by the Akademie and the expository contribution to the *Repertorium*, which were written in German, all the other memoirs of this period were written in French, thereby reaching a wider and more diverse audience.

9.1 Quadratic Residues in the Complex Field

In September 1832, Dirichlet signed off on **1832a**, the "Demonstration of a property analogous to the law of reciprocity which exists between any two prime numbers." As he explained in his introduction, it was prompted by the recently published memoir of Gauss that had contained Gauss's extension of many known number-theoretic results for real numbers to complex numbers. This, [Gauss 1832b], was the second of Gauss's planned three memoirs on residues, announcement for the first of which, [Gauss 1828], had led to Dirichlet's **1828b**. Now Dirichlet noticed that elementary properties were extended easily. But, among the more difficult ones, Gauss himself had singled out "the Fundamental Theorem" (the designation Gauss would continue to use rather than Legendre's "law of quadratic reciprocity") as one the expansion of which to complex numbers presented special difficulties:

> Despite the great simplicity of this statement, its proof belongs to the most hidden secrets of higher arithmetic, so that, at least as matters stand now, it can be carried out only by the most subtle investigations, which would exceed the limits of this memoir.[1]

Gauss's statement concerning the difficulty of the proof was just the spur Dirichlet needed to work out his own proof. He not only succeeded in this, but **1832a** would be the first of a number of memoirs Dirichlet produced as a result of following Gauss's lead in expanding number theory to include the study of complex numbers. Gauss had indicated that he would supply a proof, along with the explanation of its connection to some previously mentioned results obtained by induction, in a future third memoir on quadratic residues and complex numbers. This publication now became unnecessary.

Following a brief introduction, Dirichlet's memoir consisted of four sections. His initial step was a careful explanation of the basic properties of complex numbers. He began by reviewing, largely following Gauss, fundamental definitions and relationships. These preliminaries remind us that, despite specific earlier results of Euler, Lagrange, Laplace, and Legendre, and the work of Fourier, Cauchy, and Poisson, there was not yet an established complex function theory and that Gauss's extension of number theory to the complex field was of enormous consequence.

Dirichlet pointed out that his reciprocity proof is based on simple considerations which can be applied to other problems as well. In a footnote to his introduction, he observed that his procedure will lead to analogous theorems if instead of considering numbers of the form $t + u\sqrt{-1}$ one considers those of the more general form $t + u\sqrt{a}$, where a has no squared divisors.

The basics Dirichlet reviewed included the following:
First, noting that real integers are special cases of complex whole numbers, there was the definition of a complex whole number:

> An expression of the form $g + h\sqrt{-1}$, where g and h denote real whole numbers, not excluding zero, will be called a complex whole number.... [J]ust as every real number is divisible by ± 1, every complex number must be considered as containing the factors ± 1 and $\pm\sqrt{-1}$. A complex number will be called prime when it cannot be decomposed into two factors both different from ± 1 and $\pm\sqrt{-1}$.

To tell whether a complex number $g + h\sqrt{-1}$ is prime or not, Dirichlet first distinguished two cases according to whether the two terms g and h of the complex number are or are not both different from zero. The second of these two cases seems to subdivide, since the subsisting term can be real or the product of $\sqrt{-1}$ and a real number. But this amounts to the same thing, for if $h\sqrt{-1}$ is prime, so is h, and reciprocally. Now in order that a real number q, considered as complex, is prime, it is necessary first that it be so "from the ordinary point of view." But this is not sufficient. Aside from the sign, it must be of the form $4n + 3$; for if it had the form $4n + 1$, it was known since Euler and Fermat that there exist numbers c and d such that $4n + 1 = c^2 + d^2$;

[1] Gauss 1832b, art. 67.

therefore, it would be decomposable into the factors $c + d\sqrt{-1}$ and $c - d\sqrt{-1}$. Reciprocally, Dirichlet showed that every real prime number q which, disregarding the sign, is of the form $4n + 3$ must also be considered prime in the theory of complex numbers.

Next, Dirichlet considered the case where neither of the two terms of the expression $g + h\sqrt{-1}$ vanishes. He showed that in order that this expression represents a prime complex number, it is necessary and sufficient that $g^2 + h^2$ be a real prime number.

From this, Dirichlet deduced that there are prime numbers of two different kinds. Those of the first kind reduce to a single term and, disregarding the sign or the factor $\pm\sqrt{-1}$, are nothing but real prime numbers of the form $4n + 3$. For greater simplicity, he suggested always considering them as cleared of the factor $\sqrt{-1}$. Those of the second kind draw their origin from real prime numbers composed of two squares which, except for 2, are of the form $4n + 1$. Denoting a prime of this kind by $c + d\sqrt{-1}$, one of the pair c and d has to be even, the other one odd.

He concluded his preliminaries with the theorem that, given a complex number $A + B\sqrt{-1}$ where A and B are relatively prime, nonzero, real numbers, and $g + h\sqrt{-1}$ some complex number, there exists a real integer s such that

$$s \equiv g + h\sqrt{-1} \pmod{A + B\sqrt{-1}}.$$

After this preliminary review, in Section 2 Dirichlet set out to establish when a complex number is a quadratic residue of another. By definition, $\alpha + \beta\sqrt{-1}$ is or is not a quadratic residue of $A + B\sqrt{-1}$ depending on whether there does or does not exist an expression $x + y\sqrt{-1}$ such that $(x + y\sqrt{-1})^2 - \alpha - \beta\sqrt{-1}$ is divisible by $A + B\sqrt{-1}$.

To decide whether a complex number is or is not a quadratic residue of a composite complex number, it suffices, as in the case of real numbers, to consider the different simple factors of the divisor. So Dirichlet first assumed that the divisor $A + B\sqrt{-1}$ is a prime number. Beginning with the simplest case, he next considered a prime number q of the first kind, and proposed to determine whether $\alpha + \beta\sqrt{-1}$, assumed not to be divisible by q and β not zero, is or is not a quadratic residue of q. Going back to the expression $t + u\sqrt{-1}$, letting each of t and u take on the values $0, 1, 2, 3, \ldots, q - 1$, and disregarding both being 0, gives $q^2 - 1$ numbers, the set of which he denoted by (k). Here, too, we can distinguish two cases, depending on whether $\alpha + \beta\sqrt{-1}$ is or is not a residue of q. He began with examining the second case. There, the set (k) can be divided into groups each of which is composed of two numbers such that their product is congruent to $\alpha + \beta\sqrt{-1} \pmod{q}$. Since there are $\frac{1}{2}(q^2 - 1)$ of these, letting K denote the product of the numbers (k), this gives $(\alpha + \beta\sqrt{-1})^{\frac{1}{2}(q^2-1)}$ is congruent to $K \pmod{q}$. Showing that this holds for the other kind, he arrived at the more general result that this congruence holds with this or the opposite sign depending on whether $\alpha + \beta\sqrt{-1}$ is not or is a residue of q.

Dirichlet next noted that on setting $\alpha = 1$ and $\beta = 0$, K will be congruent to $-1 \pmod{q}$, which is analogous to Wilson's theorem.

After continuing with simple algebraic manipulations, he summarized his results in the form of the following theorems:

> Theorem I. One has $(\alpha + \beta\sqrt{-1})^{\frac{1}{2}(q^2-1)}$ is congruent to $+1$ or to $-1 \pmod q$ depending on whether $\alpha + \beta\sqrt{-1}$ is or is not a residue of q.

> Theorem II. The product of any number of factors is or is not a residue of the prime number q depending on whether among these factors there is an even or odd number of non-residues of q.

Using Legendre's symbol and referring to results of both Euler and Legendre, combining the preceding results now led Dirichlet to a third theorem:

> Theorem III. The expression $\alpha + \beta\sqrt{-1}$, which is assumed not to be divisible by the prime number q (of the first kind) is or is not a residue of q depending on whether one has $(\frac{\alpha^2+\beta^2}{q}) = +1$ or $(\frac{\alpha^2+\beta^2}{q}) = -1$.

As a corollary, by successively setting $\beta = 0$ and $\alpha = 0$, he observed that every real number α is a residue of q and similarly of every imaginary expression of the form $\beta\sqrt{-1}$.

We note in passing that this is one of the first, rare, instances where Dirichlet used the term "imaginary." Later on, perhaps influenced by careful reading of Gauss's definitions in arts. 30–31 of [Gauss 1832b] and his direct study of Euler's and Lagrange's writings, his use of the term "imaginary" became more frequent.

Considering primes of the second kind he established a fourth theorem, stated as follows:

> Theorem IV. Assume that the number $\alpha + \beta\sqrt{-1}$ is not divisible by the prime number of the second kind $A + B\sqrt{-1}$; if one sets
>
> $$A^2 + B^2 = P,$$
>
> I say that $\alpha + \beta\sqrt{-1}$ will or will not be a residue of $A + B\sqrt{-1}$, depending on whether one has
>
> $$\left(\frac{A\alpha + B\beta}{P}\right) = +1 \text{ or } \left(\frac{A\alpha + B\beta}{P}\right) = -1.$$

Letting $(\frac{A\alpha+B\beta}{P}) = \epsilon$ and $(\frac{A\alpha'+B\beta'}{P}) = \epsilon'$, the product $\epsilon\epsilon'$ quickly leads to an expression that shows Theorem II applies to primes of the second kind as well.

This concluded Dirichlet's firming up the conditions governing when a complex number is or is not a quadratic residue of some prime number.

In his fourth section, he now considered finding a simple expression for the characteristics of a prime number of which a given complex number is a residue.

First he remarked that we are permitted to limit ourselves to the case where the given number is prime; for if it is composite, it results from the previous Theorem II that its relation to any prime depends on those of its simple factors to this same prime number.

Dirichlet began with the prime number $1 + \sqrt{-1}$. By Theorem III, this number will or will not be the residue of a prime number of the first kind q according to whether one has $\left(\frac{2}{q}\right) = 1$ or $\left(\frac{2}{q}\right) = -1$.

On the other hand, one knows that the first or second of these cases will hold according to whether q, taken positively, is of the form $8n + 7$ or $8n + 3$. Therefore, $1 + \sqrt{-1}$ will be a residue of every prime of the form $8n + 7$ and non-residue of primes of the form $8n + 3$.

Next, Dirichlet turned to prime numbers of the second kind. To decide whether $1 + \sqrt{-1}$ is or is not residue of a similar number $A + B\sqrt{-1}$, by Theorem IV it is sufficient to know whether $\left(\frac{A+B}{P}\right) = +1$ or $\left(\frac{A+B}{P}\right) = -1$, where $P = A^2 + B^2$.

After a few more algebraic manipulations and decomposition into simple factors, Dirichlet now arived at the theorem

> $1 + \sqrt{-1}$ is a quadratic residue or non-residue of the prime number $A + B\sqrt{-1}$ according as to whether one has $A + B \equiv \pm 1$ or $A + B \equiv \pm 3 \pmod 8$.

Commenting that the case for $1 - \sqrt{-1}$ follows from the preceding, Dirichlet next turned to the more general primes of the form $\alpha + \beta\sqrt{-1}$ and $A + B\sqrt{-1}$.

He first looked at two primes of the second kind where the β or B are even; their relation is determined after demonstrating the theorem that the first is or is not a residue of the second if the second is or is not a residue of the first. To decide whether a complex number is or is not a quadratic residue of a composite complex number, it suffices, as in the case of real numbers, to consider the different simple factors of the divisor. There Dirichlet says let us assume that the divisor $A + B\sqrt{-1}$ is a prime number.

An analogous reciprocity is now proved for the case where the two prime numbers do not both belong to the second kind. Again, a simple manipulation using the Legendre symbol gives the same reciprocity.

Finally, the remaining third case, where both prime numbers are of the first kind, is considered. Here the reciprocity follows immediately from the previously established property of such numbers. These three cases together result in the theorem proposed at the outset of the memoir:

> Denote two complex prime numbers by $\alpha + \beta\sqrt{-1}$ and by $A + B\sqrt{-1}$ (where β and B are even and can be reduced to zero); then the first will or will not be a quadratic residue of the second according to whether the second is or is not a quadratic residue of the first.

The method Dirichlet employed is not unlike one he used repeatedly. Whereas he favored looking for a less complicated approach or more suitable "innate" concept if a problem appeared intractable or too cumbersome when treated by traditional means, yet, before looking for a new approach or concept, he frequently, as in this case, simply broke the problem down into more basic components that permitted relatively easy solutions, particularly if these could be obtained by simple algebraic means. This may leave the impression that the method owes more to Euler or Legendre than to Gauss. That is misleading, however. It is true that Gauss frequently did not reveal all his intermediate steps in so much detail as did Legendre in his textbooks. However, in addition to drawing on his well-known preference for arithmetic examples, Gauss used a similar step-by-step approach in [Gauss 1832b]; Dirichlet facilitated the same approach by resorting to algebraic manipulations and the abbreviating mechanism of the Legendre symbol.

By the mid-1830s, Dirichlet was to find a means of circumventing this still tedious step-by-step approach. The question that Dirichlet's success up to that point raises is whether that famous Gaussian motto "notions not notations" may sometimes get in the way when it leads to losing a connecting thread, so that the problem under consideration appears to be more inherently difficult than it is shown to be by an algebraic breaking down and the use of an efficient notation.

9.2 Fermat's Last Theorem for $n = 14$

The second publication of the year 1832 in Crelle's *Journal*, **1832b**, signed in October, was Dirichlet's proof of Fermat's Theorem for the case $n = 14$. It is quite short.

In publishing the solution for the case $n = 14$, Dirichlet showed that applying the approach previously used for $n = 5$, he could not succeed for the case $n = 7$, which would have been the expected next example.

In his new proof, Dirichlet referred to parts of the 1825 argument, but throughout emphasized the effectiveness of looking at common divisors in the equation. Beginning with the desired equation $t^{14} = u^{14} + v^{14}$, by setting $v = 7w$ he transformed the original equation into $t^{14} - u^{14} = 7^{14}w^{14}$ and then considered the more general equation $t^{14} - u^{14} = 2^m 7^{(1+n)} w^{14}$, with t and u having no common divisors and m and n being nonnegative. After a series of further substitutions and considerations of the existence of common divisors, he arrived at a proof by contradiction showing that neither this more general equation nor the original one can hold.

This memoir played only a minor role in the long saga of Fermat's Last Theorem. Whereas it may have encouraged Lamé and others to seek new approaches for a general proof—Lamé's successful but complicated proof for $n = 7$, published in 1839, has been described by Harold Edwards as seeming to lead to the edge of an impenetrable thicket—the next major step was only taken by Kummer in the late 1840s, after he had introduced his ideal factors.[2]

9.3 Quadratic Forms and Divisors

The first paper Dirichlet presented to the Akademie, **1835**, was read on Thursday, August 15, 1833. Titled "Investigations on the theory of quadratic forms" he preceded the substance of his findings by a lengthy historical introduction.

He began this introduction by explaining why Fermat had justifiably been considered the creator of the theory of numbers although almost none of the proofs

[2]Edwards 1977a, Chapters 3 and 4; also see Edwards 1975 and 1977b.

for his many theorems were extant. Dirichlet pointed out that Fermat's theorems dealing with the relationship of certain linear and second-degree forms are particularly noteworthy as these had been the primary motivation for the further development of the theory of numbers. He commented, as had Gauss in [Gauss 1808a], that it seems to be characteristic of the theory of numbers that most major advances had been achieved by attempts to verify individual results obtained by induction, whereas in all other branches of mathematics significant results had been achieved by finding new points of view which would allow the discoverer to obtain results not achievable by traditional methods. This, as will be seen, remained a dominant motivating factor for Dirichlet's subsequent discoveries in the theory of numbers. By emphasizing methodology over individual results, he sought to bring number theory in line with the other branches of mathematics.

Dirichlet addressed another factor of interest while explaining his belief that Fermat's claims to have proofs for his statements should be considered carefully. He pointed out that Fermat lived in a century still used to the rigor that the ancient Greeks demanded in arithmetic questions as well as geometric ones, so that it is less likely Fermat could have fooled himself into believing he had a proof than it would have been in later times where the ease of the new analytic methods occasionally allows mathematical treatments to degenerate into a mechanism that one follows without even considering the possibility that the results obtained may have to be subjected to some limitations. Again, this is a point that keeps recurring in Dirichlet's thoughts. Knowing that he is mindful of limiting conditions (though he may not always speak of them so explicitly) clarifies a number of issues that arose in discussions of his later work.

Dirichlet continued his historically oriented observations by noting that Fermat's presenting his statements as having been based on rigorous proofs rather than mere inductions contributed greatly to the further development of number theory by making it a point of honor to follow this model. In this connection, Dirichlet called attention to Euler's preoccupation with the relationship between first- and second-degree forms, as exemplified by his theorem that every prime number of the form $4n + 1$ is the sum of two squares.

Dirichlet next brought to the fore Lagrange's achievement of founding a unified theory of quadratic forms. He pointed out that when considering a quadratic form such as $t^2 + cu^2$, where c is a given positive or negative whole number, t and u indefinite whole numbers, Lagrange's methodology hinged on the observations of this form's single divisors. As spelled out later in [Legendre 1830], the *Théorie des nombres*, every such divisor is contained in a trinomial form $gt^2 + 2htu + ku^2$, whose coefficients $g, 2h, k$ are related to c by the equation $gk - h^2 = c$. Considering a specific number c, this could allow infinitely many forms for the divisor. These can be reduced to a finite number, which is not too difficult for positive c but stumped Lagrange when c is negative:

> Now although induction would seem to prove that prime numbers of forms that conform to the divisors of $t^2 \pm au^2$, could always be effectively divisors of such numbers, this proposition can only be proved rigorously with respect to prime numbers $4n + 1$ for a very small number of cases; at least all the attempts I have made at reaching this goal have been useless so far;

> so that I shall restrict myself here to report results of my researches in some particular cases where I succeeded in finding the demonstration of the proposition in question. (Lagrange *Mém. de l'Acad. de Berlin* for the year 1775, 350.)

This reference, which Dirichlet noted explicitly, is to a passage (Section 47) in Lagrange's historic *Recherches d'Arithmétiques*.[3]

Having quoted Lagrange's acknowledgment of the missing piece in his number-theoretic structure, Dirichlet continued by stating that, as Legendre had shown, the theorem whose full proof had eluded Lagrange depended on the law of reciprocity. Dirichlet again reminded the reader, however, that despite its apparent simplicity, Legendre, although using "highly perspicuous" considerations had been able to resolve the difficulties in the proof of that law only in part.

Reflecting his recent preoccupation with mathematical textbooks, Dirichlet noted that, while Gauss in the *D.A.* of 1801 had offered two proofs of the law of reciprocity—which Gauss chose to call the Fundamental Theorem[4]—among later treatises of "this great mathematician" there are several additional proofs, among which two in particular are so simple that they no longer leave anything to be desired for including this theory in an elementary book.[5]

Dirichlet now wished to present to the Akademie one of several remaining questions pertaining to the theory of quadratic forms. The open problem he addressed was to decide a priori to what quadratic forms a given number c belongs. He had discovered that this question could not be resolved by recent methods involving prime numbers and linear forms. These only showed that a prime number contained in a linear form may be involved in one of the corresponding quadratic forms but does not help to determine a priori which of these quadratic forms that may be.

Instead, Dirichlet had found that previous studies of his, having no apparent connection with this question, led him to a theorem that allowed him to determine characteristic properties of prime numbers contained in various quadratic forms and guided him to find how an induction could lead to generalizations. Although his results were not limited to the case where c is a prime number, he wished in the present memoir to restrict himself to that consideration.

His argument, in **1835**, was divided into eight sections. Noticeably, he made frequent references to the recently (1833) deceased Legendre's *Théorie des nombres* of 1830.

In the first three sections, Dirichlet introduced and defined the Legendre symbol $\left(\frac{k}{p}\right)$ and, referencing the appropriate sections of [Legendre 1830], listed its basic

[3] Lagrange 1773/1775; reissued in Lagrange's *Oeuvres* 3:789.

[4] *D.A.*, art. 131.

[5] Except for Lagrange 1775, Dirichlet did not give specific references in this discussion. The reader may wish to note the following: Legendre 1785; Gauss 1801 (reciprocity proofs 1 and 2: *D.A.* art. 131ff. and art. 262); Gauss 1808a = Gauss *Werke* 2:3 (proof 4); Gauss 1808b = Gauss *Werke* 2:9 (proof 5); Gauss 1817 = Gauss *Werke* 2:51 and 55 (proofs 6 and 7); *Analysis Residuorum* = Gauss *Werke* 2:234 (proof 3, unpublished in Dirichlet's lifetime). The numbering of Gauss's proofs in Bachmann 1872:103 differs, as Bachmann does not include proof 3. In addition, Weil 1983 contains several careful discussions concerning Euler's and Legendre's work on quadratic reciprocity.

properties, relevant to reciprocity. Considering the equation $t^2 + au^2 = pq$, with p and q primes and t and u having no common divisors, and dividing it into two classes and several subclasses, he obtained the result:

> The prime numbers p and q either are both in biquadratic reciprocity with respect to a or both in biquadratic non-reciprocity.

Given the previously stated conditions, this is equivalent to the following theorem:

> Let a be a prime of the form $8n + 1$; then all the prime numbers contained in the same quadratic divisor $4n + 1$ of $t^2 + au^2$ either are in biquadratic reciprocity with respect to a or in the opposite, i.e., in non-reciprocity.

As example of the foregoing, Dirichlet, in Section 4, used the value $n = 17$. Referring to one of the tables (IV) in [Legendre 1830], he noted that there are two quadratic divisors $4n + 1$ in this case, namely $t^2 + 17u^2$ and $2t^2 + 2tu + 9u^2$. Each of these gives one class. Setting specific values for t and u, he obtained the result that

> Every prime number of the form $4n + 1$ which is contained in the expression $t^2 + 17u^2$ will be contained simply or doubly in the same expression depending on whether it is in biquadratic reciprocity or nonreciprocity to 17.

Dirichlet used this example to remark that in all similar cases, where there are only two quadratic divisors $4n + 1$ each of which forms a class of its own, this statement will provide the characteristic properties of the prime numbers contained therein. This does not apply when a class consists of two or more forms. That case he described as being very difficult, and for that reason, he wished in the following only to engage in some investigations as to how all the quadratic divisors of the form $4n + 1$ are divided among the two classes.

He devoted Section 5 to a brief discussion of conjugate divisors. Letting a again denote a prime of the form $8n + 1$ and letting every divisor of $t^2 + au^2$ be expressed in the form $2\alpha t^2 + 2\beta tu + \gamma u^2$, where α, β, γ are odd positive numbers satisfying the equation $a = 2\alpha\gamma - \beta^2$ and the inequalities $\alpha \geqq \beta$ and $\gamma \geqq \beta$, he could now define conjugate divisors as pairs of the forms $2\alpha t^2 + 2\beta tu + \gamma u^2$ and $\alpha t^2 + 2\beta tu + 2\gamma u^2$.

This allowed him to observe that every odd number that can be represented by one of these when doubled will be contained in the other. He followed this by showing that since both are either contained in the form $4n + 1$ or $4n + 3$ either one of these forms will belong to both of the divisors. In order for a divisor to be self-conjugate, one must have $a = 2\alpha^2 - \beta^2$ and $\alpha > \beta$. From this, he concluded that there is always one and only one self-conjugate divisor to which the form $4n + 1$ or $4n + 3$ will belong, depending on whether the number α representable by it is of one or the other of these two forms. In the sixth section, Dirichlet noted that among the quadratic divisors $4n + 1$ of the form $t^2 + au^2$ that form itself is included. He then proceeded to show that this form belongs to the first class.

Next, in Section 7, Dirichlet considered the question under what conditions conjugate divisors will belong to the same, and when to opposite classes. To examine this, he drew on [Gauss 1832] and the theorem utilized in his own **1828b**, according to which "letting $a = \phi^s + \psi^2$, and assuming ψ to be even, 2 will be a biquadratic residue or non-residue of a depending on whether ψ is contained in the form $8n$ or in the form $8n + 4$." This allowed him to arrive at the following statement:

> Let $a = \phi^2 + \psi^2$; then any two conjugate divisors $4n + 1$ of the form $t^2 + au^2$ belong to the same class or to opposite classes according as to whether $\phi + \psi$ are contained in the form $8n \pm 1$ or in the form $8n \pm 5$.

In the last section of the memoir, Dirichlet proposed to seek a criterion to determine whether the self-conjugate divisor $\alpha t^2 + 2\beta tu + 2\alpha u^2$ belongs to the form $4n + 1$ or $4n + 3$. If the conjugate forms belong to different classes, the self-conjugate divisor must be of the form $4n + 3$ since the other case leads to a contradiction; the matter requires more consideration if the conjugate divisors belong to the same class. In that case, he needed to draw on the results of his preceding arguments, in Sections 5 and 7.

Applying the quoted statement of the preceding section to the problem at hand, he finally obtained the new theorem:

> Letting a be a prime number of the form $8n + 1$, if one sets $a = \phi^2 + \psi^2$, then the self-conjugate divisor of $t^2 + au^2$ belongs to the form $4n + 1$ or $4n + 3$ depending on whether $\phi + \psi$ is contained in the form $8n \pm 1$ or in the form $8n \pm 5$.

Dirichlet's methodology followed the same pattern seen in his other work of the early 1830s. This tended to have two basic components: It consisted of approaching a complicated problem by breaking it into separate parts, pursuing these step-by-step. It also involved illustrating the problem by using a simplified version of the statement to be proved by placing restrictions on certain of its elements, such as limiting them to being primes, or product of primes, or even assigning specific numerical values to them. This is reminiscent of procedures common in the work of Euler as well as Gauss, although Dirichlet's emphasis on the more general "limiting conditions," as opposed to simply providing limited examples, would become a prominent feature of his future research results, in number theory as well as classical analysis.

9.4 Existence and Uniqueness Issues

The following year, on June 19, 1834, Dirichlet read a memoir that would be published in 1836 with the title "Some new statements concerning indeterminate equations" (**1836a**). He introduced the subject of methods for solving second-degree indeterminate equations with two unknowns by immediately calling attention to the importance of distinguishing between solving an equation with numerical coefficients and establishing whether a solution exists. His opening statement contained a reminder that Lagrange's method of solving these equations leaves nothing to be desired if one actually wishes to solve such an equation numerically.

Dirichlet stressed the Lagrangian stamp in his emphasis on the two approaches—verifying the existence of a solution as opposed to solving an equation, whether in rational numbers, which leads to a straightforward solution, or with the further restriction of seeking whole-number values for the unknowns. The last case requires carrying out a series of computations. Dirichlet commented that this sharp distinction is less surprising if one recalls that the second case requires transformation of the root of the equation into a continued fraction and that one was still very much in the dark concerning the relationship of the members of such a fraction to the coefficients of the equation.

Dirichlet remarked that there are special cases, however, where there are criteria for deciding on the possibility of a solution without having to run through the transformations prescribed in the general method of solution. He noted that he had been able to add to the few known cases among these. These additions he now wished to present.

He pointed out that these special theorems are closely related to the equation $t^2 - Au^2 = 1$. He added background to the story of the equation, emphasizing that it was Fermat who placed it before contemporary mathematicians, among whom were Pell and Brouncker, whose restricted solutions became well-known through the textbooks of Euler and Wallis, and noting that Euler was responsible for calling attention to the equation's significance in solving second-degree equations.[6] Dirichlet pointed out that one had needed a rigorous proof for the equation actually being solvable for every non-square value of A, something that had been tacitly assumed but in fact had only been shown in the special cases treated by Brouncker and Pell. The matter was settled by Lagrange who applied the theory of continued fractions and "thereby laid the foundation for the complete treatment of undetermined second-degree equations."[7]

Dirichlet concluded his introduction by noting that Legendre had provided generally useful instances for which values of A permit solution of the equation. He pointed out in a footnote, that, in his first memoir on undetermined equations, *Mélanges de Turin*, vol. 4, pt. 2, p. 88,[8] Lagrange had erroneously conjectured that this equation is possible whenever A contains no odd prime factors except for those of the form $4n + 1$. Using the example when $A = 5.41$, Dirichlet showed that this condition is necessary but not sufficient. Whereas it is adequate for a large number of cases, he observed that it appears to be quite difficult to provide a complete criterion that would furnish all values of A for which the equation holds.[9]

---

[6]In none of his publications until after his return from Italy in 1845 did Dirichlet use the term "Pell's equation" but always either spelled out the equation or occasionally referred to it as "Fermat's equation" when the context is unmistakeable. Dirichlet's publications after 1845 include Pell's name more than once, presumably a sign that he had become more conscious of Euler's consistent use of the term and that it was now too firmly associated with that important equation to be disregarded. Legendre, too, had resisted the reference to Pell; his reason had a more nationalistic basis; however, for in the preface to Legendre 1808 he related it to the competition between England and France. Gauss also had resisted using the expression; see the *D.A.* art. 202.

[7]Dirichlet's account varies slightly from the summary given by Gauss in art. 202 of the *D.A.* Among other references, note Dickson 1919–23 (2005) 2: Chapter XII.

[8]Lagrange *Oeuvres* 1:671–731.

[9]For references to Lagrange's significant development and publications of this topic between 1761 and 1771, see Weil 1983:233.

Dirichlet divided the substance of his memoir into five sections, in which he gave conditions for the existence of a solution depending on the nature of A. In the first section, he described what he called Legendre's method, referring to Section 7 of the first part of [Legendre 1830]. This essentially consisted of proving the impossibility of a solution for numerous values until only one is left. Legendre could show that there is always one that is impossible, but that did not help determine that this is the only one.

In the second section, Dirichlet assumed that A is an odd prime. Applying the preceding methodology, he arrived at the following statements:

> For every prime number A of the form $4n + 1$ it is possible to have an equation $t^2 - Au^2 = -1$.

> For every prime number A of the form $8n + 7$ the equation $t^2 - Au^2 = 2$ is possible; for every prime number A of the form $8n + 3$, however, the equation $t^2 - Au^2 = -2$ is possible.

He noted that, whereas one can always tell which equation holds if A is a prime, this is not the case if A can be divided into two or more factors.

In the third section, Dirichlet treated the case where A is the product of two odd primes a and b which either both have the form $4n + 1$ or the form $4n + 3$. Using the method introduced in Section 1, he proved that

> If a and b are two prime numbers of the form $4n + 3$, then the equation $at^2 - bu^2 = \pm 1$ is always possible, the sign coinciding with that of the expression $\left(\frac{a}{b}\right) = \pm 1$.

Next, he showed that if a and b are of the form $4n + 1$, and $\left(\frac{a}{b}\right) = -1$, then the equation $t^2 - abu^2 = -1$ is always possible.

However, if $\left(\frac{a}{b}\right) = \left(\frac{b}{a}\right) = 1$, the preceding procedure does not determine which of the remaining three equations allows a solution. To find the appropriate criteria in this case, he applied the previous procedure to the first two of these equations, combining these to read

$$at^2 - bu^2 = \pm 1.$$

In the case of positive 1, we must assume that t and u are, respectively, even and uneven. Setting $u = 2^\nu h' h'' \ldots$, where $h, h', h'', \ldots$ are odd primes, it follows that

$$\left(\frac{a}{h}\right) = 1,$$

and, thanks to the law of reciprocity,[10]

$$\left(\frac{h}{a}\right) = 1.$$

[10]Writing in French, Dirichlet here used Legendre's term.

Multiplying this and the similar equations for h', h'', etc., one obtains

$$\left(\frac{h'h''\ldots}{a}\right) = 1,$$

where the number a has either the form $8n + 1$ or $8n + 5$.

By again breaking the several possibilities down into a series of cases, this consideration led Dirichlet to the theorem

> Let a and b be two prime numbers of the form $4n + 1$, for which $\left(\frac{a}{b}\right) = 1$, and at the same time $\left(\frac{a}{b}\right)_4 = -1$, and $\left(\frac{b}{a}\right)_4 = -1$; then the equation $t^2 - abu^2 = -1$ can always be solved.

He defined the notation used above as follows:

> If c is a prime number of the form $4n + 1$ and k a number not divisible by c, such that $\left(\frac{k}{c}\right) = 1$, i.e. $k^{\frac{c-1}{2}} \equiv 1 \pmod{c}$, then either $k^{\frac{c-1}{4}} \equiv +1$ or $k^{\frac{c-1}{4}} \equiv -1 \pmod{c}$. The symbol $\left(\frac{k}{c}\right)_4$ denotes that residue, whether $+1$ or -1.

As an example of the theorem, he chose $a = 5$, $b = 89$. This satisfies all the conditions, the last equation is solvable, and one obtains the smallest values $t = 4662$ and $u = 221$. In addition, he provided two examples showing that the equation at times may also be solvable if only one or even none of the two conditions involving his modified Legendre symbol hold.

This is a rare instance where Dirichlet contributed an abbreviating notation, modeled after the Legendre symbol. He would have little further use for it after completion of the next several examples, however.

In section four, Dirichlet let the reader deal with the case where $A = ab$ and one of the factors a or b is of the form $4n + 1$, the other of the form $4n + 3$, as well as with the case where $A = 2ab$, with a and b being odd prime numbers.

In addition, he considered, as the last example for this method, the case where $A = abc$, all three factors being prime numbers each of which is contained in the form $4n + 1$. Once more subdividing the possibilities into several further cases, followed by a number of fairly lengthy manipulations, he was able to provide statements and examples for these as well.

9.5 Gauss Sums

Dirichlet's Akademie discourse **1837a** read the following year, on Thursday, June 25, 1835, dealt with "A new application of definite integrals for summing finite or infinite series." Here Dirichlet turned to Gauss, specifically what later would be called Gauss sums. He introduced his topic by calling attention to the "numerous and unexpected consequences" Gauss derived from his method of solving binomial equations—the cyclotomic equations. Referring to noteworthy difficulties with certain finite series among these, Dirichlet called attention to the problem of determining the sign for their sums.

Dirichlet explained that, letting p be a prime number, depending on whether p has the form $4\mu + 1$ or $4\mu + 3$, either of the sums

$$\sum_{s=0}^{p-1} \cos s^2 \frac{2\pi}{p} \text{ or } \sum_{s=0}^{p-1} \sin s^2 \frac{2\pi}{p}$$

will equal to $\pm\sqrt{p}$; in other words, it can be either positive or negative.

The determination of the sign for such a sum can usually be obtained from the nature of the problem to be solved. That is not so in this case, however, because of the fact that the members of these series can be partly positive, partly negative, and it cannot generally be established which of these predominates. For certain known values of p, it can be determined by use of trigonometric tables; but that does not apply in general; and, Dirichlet further observed, cyclotomy has seemed to present no means of determination for results obtained by induction. He noted that Gauss had not touched the issue of the sign in the 1801 *D.A.* but had resolved it in a later publication devoted to that problem.[11] Dirichlet explained that Gauss's approach, "the idea of which is as simple as the execution is ingenious," consisted of converting such a series, even one generalized to using any integer instead of the prime p, into a product of sines whose arcs are in arithmetic progression. That allows determination of the sign, since all the negative factors appear in even numbers. However, it still entails lengthy manipulations before the answer is finally obtained. This motivated Dirichlet to look for another angle of attack that would allow him to reach the final answer without having to go through all the intermediate calculations. He remarked that it was particularly interesting to find an alternate solution since Gauss's had been the only successful one so far.

As an example of a failed attempt to obviate the ambiguity of the sign, Dirichlet mentioned [Libri 1832], which, he noted, also leads to a quadratic equation. Libri, too, attempted to use a series converted to a sine product but Dirichlet commented that, by failing to verify the equivalence of his two series, Libri missed the salient point which would make every other consideration superfluous, since the resulting product belongs to those well-known since Euler's treatment in his *Introductio in analysin infinitorum*, [Euler 1748].

Dirichlet divided this memoir into four sections. In the first, he stated the following two theorems which he considered the basis of what was to follow, and to which, he stressed, one is led by considering trigonometric series that represent an arbitrarily given function in a given interval.

I. Let c denote a constant that satisfies the double condition that $0 < c \leq \frac{\pi}{2}$, and let $f(\beta)$ be a function of β remaining continuous between $\beta = 0$ and $\beta = c$. Then the integral

$$\int_0^c f(\beta) \frac{\sin(2k+1)\beta}{\sin\beta} d\beta$$

[11] Gauss 1811. Gauss had presented it to Göttingen's Royal Society in 1808, but, as in Berlin, there was a delay in publication; it only appeared three years later.

approaches the limit $\frac{\pi}{2}f(0)$ as the positive integer k becomes infinite.

II. Let b and c be constants such that $o < b < c \leq \frac{\pi}{2}$, and let the function $f(\beta)$ be continuous from $\beta = b$ to $\beta = c$. Then the integral

$$\int_b^c f(\beta)\frac{\sin(2k+1)\beta}{\sin\beta}d\beta$$

approaches the limit zero as k increases indefinitely.

He now called attention to the fact that in **1829b** as well as in his memoir for Dove's *Repertorium*, **1837f**, he had shown how to prove such statements under the assumption that the function $f(\beta)$ does not alternately increase or decrease between the limits of integration stipulated. He next wished to show how to extend the statements to the more general case where the function has an arbitrary number of maxima or minima within the limits of integration. To do so, he now once again followed a step-wise procedure, dividing the integral into several whose limits of integration are the values of β for which the function assumes a maximum or minimum.

Next, he established the limit of

$$\int_0^c f(\beta)\frac{\sin(2k+1)\beta}{\sin\beta}d\beta$$

where c is an arbitrary constant, as k increases indefinitely. Letting $l\pi$ denote the largest multiple of π contained in c, and decomposing the integral into two, with limits ranging from 0 to $l\pi$ and from $l\pi$ to c, then proceeding with a further sequence of decompositions for each of these, and applying the statements I. and II. he was able to recombine the results to find that, as k increases indefinitely, the integral approaches the limit

$$\pi(\tfrac{1}{2}f(0) + f(\pi) + \cdots + f(l\pi)) = \tfrac{\pi}{2}f(0) + \pi\sum_{s=1}^{l} f(s\pi),$$

where again l is the largest quantity contained in $\frac{c}{\pi}$, unless c is a multiple of π, in which case the last member of the sum is divided by 2.

In his second section, Dirichlet considered the two integrals $\int_{-\infty}^{\infty}\cos(\alpha^2)d\alpha = a$, and $\int_{-\infty}^{\infty}\sin(\alpha^2)d\alpha = b$. After remarking that one need not assume the fact, already known to Euler, that the constants a and b both have the value $\sqrt{\frac{\pi}{2}}$, as this will result in the course of the analysis, he let β be a new variable and n a positive integer, then set

$$\alpha = \frac{\beta}{2}\sqrt{\frac{n}{2\pi}}.$$

This resulted in the two integrals assuming the form

$$\int_{-\infty}^{\infty} \cos \frac{n\beta^2}{8\pi} d\beta = 2a\sqrt{\frac{2\pi}{n}}$$

and

$$\int_{-\infty}^{\infty} \sin \frac{n\beta^2}{8\pi} d\beta = 2b\sqrt{\frac{2\pi}{n}}.$$

Letting k be a positive integer, considering the limits of integration as running from $-(2k+1)\pi$ to $(2k+1)\pi$, he let k increase to ∞.

After a sequence of substitutions, decompositions, and transformations, and successively replacing β by $-2k\pi + \gamma, -2(k-1)\pi + \gamma, \cdots, 2(k-1)\pi + \gamma, 2k\pi + \gamma$, all the new integrals will have the limits $-\pi$ and $+\pi$; this allowed him to obtain two new integrals

$$\int_{-\pi}^{+\pi} d\gamma \sum_{h=-k}^{k} \cos \frac{n}{8\pi}(\gamma + 2h\pi)^2 \text{ and}$$

$$\int_{-\pi}^{+\pi} d\gamma \sum_{h=-k}^{k} \sin \frac{n}{8\pi}(\gamma + 2h\pi)^2.$$

Commenting that n can have one of the four forms 4μ, $4\mu+1$, $4\mu+2$, $4\mu+3$, he next proceeded to deal with the first case, where n is divisible by 4. Dealing with the sums under the integrals, he again performed a sequence of lengthy manipulations, ingenious though easy to follow, as the result of which he arrived at the two sums

$$1 + \cos 1^2 \frac{2\pi}{n} + \cos 2^2 \frac{2\pi}{n} + \cdots + \cos(n-1)^2 \frac{2\pi}{n} = \sqrt{n}$$

and

$$\sin 1^2 \frac{2\pi}{n} + \sin 2^2 \frac{2\pi}{n} + \cdots + \sin(n-1)^2 \frac{2\pi}{n} = \sqrt{n}.$$

In the third section of the memoir, Dirichlet used a modified method for the cases where n is not divisible by 4 but is contained in one of the forms $4\mu+1$, $4\mu+2$, $4\mu+3$. Again using a lengthy procedure which would produce generalized results, he finally arrived at the following sums:

$$\sum \cos \frac{2i^2\pi}{n} = \sqrt{n}, \ \sum \sin \frac{2i^2\pi}{n} = \sqrt{n}, \ \text{for } n = 4\mu,$$

$$\sum \cos \frac{2i^2\pi}{n} = \sqrt{n}, \ \sum \sin \frac{2i^2\pi}{n} = 0, \ \text{for } n = 4\mu + 1,$$

$$\sum \cos \frac{2i^2\pi}{n} = 0, \ \sum \sin \frac{2i^2\pi}{n} = 0, \ \text{for } n = 4\mu + 2,$$

$$\sum \cos \frac{2i^2\pi}{n} = 0, \ \sum \sin \frac{2i^2\pi}{n} = \sqrt{n}, \ \text{for } n = 4\mu + 3,$$

where the summation extends from $i = 0$ to $i = n - 1$.

In the last, fourth, section of the memoir Dirichlet demonstrated how (following Gauss's *D.A.*, arts. 356 and 357, and pointing to Section IV of the *D.A.*) one can use the four sets of sums just obtained to prove the Fundamental Theorem. Dirichlet's new approach consisted of considering the sum

$$M = \sum_{s=0}^{p-1} e^{s^2 \frac{2q\pi}{p}\sqrt{-1}},$$

where e, as usual, denotes the base of the natural logarithms.

After a new sequence of easy, but lengthy, substitutions and algebraic manipulations, accompanied by the breaking down into a number of cases depending on the residual relationship of p to q, he was able finally to determine the conditions under which the sign of the sum is positive or negative. In particular, he noted that if both p and q have the form $4\mu + 3$, the sign is negative, whereas it is positive otherwise.[12]

A French version (**1837f**) of the memoir on Gauss sums was included in Crelle's *Journal* for 1837.[13] Understandably, Dirichlet here referred to Gauss's "Fundamental Theorem" as the "Law of Reciprocity."

9.6 Eulerian Integrals

In 1836, Dirichlet published a short article on Eulerian integrals **1836c** in Crelle's *Journal*. His object was the demonstration of the equation

$$\Gamma(a)\Gamma\left(a + \frac{1}{n}\right)\Gamma\left(a + \frac{2}{n}\right)\dots\Gamma\left(a + \frac{n-1}{n}\right) = (2\pi)^{\frac{n-1}{2}} n^{\frac{1}{2}-na}\Gamma(na)$$

without recourse to infinite developments.

[12]The influence of this memoir is suggested in Chapter 2 of [Patterson 2010]. Patterson, commenting on the statement in [Davenport 2000] that Dirichlet's method "is probably the most satisfactory of all that are known," added that it is also the one least frequently reproduced. This is not surprising if one follows the entire long argument, which has been considerably condensed in our lengthy outline.

[13]Although described as an "excerpt," **1837f** is actually an only slightly modified summary of the memoir **1837a** printed in the *Abhandlungen*.

As Dirichlet noted at the outset, Legendre had coined the expression Eulerian integral in his *Exercises* of the integral calculus and discussed them in his treatise on elliptic functions. Dirichlet used the following definition:

The integrals that Legendre called Eulerian of the first and second kind are those where the equations

$$(1) \qquad \int_0^1 (1-x)^{a-1} x^{b-1} dx = \int_0^\infty \frac{y^{a-1} da}{(1+y)^{a+b}} = \left(\frac{b}{a}\right)$$

$$(2) \qquad \int_0^1 \left(\log \frac{1}{x}\right)^{a-1} dx = \int_0^\infty e^{-y} y^{a-1} dy = \Gamma(a),$$

in which the constants a and b, or at least their real parts, must be assumed to be positive in order that the integrals do not become infinite. Dirichlet noted that Euler had shown the connection between the two equations can be expressed by the very simple relationship

$$(3) \qquad \left(\frac{b}{a}\right) = \frac{\Gamma(a)\Gamma(b)}{\Gamma(a+b)}.$$

$$\Gamma(a)\Gamma(1-a) = \int_0^\infty \frac{y^{a-1} dy}{1+y}.$$

Dirichlet further observed that, as Euler and subsequent authors had shown using different approaches, this last integral has the very simple value $\frac{\pi}{\sin a\pi}$, so that the last equation becomes

$$(4) \qquad \Gamma(a)\Gamma(1-a) = \frac{\pi}{\sin a\pi}.$$

Referring to work by Gauss, Legendre, Poisson, and Jacobi, as well as Cauchy and Crelle, Dirichlet suggested that the diversity of their procedures called for a more uniform approach.

9.7 Efficacy of Least Squares

The memoir **1836b** Dirichlet read before the plenum of the Akademie on July 28, 1836, was related to two courses he gave in the summer term of 1836 pertaining to probability theory, and one, given in the winter term 1836/37, on the figure and movement of celestial bodies. In addition to recalling his mentor Lacroix's explication of probability theory, these led to his taking a close look at Laplace's work on probability and on celestial mechanics.

According to **1836b**, the Akademie's *Bericht*, the presentation carried the lengthy title "On the question how far the method of least squares can be considered the most advantageous means of determining unknown elements among all linear combinations of the equations of conditions in the case of very numerous observations." The issue was one of concern to astronomers as well as geodesists, physicists, and other scientists. It had been discussed by Lagrange, Laplace, most recently and frequently by Encke, and others. Dirichlet's analysis was outlined in the *Bericht* but not published in the Akademie's related *Abhandlungen*.

Dirichlet observed that Laplace in his work on the analytic theory of probabilities assumed that the various "factor systems" among which one needs to choose are independent of the constants in the equations. Without this assumption, one can provide systems that are quite different from that corresponding to the method of least squares but generally appear to provide an equal degree of precision. As the simplest example of such an alternative method, Dirichlet cited that of sorting the values of a large, odd number of these observations according to size, and taking the median. In the brief report he stated that, comparing the limits within which the error for the value determined thereby lies for a given probability with those corresponding to the arithmetic mean, which, for the case at hand, the method of least squares approaches, the result will be that, with equal probability, the error limits for both methods will correspond to one another like the constants $\frac{1}{\sqrt{2}f(0)}$ and $2\sqrt{\int_0^a x^2 f(x)dx}$. The function $f(x)$, for which $f(-x) = f(x)$, expresses the law of the errors of observation, which are taken between $-a$ and a. He concluded that it is clear that, as long as one makes no assumptions concerning $f(x)$, one cannot decide which of those constants is larger, and so it remains undecided whether the arithmetic mean or the other procedure is preferable.

In an expanded note on the subject, including historical references to Lagrange's "Sur l'utilité de la méthode de prendre le milieu," to Laplace, and to Encke, Dirichlet concluded that "the relationship is that of two constants both of which depend on the unknown error law of the observations; the constant corresponding to the arithmetic mean is expressed by an integral extending over the entire extent of the error curve whereas the other one is merely determined by the ordinate at the origin of the coordinates."[14]

For some related details, see our Chap. 16 and the noteworthy, extended discussion of Dirichlet's studies of probability theory in [Fischer 1994]. The short report **1836b** is the only publication by Dirichlet on the subject of error theory.

Fischer described **1836b** as being an example of "Dirichlet's total neglect of the practical aspects of error theory." This characterization may be moderated by noting that Encke, in lectures and written statements, in addition to discussing the entire issue of optimal grouping of observations, had for some time treated the significance of the method of least squares for applications, as had his teacher Gauss, who claimed its discovery, applied it, and gave occasional lectures on the subject. It appears appropriate to suggest that Dirichlet here consciously supplemented Encke's treatment by

[14] Werke 2:351.

wishing to focus on the relative merits of the two approaches (taking the median vs. an arithmetic mean or the method of least squares) without duplicating Encke's efforts of dealing with specific applications. This need not imply that Dirichlet ignored such applications.

9.8 Primes in Arithmetic Progressions

A memoir read before the plenum of the Akademie on Thursday, July 27, 1837, marks a milestone in the progress of Dirichlet's research and in the history of number theory. The Akademie's *Bericht* **1837c** simply listed it as "A proof of a statement concerning the arithmetic progression." The full memoir **1839a** was titled more explicitly as "Proof of the statement that every unlimited arithmetic progression, whose first member and difference are integers without a common factor, contains infinitely many prime numbers." It is often regarded as constituting the beginning of systematic analytic number theory and, as we shall see, is frequently mentioned by Dirichlet himself for his introduction therein of his L-series. His being described as the founder of analytic number theory, sometimes even its father, is a questionable patrimony, however, unless one wishes to disregard the fundamental concepts and results of Euler, Lagrange, and Gauss, and to consider only the cohesive approach Dirichlet established while building on their concepts and methodology employed in this 1837 discourse on arithmetic progressions.[15]

1837c

This abstract in the Akademie's *Bericht* begins by stating that up to that time, there was no rigorous proof of the statement that "every arithmetic progression whose first member and difference have no common factor, contains infinitely many prime numbers." Dirichlet stressed that this proposition is not without importance for higher arithmetic, "not only because it can be used as a lemma in a variety of investigations, but also because it can be regarded as the complement of one of the most beautiful theories of this part of science, namely, the doctrine of linear forms of simple divisors of quadratic expressions."[16]

As an example, he noted that when we deduce from "the fundamental statement of this doctrine, that is, from the so-called law of reciprocity," that the expression $x^2 + 7$ has all prime numbers of the three forms $7n + 1, 7n + 2$, and $7n + 4$ and only

---

[15]The full proof in **1839a**, while rigorous, is very long and, because of its detail, appears complex. We provide a greatly abbreviated outline. The reader interested in proof details may wish, in addition to consulting the memoir itself and Dirichlet's later expansion of the theorem to quadratic forms, to take note of subsequent smoother but equally rigorous presentations of the 1837 argument. Supplement VI in Dedekind's editions of Dirichlet's Lectures on Number Theory, except for some reorganization and Dedekind's customary increased clarity, comes closest to Dirichlet's 1837 proof; see Chapter 16.3. Among more recent discussion and proofs, Chapter 1 of [Davenport 2000] also includes an indication of the different approaches Dirichlet took to the problem.

[16]*Werke* 1:309.

those as divisors, it still remains undecided how these simple divisors are distributed among those forms. As long as the above statement has not been proved, one could imagine that one or two of the forms contain no prime numbers at all.

Dirichlet proposed in this memoir to precede a later more general proof by here giving only an indication of the proof for the case that the difference of the progression is an odd prime number p. He sketched a brief introductory summary of the proof in the Report, noting there, as he would later frequently repeat, that his proof presents a certain analogy with the content of Chapter XV in the first volume of Euler's *Introductio in analysin infinitorum*, [Euler 1748].

1839a

Dirichlet began the published version of the memoir by observing that careful observation of the sequence of prime numbers can lead to a multitude of statements whose general applicability, by continued induction, can be raised to any arbitrary degree of probability. Finding a proof for such a statement that satisfies all the demands of rigor is tied to the greatest difficulties, however. The statement that "every unlimited arithmetic series, whose first member and difference have no common factor, contains infinitely many prime numbers" is a noteworthy example. Dirichlet pointed out that there was no adequate proof for this simple statement. As far as he knew, Legendre was the only mathematician who had attempted to justify it. Dirichlet noted that the statement can be useful because of numerous applications to be derived from it, and he thought that, aside from being attracted by the difficulty of a proof, Legendre would have been particularly interested in finding one because he had used the statement as a lemma at an earlier date.

Dirichlet told how, having noted that Legendre's induction did not provide a satisfactory proof, he had attempted to follow Legendre's example of considering the largest number of successive members of an arithmetic series which could be divided by given prime numbers, but he found that this got him nowhere. It was only after leaving Legendre's procedure that he arrived at a rigorous proof, which, however, is not purely arithmetic, but rests in part on the consideration of variable quantities. Dirichlet concluded his introduction by commenting that because of the novelty of the principles he is applying, he should prefix the proof of the theorem in its full generality by treating the special case where the difference between successive members of the progression is an odd prime number.

Dirichlet divided his complete memoir into eleven sections. Initially, he reviewed some properties of prime numbers, primitive roots, and residues, calling attention to Gauss's definition of the index of a congruence and referring to Legendre's symbol.

He let p be an odd prime number, q a prime number different from p, ω any root of the equation $\omega^{(p-1)} - 1 = 0$, and formed the geometric series

$$\frac{1}{1-\omega^{\gamma}\frac{1}{q^s}} = 1 + \omega^{\gamma}\frac{1}{q^s} + \omega^{2\gamma}\frac{1}{q^{2s}} + \omega^{3\gamma}\frac{1}{q^{3s}} + \cdots. \tag{2}$$

Next, he obtained the equation

$$\prod \frac{1}{1-\omega^{\gamma}\frac{1}{q^s}} = \sum \omega^{\gamma}\frac{1}{n^s} = L, \tag{3}$$

where the multiplication extends over all primes except for p, and n is any number not divisible by p.

Explaining that this equation represents $p-1$ distinct equations, formed upon replacing ω by its $p-1$ values, he denoted the resulting series by $L_0, L_1, L_2, \ldots, L_{p-2}$.

Before proceeding, he justified the condition that $s > 1$. Then, having set $s = 1+\rho$, he next proved that, as ρ, which is positive, becomes infinitely small, the finite limit approached by $\sum \omega^{\gamma}\frac{1}{n^{1+\rho}}$ is not zero provided that ω does not equal 1.

In Section 5 of the memoir, he took the logarithm of the left-hand side of equation (3); this allowed him to prove that, as ρ becomes infinitely small, if m is positive and neither equal to 0 nor to $\frac{1}{2}(p-1)$, the limiting value of L_m, will be finite and nonzero; L_0 will become infinite. He then showed that, as ρ becomes infinitely small, if $\omega \neq 1$, $\log L$ will approach a finite limit, but if $\omega = 1$ it will become infinitely large.

Next he proved the theorem for the case that the difference of the series equals a prime, and that the first member is not divisible by p. Before extending the preceding to an arithmetic series whose difference is a composite number, Dirichlet reviewed a number of results from Gauss's Section III of the *D.A.* on power residues. After this, he set out to prove the theorem in all generality. He divided the L-series into three classes. The first consists of just one series, where each of the previous roots of unity equals 1. The second contains all the series where there are only real roots. The third class includes all the rest, that is, all series containing at least one imaginary root. His discussion resulted in the conclusion that, as ρ becomes infinitely small, all the series of the second and third class approach the limit

$$\sum \theta^{\alpha}\phi^{\beta}\omega^{\gamma}\omega'^{\gamma'}\cdots\frac{1}{n} = \int_0^1 \frac{\sum \theta^{\alpha}\phi^{\beta}\omega^{\gamma}\omega'^{\gamma'}\ldots x^{n-1}}{1-x^k}dx. \tag{14}$$

As he noted at the end of this section, it remained to be proved that this limit is different from zero. He accomplished this in the final two sections by means of another sequence that included taking logarithms, and performing a variety of summations, reductions, and substitutions. This lengthy activity led to the following expression (labeled 16):

$$\sum (\pm 1)^{\alpha}(\pm 1)^{\beta}(\pm 1)^{\gamma}(\pm 1)^{\gamma'}\cdots\frac{1}{n},$$

showing that for any one combination of the roots ω of the form $\pm 1, \pm 1, \pm 1, \ldots$, where one needs to exclude the possibility of $+1, +1, +1, \ldots$, the sum (16):

$$\sum (\pm 1)^{\alpha}(\pm 1)^{\beta}(\pm 1)^{\gamma}(\pm 1)^{\gamma'}\cdots\frac{1}{n},$$

where $\alpha, \beta, \gamma, \gamma', \ldots$, designate the system of indices for n, and n all positive integers, not divisible by any of the prime numbers 2, p, p', p'', $\cdots$, following in order of size, has a nonzero value.

Dirichlet concluded the memoir by remarking that his first proof for this nonvanishing of the limit (14) that he had presented to the Akademie was too indirect and complicated. Since then, he had found another, shorter path to a proof. He noted that the principles he had used could be extended to numerous other applications which appeared to have no connection with the problem at hand. In particular, he pointed out that these principles can be applied to solve "the very interesting problem" of determining the number of distinct quadratic forms that correspond to an arbitrary positive or negative determinant. Stressing that this is not the final form of his result, he called attention to the fact that this number, [the class number], can be represented as the product of two factors, the first of which is a very simple finite function of the determinant, while the other consists of a series which coincides with (16). This shows that (16) cannot vanish, since this would imply that the determinant which is always greater than or equal to 1, would also be zero. For that reason, he wished to omit his former proof of this property of (16) but instead refer to his forthcoming investigations concerning the number of quadratic forms, which would contain the statement necessary for completing the proof of the present memoir as a corollary.

An added footnote called attention to **1838b** described as a "provisional [*vorläufige*] note on the subject that had appeared in the meantime, i.e., between his presentation in 1837 and this publication in 1839.

9.9 The *Repertorium* Report on Arbitrary Functions

Our preceding Chap. 8 contains a brief explanation of the purpose and nature of the revised *Repertorium der Physik*. Dove's preface to his first volume of the *Repertorium* introduced Dirichlet's contribution with the following remark:

> The numerous applications which the representation of entirely arbitrary functions by sine and cosine series have found recently in the analytic treatment of physical problems require, even if they were to be reproduced merely through their results, that the mathematical considerations from which they stem be set forth. This is done in the... work by Professor Dirichlet and therefore can be viewed as an introduction to later reports.

Dirichlet's memoir **1837d**, titled "Concerning the representation of arbitrary functions by sine and cosine series," opened with a short statement essentially echoing Dove's preface. He pointed out that the noteworthy series which represent functions in a given interval that within this interval or in separate portions of the interval either follow no law or different laws in different portions of the interval, since Fourier's founding of mathematical heat theory has received many applications in the analytic treatment of physical problems. Therefore, it seemed appropriate to introduce the excerpts from the most recent works on certain parts of mathematical physics intended for the following volumes [of the *Repertorium*] with the development of some of the most important of these series.

The memoir is divided into six sections. In the first, Dirichlet defined a continuous function:

> Consider two fixed values a and b, and a variable quantity x which is to assume all values lying between a and b. If to each x there corresponds a single, finite y such that while x traverses the interval from a to b continuously, $y = f(x)$ also varies, then y is called a continuous function of x for this interval.[17]

Dirichlet stated that it is not necessary that y depend on x according to the same law throughout this interval, and that it is not even necessary to think of a dependence that can be expressed by mathematical operations. Considering this geometrically, he suggested that if x and y are treated as abscissa and ordinate, a continuous function appears as a connected [*zusammenhängende*] curve, of which to every abscissa contained between a and b corresponds only one point. He emphasized that this definition does not prescribe a common law for single parts of the curve; one can think of it as combined of different parts or as drawn without following any law. Therefore, such a function can only be considered as completely determined in an interval if it is either given graphically for its entire extent or is subject to mathematical laws valid for the separate parts of the interval. He added that as long as one has only determined the function for one part of the interval, the manner of its continuation over the rest of the interval remains totally arbitrary. At the end of this first section, he introduced the definite integral as limit of a sum but also showed under what conditions it can be treated as a surface area.

In the second section, Dirichlet summarized various well-known properties of the definite integral known since relevant work of Lagrange, Lacroix, Poisson, and Cauchy. Because it was understood that this memoir was an expository review of the state of the art rather than a presentation of new results, he did not mention anyone by name, however, except for having referred to Fourier at the outset of this work. His thereby stressing the significance of Fourier's work most likely was prompted by the importance for physicists of Fourier's *Analytical Theory of Heat* and of Dirichlet's awareness that questions about its validity raised in previous decades among mathematicians, not only by Biot but also by the prodigious Poisson, still lingered.

In the next section, he observed that various problems in mathematical physics require representation by an infinite sine series of a function given arbitrarily for the interval from 0 to π. Dirichlet devoted this entire third section to an introduction of trigonometric series, their coefficients and transformations. At first, he explained why it appears that the "most natural way" of attaining the required series expansion is the so-called transition from the finite to the infinite. His detailed discussion involved determination of the coefficients for a finite series, setting it equal to the function, expanding it, noting the effect on a specific coefficient of the unlimited increase of the terms of the series, notably its transition to a definite integral; he then outlined the difference between dealing with a sine, a cosine, or a combined series, and the nature of the respective coefficients.

---

[17] *Werke* 1:135.

In Section 4, Dirichlet cautioned the reader that, as convincing as the procedure may be whereby he arrived at the series discussed in the preceding section, this cannot be taken as a rigorous proof for the validity of these series. This hint for the need of additional restrictions also served as an opening to an invitation for the intended readership of physicists to acquaint themselves with a variety of series and limiting conditions needed to assure convergence of the chosen series and existence of a proper representation within the original interval.[18]

This discussion led to a converging cosine series which, when summing its $2n + 1$ terms, resulted in the integral

$$\frac{1}{\pi}\int_{-\pi}^{+\pi} d\alpha\phi(\alpha)\frac{\sin(2n+1)\frac{\alpha-x}{2}}{2\sin\frac{\alpha-x}{2}}.$$

For the series to converge to the value $\phi(x)$, the difference between this integral and $\phi(x)$ must become less than any considered small positive quantity as n increases without limit.

As was his custom, to prove that statement, he preceded the study of this general integral by noting special cases. He chose two obvious ones. Taking these up in his fifth section, for the first one, he considered the integral

$$\int_{0}^{\frac{\pi}{2}} \frac{\sin(2n+1)\beta}{\sin\beta} d\beta,$$

where n is positive; for the second one, he dealt with the integral

$$\int_{0}^{h} \frac{\sin k\beta}{\sin\beta} f(\beta)d\beta = \frac{\pi}{2}.$$

Combining the results of his fairly lengthy discussion, Dirichlet arrived at two theorems. The first theorem (labeled 17) states that

> If $f(\beta)$ is a continuous function of β, which, while β increases from 0 to h, where $(0 < h \leq \pi/2)$, will never alternate from decreasing to increasing and conversely, then, if one attributes to n constantly increasing positive values, the integral
>
> $$\int_{0}^{h} \frac{\sin(2n+1)\beta}{\sin\beta} f(\beta)d\beta$$
>
> will differ less than any given quantity from $\frac{\pi}{2} f(0)$.

[18]For a more detailed treatment of trigonometric (Fourier) series with related nineteenth-century references, Whittaker & Watson (1927) 1962, Chapter IX, is still useful. Among the numerous nineteenth-century memoirs, notably Riemann 1867, attempting to refine the concept of representing an arbitrary function, note Lipschitz 1864 and du Bois-Reymond 1873. A sound early overview was provided by Sachse 1880; it was subsequently superseded by Dugac 1981.

The second theorem (labeled 18) states that

> If g and h are constants satisfying the conditions that $g > 0$, $\frac{\pi}{2} \geq h > g$, and if the function $f(\beta)$, with β increasing from g to h, will never change from decreasing or increasing, or conversely [i.e., if it will be monotonic, so that it will have no maximum or minimum between g and h], then the integral
>
> $$\int_g^h \frac{\sin(2n+1)\beta}{\sin\beta} f(\beta)d\beta$$
>
> will become equal to 0 for an infinitely large n.[19]

Another detailed discussion, drawing on the prior results, leads to the expression

$$(20) \qquad \begin{cases} \frac{1}{2}b_0 + b_1\cos x + b_2\cos 2x + \cdots + b_m\cos mx + \cdots \\ \quad + a_1\sin x + a_2\sin 2x + \cdots + a_m\sin mx + \cdots, \end{cases}$$

where the coefficients are determined by the equations

$$b_m = \frac{1}{\pi}\int_{-\pi}^{+\pi} d\beta\phi(\beta)\cos m\beta, \quad a_m = \frac{1}{\pi}\int_{-\pi}^{+\pi} d\beta\phi(\beta)\sin m\beta.$$

Dirichlet pointed out that

> it emerges rigorously from the foregoing that this series is always convergent, i.e. that there is always a certain value from which the sum of the $2n+1$ first terms of the series, when one thinks of n as increasing beyond all limits, at the end will always differ by less than any assignable quantity, and that this value or the sum of the infinite series, when x lies between $-\pi$ and π, will be represented by $\frac{1}{2}[\phi(x+0)+\phi(x-0)]$, but by $\frac{1}{2}[\phi(\pi-0)+\phi(-\pi+0)]$ for $x=\pi$ and $x=-\pi$.

He briefly provided several examples of series that can be considered as special cases of (20), noting properties that only thereby become obvious, and concluded this extended survey of trigonometry, trigonometric series, and their use in representing "arbitrary" functions as follows:

> If one continues the function $\phi(x)$, arbitrary from 0 to π, according to the equation $\phi(-x) = \phi(x)$ it will be clear that for $x = 0$ there will be no discontinuity and that $\phi(-\pi) = \phi(\pi)$. Therefore the series (20) will be $\phi(0)$ for $x = 0$ and $\phi(\pi)$ for $x = \pi$.... [Elementary trigonometric operations on the coefficients b_m and a_m will produce]
>
> $$b_m = \frac{2}{\pi}\int_0^{\pi} d\beta\phi(\beta)\cos m\beta, \quad a_m = 0.$$
>
> Therefore the function $\phi(x)$, given arbitrarily from $x = 0$ to $x = \pi$ will be represented by the series:
>
> $$\frac{1}{2}b_0 + b_1\cos x + b_2\cos 2x + \cdots + b_m\cos mx + \cdots,$$
>
> which will still be valid for the values 0 and π that bound the interval.[20]

[19] *Werke* 1:155.

[20] *Werke* 1:159–60.

Dirichlet added that it is understood ["es versteht sich von selbst"] that if there is a discontinuity for $\phi(x)$ between 0 and π, then for every such value of x the series expresses half the sum of the corresponding values of $\phi(x)$. He noted that, upon using a procedure similar to the preceding one for the cosine series, the corresponding sine series generally is no longer valid between $x = 0$ and $x = \pi$; he commented that this is self-evident since that series will vanish for the values named, regardless of its coefficients.

On reading the full elaborate discussion in the *Repertorium der Physik*, it appears that Dirichlet assumed its readers to be mostly unfamiliar with basic concepts of Fourier series, continuity, and convergence criteria, while having had some elementary exposure to integrals and inequalities.

9.10 Series Expansions and Spherical Functions

Dirichlet's study of Laplace's celestial mechanics, mentioned in our discussion of **1836b** led to a memoir (**1837e**) in Crelle's *Journal* dealing with "the series whose general term depends on two angles, and which serve to express arbitrary functions between given limits."

Dirichlet introduced his subject with the explanation that the series under consideration are ordered according to the P-functions first used by Legendre in his work on the attraction of spheroids and the figure of planets. He observed that, aside from a number of other remarkable properties which these functions have, the series based on them can be used to represent arbitrary functions between given limits. Yet as Laplace had noted in the *Mécanique celeste* (vol. 2), this had not been rigorously established in the context of the theory of the attraction of spheroids. For that reason, Dirichlet stated that a general proof independent of this theory seemed desirable.[21] He proceeded to provide a rigorous demonstration, drawing in part on his convergence proof given in **1829b**, offering a geometric illustration, but also, in an addition to the memoir,[22] supplying alternate sufficiency conditions for this and similar proofs. In this addition, he took note of the following two theorems:

> Let the function $f(\beta)$ remain finite from $\beta = 0$ to $\beta = h$ (where $0 < h \leq \frac{1}{2}\pi$); then the integral $\int_0^h f(\beta)\frac{\sin k\beta}{\sin\beta}d\beta$ will converge to $\frac{1}{2}\pi f(0)$, if the positive quantity k becomes infinite.

and

> Let the function $f(\beta)$ remain finite from $\beta = g$ to $\beta = h$ (where $0 < g < h \leq \frac{1}{2}\pi$); then the integral $\int_0^h f(\beta)\frac{\sin k\beta}{\sin\beta}d\beta$ will vanish for $k = \infty$.

In discussing the relevance of these theorems with respect to a change of convergence conditions, Dirichlet led the way to weakening the conditions for applying his earlier convergence proofs. His explanation provides one of the numerous

[21] He also reviewed an approach given by Poisson, noting its insufficiency.

[22] *Werke* 1:305–6.

examples of the (not always explicit) distinction drawn by him in his lectures and publications between treating a mathematical demonstration with the restrictions imposed by applications to real-world problems and a more general purely abstract proof. This would be spelled out more explicitly by Edmund Heine, who discussed Dirichlet's proof in his *Handbuch der Kugelfunctionen*, noting the historical background in the introduction to the first edition of his *Handbuch*, [Heine1861], and devoting the fifth chapter of that work's second part to Dirichlet's proof. Heine referred to that proof of the necessary and sufficient condition for the expansion of a function of two variables according to spherical functions as the "foundation for the applications of physical problems."[23]

9.11 Pell's Equation and Circular Functions

In a memoir, **1837g**, appearing in the third (of four) issues of Crelle's *Journal* for 1837 titled "On the manner of solving the equation $t^2 - pu^2 = 1$ by means of circular functions," Dirichlet called attention to the short Akademie *Bericht* for the preceding July (**1837c**) of his work on primes in arithmetic progression. He noted that his study leading to the proof of the theorem established there had caused him to observe an unexpected rapport between two branches of number theory which until then had appeared to have nothing in common.

Dirichlet was referring to two approaches to the solution of Pell's equation, still without referring to Pell by name but simply stating the form of the equation. He reviewed the fact that the equation $t^2 - pu^2 = 1$, where p is a positive non-square integer, can always be solved in whole numbers, and that this fundamental proposition in the theory of second-degree indeterminate equations was deduced by Lagrange from the consideration of the periodic continued fraction which results from the expansion of the radical $\sqrt{p}$. He now commented that it is remarkable that the solution of the preceding equation can also be attached to the theory of binomial (cyclotomic) equations, knowledge of which is owed to Gauss. It results from this theory not only that the equation $t^2 - pu^2 = 1$ is always solvable, but one can even deduce from it general formulas which express the unknowns t and u as circular functions.

Although this second manner of treating the equation in question is applicable in every case, Dirichlet wished to limit himself in this note to developing the one where p is a prime number, since this case is sufficient to make known the *esprit* of the method. He cautioned the reader that the manner of solving the problem he was about to pursue is far less appropriate for numerical calculation than that which derives from the use of continued fractions. He stressed that his new method of solving the

[23]Heine 1861:266. Aside from studying the entire discussion in *Werke 1:285–306*, the reader wishing to work through the details of Dirichlet's rather lengthy argument may wish to consult the references in Heine 1861 just listed, and note several remarks, with allusions to set theory, in Dauben 1979:10–11.

equation $t^2 - pu^2 = 1$ is to be viewed strictly from a theoretical point of view and as an example of the *rapprochement* between two branches of the science of numbers.

Letting p be an odd prime number, Dirichlet began by considering the equation

$$\frac{x^p - 1}{x - 1} = X = 0.$$

He noted that the roots of this equation can be expressed as $e^{m(\frac{2\pi}{p}\sqrt{-1})}$ where m is any integer in the sequence $1, 2, 3, \ldots, p - 1$. Among these integers are $\frac{1}{2}(p - 1)$ residues and an equal number of non-residues of p. Drawing on article 357 of Gauss's *D.A.*, he next engaged once again in a series of easily followed algebraic manipulations and of splitting the problem into a number of cases depending on the nature of p. Gauss had previously (*D.A.*, art. 124) stated without proof the theorem he proved in art. 357. In art. 124, which dealt with residues of ± 7, Gauss had noted that this originated with Lagrange in 1775 who did not carry out his procedure for $p > 7$.[24]

Following this approach with expanded manipulations, Dirichlet found that his proof of the theorem in question becomes a corollary of Gauss's theorem of the *D.A.*, art. 356, according to which the polynomial $4X$ can always be put into the form $Y^2 \mp pZ^2$. He noted that, following Gauss's method, the result can be easily extended to a non-prime value of p; he then concluded the memoir with a decomposition of two polynomials Y and Z and a numerical example.[25]

9.12 Asymptotic Laws in Number Theory

On Thursday, February 8, 1838, Dirichlet read a memoir "On the determination of asymptotic laws in number theory" to the Berlin Akademie. Only a brief notice **1838a** appeared in the Akademie's *Bericht.*

He introduced the topic by reminding his audience of a well-known analytic phenomenon: Functions that appear more composite as the independent variable increases often vary with increasing regularity despite the apparently never-ending growing complication involved, so that there is a simple expression which comes closer and closer to such a function and describes its course similarly as one curve represents another of which it is the asymptote. On the basis of this analogy to geometry, one can call such an expression the asymptotic law of the more complicated function, provided one interprets the word "asymptotic" more generally and refers to the quotient of both functions, which is to be regarded as indefinitely approaching unity, whereas their difference does not necessarily decrease infinitely.

Dirichlet noted that the oldest example of such an asymptotic law is provided by the expression

[24] Lagrange *Oeuvres* 3:788; Lagrange later provided a full proof.

[25] *Werke* 1:349–50.

$$\frac{2^{2n}}{\sqrt{n\pi}}$$

which Stirling had derived, from Wallis's infinite product for π, to approximately determine the middle binomial coefficient of a very high even power. Subsequent studies provided quite a few similar results, which had become especially important for probability calculations.

Dirichlet next stressed, as he had done in **1836b**, that the existence of such asymptotic laws need not be restricted to analytic functions; they can even occur when there is no analytic expression at all, as is usually the case when considering functions pertaining to properties of numbers. As examples of such functions, he called attention to Legendre's formula, found by induction, which approximates the number of primes not exceeding a given limit. He also noted that the *D.A.* contains a number of similar expressions belonging to the theory of quadratic forms that represent the mean number of classes and orders of such forms as function of the determinant. He added, however, that so far no proof was known either for these or for Legendre's formula.

He explained that the purpose of the memoir he was presenting to the Akademie was to develop several methods that often can be used successfully in investigations of this sort and the application of which will result in Legendre's formula as well as in some of those reported by Gauss. In the extract, he wished to limit himself to providing only one example by dealing with a problem not treated so far that relates to the theory of divisors.

Let b_n denote the number of divisors of n, including 1 and n. Then b_n will be an irregularly progressing function of n, which, although increasing beyond every limit as n increases, nevertheless will assume small values such as 2, 3 infinitely often. If instead of considering this function, one considers its mean value as defined in the *D.A.* [art. 301], then the irregularity disappears, and this mean value will be capable of an asymptotic law.

To determine this, Dirichlet suggested considering the infinite series

$$b_1\rho + b_2\rho^2 + \cdots + b_n\rho^n + \cdots = f(\rho),$$

which, he remarked, as Lambert had already noted, can also be represented as follows:

$$\frac{\rho}{1-\rho} + \frac{\rho^2}{1-\rho^2} + \cdots + \frac{\rho^m}{1-\rho^m} + \cdots = f(\rho).$$

The sum of the series will be finite as long as ρ is a proper fraction; it grows indefinitely while ρ (considered > 0) approaches unity. Setting $\rho = e^{-\alpha}$ and representing the series by a definite integral, then, for indefinitely small positive values of α, the series can be represented as

$$\frac{1}{\alpha}\log\left(\frac{1}{\alpha}\right) + \frac{C}{\alpha},$$

where C is Euler's constant.

Dirichlet now observed that there is a necessary connection between this formula, which expresses the rapidity whereby the function $b_1e^{-\alpha} + b_2e^{-2\alpha} + \cdots + b_ne^{-n\alpha} + \cdots$ increases and is to be regarded as its asymptotic law for decreasing values of α, and the mean value of the general coefficient b_n. Using the properties of the gamma function $\Gamma(k)$ results in obtaining $\log n + 2C$ for the asymptotic law of b_n.

Summing this expression from $n = 1$ to $n = n$, Dirichlet obtained the formula

$$(n + \tfrac{1}{2}) \log n + n + 2Cn$$

for the asymptotic law of the sum $b_1 + b_2 + \cdots + b_n$, which grants a considerable approximation, as Dirichlet illustrated by setting $n = 100$, which produces 482 for the sum of the b_i, versus 478.2 for the approximation; or, setting $n = 200$, gives the respective values of 1098 and 1093.2.

He noted that if one wished to determine the mean sum of the divisors rather than the median number, then, instead of using Lambert's sum, one should use

$$\frac{\rho}{(1-\rho)^2} + \frac{\rho^2}{(1-\rho^2)^2} + \cdots + \frac{\rho^m}{(1-\rho^m)^2} + \cdots.$$

If this is expanded in powers of ρ, its general term $c_n\rho^n$ will have the sum of the divisors of n as coefficient. Similarly, the mean value of this coefficient will give the asymptotic expression $\frac{1}{6}\pi^2 n - \frac{1}{2}$.

A decade would pass before his next report on the topic appeared in the publications of the Akademie, in **1851b**.[26]

9.13 Infinite Series and Number Theory

Dirichlet's influential memoir **1838b**, titled "On the use of infinite series in the theory of numbers," signed "Berlin, May 1838," was one of those highlighting the use of his L-series, thereby once again exemplifying the application of his frequently repeated guideline that, when confronted by a task that is either unattainable or too cumbersome for resolution by previously employed techniques, one should seek a concept or approach that is closer to the "innate nature" of the topic. It also provided his first sketch of the class number formula discussed in greater detail the following year.

Dirichlet noted that one argument which needed to be replaced was that used originally in Legendre's attempt to demonstrate his statement concerning the number of primes in an arithmetic interval. Commenting on the flaw in Legendre's proof of the theorem on arithmetic progressions, while reminding the reader of his own valid proof of 1837 (**1839a**), he remarked, as he had in 1837, that the theorem presents

---

[26] See Sect. 13.5.

itself, so to speak, by itself, but that its rigorous demonstration is subject to great difficulties. As Dirichlet now pointed out once more, the statement had been used by "the illustrious Legendre" so frequently as a lemma that it led to a considerable number of new theorems, including the law of reciprocity. Yet Legendre had simply based the statement on an induction which, Dirichlet noted, is perhaps no less difficult to prove than the proposition that the author deduced from it. He reiterated that a valid proof had been especially desirable because of the numerous applications of which the proposition is susceptible, but that, although Legendre had devoted himself to proving the proposition, his "proof," which is very ingenious, is incomplete.

Dirichlet explained that because his initial efforts to complete the investigations of Legendre were unsuccessful, he had to have recourse to entirely different means and succeeded in establishing the proposition in question by depending on the properties of the infinite series which he had described in 1837 as being so greatly analogous to those that Euler considered in Chapter XV of his *Introduction to the Analysis of the Infinite*.[27] Dirichlet added that, since writing the 1837 memoir containing his proof, which was just then still in press, he had continued to deepen the exploration of the properties of these series. This led him to realize that his series provide "a very fruitful method of indeterminate analysis, which is applicable to very varied questions."

Dirichlet ended his introductory remarks by explaining that he was now offering some new applications of this type of analysis which would be amplified when he could work out a more extended work on the material.[28] He summarized the importance of his approach in the concluding statement of his introduction:

> The method which I use seems to me above all to merit some attention by the liaison that it establishes between infinitesimal Analysis and transcendental Arithmetic, and I hope that by this rapport it could even interest some geometers who do not occupy themselves especially with questions related to the properties of numbers.[29]

Dirichlet began the substance of his memoir by recalling that, given a positive prime number q equal to $4\nu + 3$, there will be two kinds of positive, odd prime numbers different from q. Designating the first kind by f, he reminded the reader that

$$\left(\frac{-q}{f}\right) = \left(\frac{f}{q}\right) = 1;$$

and the second kind designated by g gives

$$\left(\frac{-q}{g}\right) = \left(\frac{g}{q}\right) = -1.$$

---

[27]Euler 1748.

[28]His promised extension is **1839–40**; see Chap. 11 for this and the later **1842b**.

[29]**1838b**; see *Werke* 1:360.

After deriving three product formulas, he discussed quadratic forms of which $-q$ is the determinant. He began by considering two such forms $ax^2 + 2bxy + cy^2$ and $a'x^2 + 2b'xy + c'y^2$, where the outer coefficients are positive and never simultaneously even, noting that these constitute what Gauss called *genus positivum proprie primitivum.* Dirichlet proceeded by explaining the difference between the classifications of Gauss and Lagrange and noted that he wished to follow Gauss's classification of forms where two forms are considered as different if they only present a Gaussian "improper equivalence."

He pointed out that Lagrange, the first to show that for a given determinant there is only a finite number of different forms, considered two expressions as equivalent when it is possible to transform one to the other by a substitution of the form $x = \alpha x' + \beta y'$ and $y = \gamma x' + \delta y'$, where $\alpha, \beta, \gamma, \delta$ are, in the ordinary manner of considering quadratic forms, integers such that $\alpha\delta - \beta\gamma = \pm 1$. Dirichlet noted that this condition is sufficient for these two forms to represent the same numbers. Nevertheless, he pointed out that there is an advantage in only considering the two completely equivalent when there is a transformation of one to the other for which $\alpha\delta - \beta\gamma = +1$. He stressed that by adopting this notion of a proper equivalence one noticeably simplifies a large number of investigations, while preserving the conciseness in many of the statements that, lacking this, we would find overloaded with restrictions. He commented that, in the ordinary manner of considering quadratic forms, there are even theorems which appear isolated and restricted to determinants that satisfy certain conditions whereas, in considering the matter from Gauss's point of view, they only appear as special cases of general properties common to forms of the same determinant. As an example, he suggested comparing the theorems in [Legendre 1830], part 4, Section VI, and in Gauss's *D.A.*, art. 252.

Dirichlet next commented that it is easy to pass from Legendre's classification to that of Gauss, noting the significance of the middle coefficients' signs and the question whether they are even.

Looking at any one of the forms

$$ax^2 + 2bxy + cy^2, a'x^2 + 2b'xy + c'y^2, \ldots, \tag{4}$$

he suggested letting the indeterminates x and y assume relatively prime positive or negative values such that the corresponding value m is odd and non-divisible by q. Then m will have only prime divisors of the kind f. Reciprocally, a number m which has only such divisors can always be expressed by one or more of the forms (4). He supported his argument by referring to the *D.A.*'s articles 180.I, 155, 156, and 105. This led to the equation

$$2\sum \frac{1}{n^s} \cdot \sum \left(\frac{n}{q}\right) \frac{1}{n^s} = \sum \frac{1}{n^2 s} \cdot \sum \frac{1}{(ax^2 + 2bxy + cy^2)^s} + \cdots. \tag{5}$$

This double summation could not be effected if the variable s remains indeterminate, but Dirichlet stressed that it is extremely simple when s surpasses unity by infinitely little. Setting $s = 1 + \rho$, where ρ is a positive infinitesimal, and expressing the double series by a definite integral, he reiterated that it is very easy, especially if using geometric considerations, to find that the value of the series is

$$\frac{(q-1)}{2q\sqrt{q}} \cdot \frac{\pi}{\rho}.$$

Letting h denote the number of forms (4), leads to the conclusion that the right-hand side of (5) is equivalent to

$$h\frac{(q-1)}{2q\sqrt{q}} \cdot \frac{\pi}{\rho},$$

where ρ is always considered to be infinitely small.

By a sequence of steps involving the separate treatment of factors on both sides of (5), taking limits, and following this with a number of algebraic manipulations, he arrived at the following formulation for h:

$$h = \frac{2\sqrt{q}}{\pi} \sum \left(\frac{n}{q}\right)\frac{1}{n} = \frac{2\sqrt{q}}{\pi} S.$$

He commented that to obtain h, one now needs to determine S. Here he again reverted to Gauss's formulas. Letting a and b denote the quadratic residues and non-residues of q less than q, and by n any integer not divisible by q, he found

$$\sum \sin \frac{2an\pi}{q} - \sum \sin \frac{2bn\pi}{q} = \left(\frac{n}{q}\right)\sqrt{q},$$

where the summations extend over all values of a or b.

At this point, Dirichlet added a footnote indicating that this theorem and those he was about to utilize were stated in the *D.A.*'s art. 256, but that Gauss, "the illustrious author," only later, in a special memoir, [Gauss 1811], gave the complete proof, which presented great difficulties because of the ambiguity of the sign. Dirichlet added that he had given another proof, founded on entirely different principles, calling attention to his memoir of 1835, **1837a, f**, and to the importance of removing the ambiguity of the sign because of the numerous applications involved.

Utilizing the Legendre symbol and performing a few substitutions and transformations in the preceding equations, he arrived at the following results for S and h:

$$S = \left[1 - \left(\frac{2}{q}\right)\frac{1}{2}\right]\frac{\pi}{q\sqrt{q}}\left(\sum b - \sum a\right)$$

and therefore

$$h = 2\left[1 - \left(\frac{2}{q}\right)\frac{1}{2}\right]\frac{\sum b - \sum a}{q}.$$

Dirichlet noted that the prime number q can present two cases. If it has the form $8\nu + 3$ then $\left(\frac{2}{q}\right) = -1$, and if it is included in the form $8\nu + 7$ then $\left(\frac{2}{q}\right) = +1$. That means that, for the number h of different quadratic forms whose determinant is $-q$, $h = 3\frac{\sum b - \sum a}{q}$ when $q = 8\nu + 3$ and $h = \frac{\sum b - \sum a}{q}$ when $q = 8\nu + 7$.

He did not fail to observe that this twofold result coincides with "the elegant theorem" that Jacobi had stated several years before. In another footnote, Dirichlet provided precise references to [Jacobi 1832] as well as to [Jacobi 1837], to art. 306.X in the *D.A.*, and especially to Gauss's relevant note at the end of the *D.A.*, where Gauss had announced further, so far unpublished, investigations on the topic. In addition, in conjunction with reconciling the double result, Dirichlet again commented on the difference between Gauss's, Legendre's, and Jacobi's usage of terms.

In a series of further manipulations, based on the preceding formula of Gauss from the *D.A.* and his own related work, Dirichlet finally arrived at the following series

$$\sqrt{p}.S = \sum\sum(-1)^{\frac{n-1}{2}}\frac{1}{n}\cos\frac{2an\pi}{p} - \sum\sum(-1)^{\frac{n-1}{2}}\frac{1}{n}\cos\frac{2bn\pi}{p}.$$

He pointed out that it is now permissible no longer to exclude the values of n that are divisible by p, because one thereby simply introduces terms of opposite signs in the two series. He noted that the summation can be effected with help of the known formula given in Fourier's heat theory of 1822, namely

$$\phi(z) = \sum(-1)^{\frac{n-1}{2}}\frac{\cos nz}{n},$$

where $\phi(z)$ is a discontinuous function of z, which has the values $\frac{1}{4}\pi$ when z lies between 0 and $\frac{1}{2}\pi$; $= -\frac{1}{4}\pi$ when z is between $\frac{1}{2}\pi$ and $\frac{3}{2}\pi$; or $\frac{1}{4}\pi$ when z is between $\frac{3}{2}\pi$ and 2π.

Letting A, A', A'' denote the numbers of values of a contained between 0 and $\frac{1}{4}p$, $\frac{1}{4}p$ and $\frac{3}{4}p$, as well as $\frac{3}{4}p$ and p, and letting B, B', B'' be the analogous numbers with regard to the values of b, the expression for S becomes

$$S = \frac{\pi}{4\sqrt{p}}(A - A' + A'' - B + B' - B'');$$

by known properties of the prime number these are combined to produce

$$S = \frac{\pi}{\sqrt{p}}(A - B).$$

This gave him the very simple formula for determining the number h of quadratic forms having determinant $-p$:

$$h = 2(A - B) = 4A - \frac{1}{2}(p - 1).$$

As he remarked, one knows that these forms are of two kinds [or genres], one only representing the odd numbers $4\nu + 1$, and the other, the odd numbers $4\nu + 3$.

After another short digression concerning the differences in the terminology and classification of Gauss and Legendre, Dirichlet observed that, by an analysis similar to that just used for the determinants $-q$ and $-p$, one can obtain the number of quadratic forms the determinant of which is any positive or negative number, prime or composite. This led him to the following theorems:

> Let q be a prime number of the form $4\nu + 3$ and let A and B respectively denote how many quadratic residues and non-residues of q there are between the limits $\frac{1}{8}q$ and $\frac{3}{8}q$. Given this, the number of forms whose determinant is $-2q$ will be expressed by $2(A - B)$, where these forms are equally apportioned between the two genres which exist for this case.
>
> Let p be a prime number $4\nu + 1$, let A and B respectively be the number of quadratic residues and non-residues comprised between 0 and $\frac{1}{8}p$; similarly, let A' and B' be the number of residues and non-residues that fall between $\frac{3}{8}p$ and $\frac{1}{2}p$. Given this, the number of quadratic forms having the determinant $-2p$ will be expressed by $2(A - B - A' + B')$. These forms will be equally distributed between the two genres.
>
> Letting the letters p and q retain the preceding significance, generally denote by a the numbers less than and prime to pq which are such that $\left(\frac{a}{p}\right) = \left(\frac{a}{q}\right)$ and by b the analogous numbers which fulfill the condition $\left(\frac{b}{p}\right) = -\left(\frac{b}{q}\right)$. Given this, the expression for the number of quadratic forms whose determinant is $-pq$ will be:
>
> $$\frac{\sum b - \sum a}{pq} \quad \text{or} \quad 3\frac{\sum b - \sum a}{pq},$$
>
> depending on whether $pq \equiv 7$ or $\equiv 3 \pmod 8$; these forms will be equally distributed between the two genres.[30]

Adding that the last theorem can be stated more simply, like the one given above for the determinant $-q$, he closed the listing of the three theorems with an "And so on."

Dirichlet next observed that when the determinant is a positive number D, the analysis to recognize the number of different quadratic forms requires special attention because of some new conditions to which one must subject one of the earlier double summations. These conditions, added to the ones holding for negative determinants, consist in (1) that the values of x and y must be so chosen that the trinomial $ax^2 + 2bxy + cy^2$ will be positive; and (2) that one only needs to use a single infinite system of values for x and y where both are obtained by the formulas $x' = xt - (bx + cy)u$ and $y' = yt + (ax + by)u$, where t and u denote all the positive or negative numbers satisfying the equation $t^2 - Du^2 = 1$.

Because of this, the expression for the double series now includes the factor $\log(T + U\sqrt{D})$, where T and U are the smallest numbers, except for 0 and 1, which solve the equation. Dirichlet here noted that the logarithm plays the same role in this case as does π in the case of negative determinants. On the other hand, there is a series which includes the other side of the equation and will also be expressed by a logarithm

[30] *Werke* 1:368–69.

such as $\log(T' + U'\sqrt{D})$, where T' and U' are the solution of $t^2 - Du^2 = 1$, as follows from the application of circular functions; for this, he referred to **1837g**.

He added that, thanks to a known theorem, $T' + U'\sqrt{D} = (T + U\sqrt{D})^\lambda$, where λ is an integer, and therefore

$$\lambda = \frac{\log(T' + U'\sqrt{D})}{\log(T + U\sqrt{D})}.$$

Hence, the number of forms depends on this integer λ, but the expression of the dependence presents a slight difference, depending on the different forms "of which the determinant D is susceptible."

Dirichlet remarked that the analysis so rapidly indicated here has an advantage beyond determining the number of forms. This is the fact that it simplifies very important theories which, although already known, had only been established by very complicated methods. Among these, he singled out those that are summed up in the *D.A.*'s articles 252, 261, and 287.III, noting especially that the last-named so far brought together a large number of very extended investigations. Dirichlet referred the reader to the end of art. 287; he did not quote it, but presumably was gratified to find that Gauss there himself had declared that "these statements, if we are not very much mistaken, count among the most beautiful in the theory of binary forms, especially because, although they are of utmost simplicity by nature, they nevertheless lie so hidden that one cannot bring a rigorous proof without the support of so many other investigations."[31]

Dirichlet added that this same theorem can also be obtained by a very simple combination of the law of reciprocity with the proposition concerning an arithmetic progression, but, he commented, this way of proving it is not essentially different from the one he had just sketched, especially if, in order to prove that every arithmetic progression contains an infinite number of primes, one has recourse to series, as he had done.

He also noted that the theorems determining the number of quadratic forms implicitly involve a large number of propositions which can be expressed independently of the theory of these forms and which would perhaps be very difficult to prove without the combined help of his series and Gauss's formulas. He provided two examples, one referring to an 1831 memoir by Cauchy in which Cauchy had discussed the relative number of residues and non-residues for primes of the form $4\nu + 3$, and the other related to his own and Jacobi's studies of a solution of the equation $t^2 - Du^2 = 1$ using circular functions. For the latter, he gave an example where the determinant D is a prime number p, which appears straightforward but requires special attention in determining the sign of the product function involved.

Dirichlet added to this discussion a mention of the manner in which his series can be used to derive the "limiting expressions" [the asymptotic laws] of the mean values of certain very irregular functions pertaining to the properties of numbers. Here he referred to article 301 in the *D.A.* for Gauss's definition of the mean values

[31] *D.A.* art. 287, the penultimate paragraph.

of functions of this kind. He then pointed to his own memoir **1838a**, explaining that he had there established principles that make it possible to pin down various useful asymptotic laws. In particular, Dirichlet pointed to his using these to resolve issues similar to that of finding the approximation to the number of primes below a given large limit for which Legendre had provided the formula that Dirichlet had recently proved.[32]

As an example of the results found, Dirichlet noted that the very simple expression

$$\log n + 2C,$$

where C denotes Euler's constant,[33] expresses with increasing exactness, as n becomes larger, the median value of the number of divisors of the integer n, and that, likewise, one has for the mean sum of these same divisors

$$\frac{1}{6}\pi^2 n - \frac{1}{2}.$$

Recalling that these results had been obtained previously from Lambert's and another analogous series, Dirichlet pointed out that they could also be deduced from the series he had just discussed. He then produced the following new result derived from them:

> Let $f(n)$ be the function of n that indicates in how many ways the number n can be decomposed into two relatively prime factors. It is known that one has $f(n) = 2^\lambda$, where λ designates the number of equal prime divisors of n. Given this, one easily finds
>
> $$\sum \frac{f(n)}{n^s} = \frac{\phi^2(s)}{\phi(2s)},$$
>
> where $\phi(s) = 1 + \frac{1}{2^s} + \frac{1}{3^s} + \cdots$, and the sum is extended over all integral values of n, beginning with $n = 1$. Setting, as previously, $s = 1 + \rho$, and expanding the right-hand side according to ascending powers of ρ, it becomes
>
> $$\sum \frac{f(n)}{n^{1+\rho}} = \frac{6}{\pi^2}\left[\frac{1}{\rho^2} + \left(\frac{12C'}{\pi^2} + 2C\right)\frac{1}{\rho} + \cdots\right].$$
>
> The constant C again denotes Euler's Constant, and C' designates the series:
>
> $$\frac{\log 2}{2^2} + \frac{\log 3}{3^2} + \frac{\log 4}{4^2} + \cdots.$$
>
> By the method shown in **1838a**, it follows that the asymptotic formula [the *expression-limite*] of the mean value of $f(n)$ is[34]

[32]When editing this memoir for the *Werke*, Kronecker called attention to Dirichlet's having added a handwritten note in the copy of the memoir he sent to Gauss, pointing out that Legendre's proof is only exact in the first term, the true "expression-limite" being $\sum \frac{1}{\log(n)}$. See Dirichlet's *Werke* 1:372n.

[33]Dirichlet referred to Euler's *Differential Calculus* p. 444, for the constant. [See *Institutiones Calculi Differentialis*, Opera Omnia Ser. 1, vol. 10.]

[34]See *Werke* 1:373.

$$\frac{6}{\pi^2}\left(\log n + \frac{12C'}{\pi^2} + 2C\right).$$

Finally, Dirichlet observed that by the same kind of analysis he could find the formulas presented in arts. 301ff. of Gauss's "beautiful work":

> Suppose, for example, that it is a question of obtaining the mean value of the number of *genera* for the determinant $-n$, a number which we shall denote by $F(n)$. If one compares art. 231 [of the *D.A.*] where all the complete characters assignable a priori are enumerated, with arts. 261 and 287, where the illustrious author showed that only half of these characters correspond to really existing *genera*, one will easily find [...] five equations, [...] which [on appropriate summing and substituting], result in the asymptotic formula of the mean value of the number of *genera* for a determinant $-n$ as
>
> $$\frac{4}{\pi^2}\left(\log n + \frac{12C'}{\pi^2} + 2C - \frac{1}{6}\log 2\right),$$
>
> which coincides with the result of M. Gauss.[35]

This memoir, as explicitly as in any of his writings, brought out Dirichlet's close study of Gauss's work, his singling out problems that Gauss had cited as involving proofs that are particularly difficult, and Dirichlet's repeated demonstrations of the relative ease whereby such questions could be resolved with help of series like his L-series or related ones, even at one point including mention of Fourier's use of a discontinuous function.

9.14 The New Method: Using a Discontinuity Factor

On February 14, 1839, a day after his birthday, Dirichlet offered to the Akademie an innovation that had given him special pleasure. It was the introduction of a discontinuity factor for dealing with multiple integrals.

Of four related publications, three were summaries. Two identical French versions (**1839b** and **1839d**) were translations of the summary **1839c** provided in the Berlin Akademie's *Bericht*; they had appeared in the Paris Académie's *Comptes rendus* and Liouville's *Journal*. With the usual delay, the publication of the full memoir (**1841a**) only occurred in the Berlin Akademie's *Abhandlungen* two years later.

Having mentioned earlier, in **1839b–d**, that he was led to his new method while studying some questions pertaining to mathematical physics, Dirichlet introduced the subject in the *Abhandlungen* by remarking on the known complexity of determining a multiple integral or the reduction of such an integral to one of lower order. He noted that this generally is one of the more difficult problems, namely when the limits of integration for the several variables are not constant but depend on one another, so that the extent of the integration is expressed by one or more inequalities which, at the same time, contain several of the variables.

[35] *Werke* 1:373–74.

Dirichlet related that, in dealing with several problems resulting in the determination of a class of multiple integrals of an undetermined order, he had arrived at the method here treated, which not only provides the values of the integrals in question, but can also be applied to many other diverse integrals. He commented that, since the method "combines such a high degree of simplicity with this fruitfulness," it is astonishing it would not have been applied earlier to similar investigations. The principle of this way of treating those multiple integrals where one cannot simply divide the individual integrations between constant bounds is based on the possibility of expressing discontinuous functions by definite integrals. He cited the example of the expression

$$\frac{2}{\pi}\int_0^\infty \frac{\sin\phi}{\phi}\cos g\phi d\phi,$$

which equals 1 as long as the constant g has an absolute value less than 1 ["is less than 1 aside from its sign"], but vanishes when g exceeds 1. He took a threefold integral as an illustration, observing, however, that he only used the one of order three because using three variables allows a geometric interpretation which makes it easier to visualize the nature of the problem. If this integral is to be taken over a defined space, such as one bounded by an ellipsoid, one must only note the following: If α, β, γ designate the main semi-axes of the bounding surface and coincide in direction with the coordinate axes, then the expression

$$\left(\frac{x}{\alpha}\right)^2 + \left(\frac{y}{\beta}\right)^2 + \left(\frac{z}{\gamma}\right)^2$$

is less than or greater than unity depending on whether the point (x, y, z) lies within or outside the space, because the integral

$$\frac{2}{\pi}\int_0^\infty \left(\frac{\sin\phi}{\phi}\right)\cos\left[\left(\frac{x}{\alpha}\right)^2 + \left(\frac{y}{\beta}\right)^2 + \left(\frac{x}{\gamma}\right)^2\right]\phi d\phi$$

equals unity inside but vanishes outside.

For that reason, upon multiplying the given differential expression $Pdxdydz$, where P is a function of x, y, z, by this integral [the discontinuity factor] one no longer needs to pay attention to the original limits upon integrating; in other words, one can simply carry out the integration from $-\infty$ to ∞, since the discontinuity factor will cause the elements outside the intended area to disappear.

Dirichlet concluded his introduction by again observing how surprising it is that such an easy transformation can simplify the most difficult integrations and make unnecessary the complicated calculations or other artifices that have been used for certain problems which now can be solved merely with the aid of some definite integrals.

In his presentation to the Akademie, Dirichlet divided the memoir into five sections. In the first one, he dealt with difficulties that may arise in his application; in the next four, he gave examples of four successful uses of a discontinuity factor.

First, he considered an integral $\int Pdxdy\ldots$, where P represents an arbitrary function of the variables $x, y, \ldots$, whose extent is determined by a variety of conditions of inequality. He pointed out that this can involve two essentially different cases, similar to those encountered when one considers infinite series.

He was referring to the distinction found when considering the absolute value of the function. Depending on whether this gives a finite value or becomes indefinitely large, the first case presents no difficulties, but in the second, there may be instances where the new method cannot be applied. He minimized the significance of this potential problem, however, remarking on techniques of avoiding an analogous issue that Cauchy and Poisson had employed, but noting that with some practice one can often tell by inspection whether the procedure is applicable. After this disclaimer, he proceeded to provide several successful examples of the method.

The first of these examples illustrating the usefulness of a discontinuity factor involved the evaluation of the integral

$$\int\int\ldots e^{-k(x+y+\cdots)}x^{a-1}dx.y^{b-1}dy\ldots,$$

where $x + y + \cdots < 1$.

Following some cautionary observations concerning treatment of the triple integral obtained upon multiplication with the [discontinuity] factor

$$\frac{2}{\pi}\int_0^\infty \frac{\sin\phi}{\phi}\cos\sigma\phi d\phi,$$

where $\sigma = x + y + \cdots$, by a carefully outlined sequence of operations, Dirichlet obtained an equation depending simply on gamma functions and Eulerian integrals. Illustrating the usefulness of the result by restricting it to three dimensions, he noted that one can thereby determine the content, center of gravity, and momentum of inertia for a large assortment of solids by Eulerian integrals.

He added two additional, more complicated, examples before, in the fifth section of the memoir, applying his method to the attraction of a homogeneous ellipsoid, "which problem, as is well-known, has occupied mathematicians more than any other of the integral calculus." He had discussed this example in the Akademie's *Bericht* as follows:

First he noted that one had traditionally treated the problem by considering an outer point separately from an inner one, which is handled more easily. If the case of the outer point could not be led back to that of the inner one, so that they had to be solved independently, this was done by using different methods. His new method made this separate treatment unnecessary. He also added that it no longer requires the assumption that the attraction is inversely proportional to the square of the distance. Furthermore, one no longer needs to assume that the density of the attracting mass is constant; instead, it can be expressed by any rational entire function of the three coordinates x, y, and z. For simplicity, he assumed, however, that it is constant and equal to 1 in an example he provided.

While the memoir was in print, Dirichlet added a sixth section in which he took note of more problems that can be handled by his new method. In particular, he pointed to examples of two attracting bodies with both masses extended. It appears that his attention was drawn to these additional examples by his study of Gauss's memoir on "General Principles of the Theory of the Figure of Fluids in Equilibrium" [Gauss 1832a]. In dealing with such cases, one frequently needs to reduce a sixfold integral to one of lower order. This can usually be accomplished with customary methods if one simply wishes to reduce the integral to a fourfold one, as Gauss had shown, but, after that, further success is frequently out of reach. Using a discontinuity factor, however, Dirichlet found that he could reduce the sixfold integral to a double one without difficulty. He noted that among the various techniques in play for accomplishing this, the simplest appeared to be the use of a system of oblique coordinates.[36]

Dirichlet would treat the examples of use of his discontinuity factor in greater detail in his lectures on multiple integrals that were given regularly after the publication of this memoir in 1841.[37] Not surprisingly, his discontinuity factor would continue to appear in the successors to his unpublished lectures on differential equations as well.[38]

9.15 Observations

The content and references in the memoirs Dirichlet wrote during the productive period between 1832 and 1839 reveal the extent to which his previous studies influenced his new methodology. Above all, they illustrate his constant reliance on and expansion of ideas gleaned from his reading of Gauss's work. He appears also to have familiarized himself more thoroughly with those of Lagrange's memoirs that the latter had produced during his time in Berlin when he created most of his work in "algebraic analysis" (including both algebra and number theory). Dirichlet stressed Lagrange's having based his contributions to the subject on rigorous arguments—a characteristic that Gauss had mentioned repeatedly. Additionally, Dirichlet was familiar with some of the most useful techniques in analysis employed by his contemporaries Cauchy, Poisson, and the older Legendre, just as he was aware of how much they differed in the degree of rigor whereby they substantiated their results.

We attribute his willingness to use geometric examples as a result not only of the models provided by Gauss, but of his early exposure to such concepts through the

[36] He based this remark on a statement of Monge later proved by Chasles. Dirichlet may have learned of Monge's proposition while studying with Lacroix and Hachette in the 1820s.

[37] See Dirichlet–Arendt 1904, Section 8; also, Dirichlet–Meyer 1871: art. 173ff.

[38] See, for example, the succinct reference in Courant–Hilbert 1931:69.

teaching of Georg Ohm, Lacroix, and Hachette, the latter two followers of Gaspard Monge, as noted in Chap. 2.

When Dirichlet sought his "more natural" approach to solving the more intractable problems in number theory and found answers in earlier analytical techniques, he was going back also to the topics Lacroix had taught. Lacroix's Treatise on the Calculus deals extensively with Euler's Introduction to the Integral Calculus, Euler's summation and product formulas, Euler's applying the integral to the summation of series, and more. Likewise, Dirichlet's other eighteenth-century references, such as those to Landen, Lambert, and Stirling, were first brought to his attention through the writings—and in some cases conversations—of Gauss, Lacroix, and others. He merged these concepts, particularly in his applying techniques of classical analysis to number theory, resulting in his introduction of L-series and a new methodology that since his time we associate with analytic number theory.[39]

The memoirs on number theory Dirichlet wrote between 1832 and 1839 fall into several groups. The earlier ones show the advantage of his combining Gauss's abstract insights, frequently explicated by arithmetic examples in the *D.A.*, with Legendre's more practically oriented, but many times incomplete, inductions, often algebraically grounded, and, though facilitated by Legendre's notation, on occasion lengthy and cumbersome, as were some of Gauss's more rigorous proofs.

Overlapping with this is another group that shows how thoroughly Dirichlet was familiarizing himself with many of Lagrange's memoirs, especially those that Gauss had singled out in the *D.A.* Most of these were ones Lagrange had produced during his time in Berlin when he created the greater part of his work in "algebraic analysis" and the *Analytic Mechanics*. Dirichlet's growing number of examples pertinent to divisor theory appear to derive from his Lagrangian studies as well.

It seems fair to suggest that Dirichlet's interest in Lagrange was prompted not only by the high esteem in which Lagrange was held for both rigor and discovery, but also by Dirichlet's wishing to show himself a worthy member of the line of succession to Lagrange in the Physical-Mathematical Class of the Berlin Akademie, to which he had been elected in 1832.

By the mid-thirties, continuing his ongoing study of Gauss's *Disquisitiones Arithmeticae* and the subsequent, consequential expansion of number theory to the complex domain, Dirichlet focused not only on the difficult last part of the Fifth Section, but also on the Seventh Section dealing with circle division (cyclotomy). This meant that he needed to return to the consideration of trigonometric series. Both he and his friend Jacobi taught related courses at this time, and their publications on number theory frequently have similar references to the *D.A.*

It was in this period that Dirichlet gave to the world the Dirichlet series (his L-series) and used them to open the door to analytic number theory, supporting his belief in the fundamental unity of the various parts of mathematics. He reinforced this belief largely by showing how number theory could utilize algebraic, analytic, and approximation techniques and concepts from the other branches. It also enabled

[39]This merging exemplifies that combinatorial activity of which the psychologist Jerome Bruner wrote while describing creativity. See Bruner 1962.

him to minimize the lengthy, step-wise, often algebraic, approach he had needed in his earlier memoirs.

It is in the mid-thirties, too, that he produced several studies stressing applications of definite integrals. Culminating this area of study, he ended the decade, early in 1839, by offering his "New Method": the uses of a discontinuity factor for dealing with multiple integrals. As time went on, he would continue to refer to its usefulness.

By 1838, he was ready to elaborate on his studies of class numbers. The early formulation of his class number theorem appearing in **1838b** would be followed by his generalized approach described in 1839 (**1839/40**) and by the achievement of **1842b**, both discussed in Chap. 11.

Chapter 10
Expanding Interactions

Once again contrast marked the passing of time: The weekend following the burial of Gans on May 8, 1839, the king had signed Dirichlet's appointment as Ordinary Professor Designate, and Humboldt could report to Dirichlet a few days later that Altenstein already had the appointment back from the court. The faculty was informed within a month. The appointment had become effective on May 11.

10.1 Professor Designate

Although now holding the rank of ordinary professor, as continuing "designate" Dirichlet still could not act as official doctoral adviser and had no voting rights in faculty meetings. Under special circumstances, he could serve as consultant in the examination for the doctorate or, as in Kronecker's case, even be recognized as an adviser in awarding the doctoral degree.

In the spring of 1839, Dirichlet attended to Akademie affairs as well as to his lecture for the summer term, which had begun on April 27 and dealt with the application of the integral calculus to the attraction of ellipsoids and other problems. On July 2, he requested permission to go to Paris for a short visit with mathematical colleagues there and to close the term by July 10.

10.2 Paris

The brief visit in Paris appeared congenial. Arago welcomed him warmly; Cauchy invited him to dinner. There were friendly exchanges with Gabriel Lamé, who had been teaching at the Ecole Polytechnique but was now additionally occupied with engineering tasks, particularly the recently constructed railway connections between Paris and Versailles and between Paris and Saint Germain. It was a year after

U. C. Merzbach, *Dirichlet*, https://doi.org/10.1007/978-3-030-01073-7_10

Dirichlet's visit that Lamé's proof of Fermat's Last Theorem for the case $n = 7$ (Lamé 1840) would be published.

In addition, Dirichlet was well received by Gustav Eichthal and Olinde Rodrigues, among others. Both Rodrigues and Eichthal were acquainted with Dirichlet's in-laws. Rodrigues's father had been affiliated with the banking house Fould in Paris, with which Dirichlet's father-in-law had had long-standing connections. Olinde Rodrigues had become a mathematician and an influential follower of Saint-Simon. One of his best-known contributions to mathematics, a memoir on transformation groups, was published in 1840, the year after Dirichlet's visit.[1] Eichthal, especially familiar to Felix and other family members of Dirichlet's generation, also belonged to a well-known banking family; a sociologist, he, too, was affiliated with the Saint-Simonians. Both Eichthal and Rodrigues represented the banking and industrially oriented branch of the followers of Saint-Simon.[2] It is possible that their acquaintance with Dirichlet went back to the 1820s; Rodrigues had met and cared for Saint-Simon in 1823, the year Dirichlet had entered the Foy household; Maximilien Foy and his wife had been early Saint-Simonians.

Liouville

Perhaps the most significant encounter in terms of establishing a close relationship with a fellow mathematician was Dirichlet's meeting Liouville. They had corresponded before; but their becoming personally acquainted would mark the beginning of a steadier correspondence between them and of regular future meetings, including numerous visits by the bilingual Dirichlet to the Liouvilles' residences in Paris and Toul. In addition, it led to an increasing number of contributions (in French) to Liouville's *Journal de mathématiques pures et appliquées* from Dirichlet and other German mathematicians.

There was a "downside" to the close relationship with Liouville. This had to do with the presence in Paris of the charming Guillaume (Guglielmo) Libri, a native of Florence, affiliated with the University of Pisa, who had first appeared in Paris in 1824, before coming back as a political refugee in 1830. He had ingratiated himself with members of the Académie, including François Arago, and with research mathematicians like Sophie Germain, by his apparent mathematical knowledge, his willingness to share the content of rare mathematical manuscripts, and his reputed political courage. He had become a French citizen and member of the Académie in 1833, through Arago had obtained teaching positions at the Collège des Sciences and the Faculté de Paris, and by 1835 was serving on review committees of the Académie. Only then had he begun to arouse the suspicions of Liouville, with whom he had been friendly before. By 1837, Liouville discovered that Libri had published and claimed for himself a method for reducing linear differential equations that had been sent by d'Alembert to Lagrange and published in the *Miscellanea Taurinensis*.

---

[1] Gray, J. J. 1979/80.

[2] Jacoud 2010.

As Liouville would remark, "I do not understand how this passage has escaped Mr. Libri who has for a long time taken an interest in the history of mathematics."[3]

In 1838, Liouville found an opportunity to quote Dirichlet's criticism of Libri for having applied a transformation, without giving the procedure for finding it, which reduces the problem in question to something Euler had done before. On top of this, Liouville the following year published a comment in which he expressed "surprise when I discovered that the formulas published by Mr. Libri are incorrect and that the general principle on which they are based is inadmissible."[4] Liouville sent his findings to the Académie, where Arago had succeeded Fourier as Secrétaire Perpétuel of the Mathematical Sciences Class. Arago analyzed the criticisms before the group in February 1838. Libri indicated he would have a future response, which, however, did not appear. From then on, matters got worse. Accusations went back and forth for years, including the time of Dirichlet's short 1839 visit in Paris. During this same period, and for some time thereafter, Libri managed to best Liouville as well as Cauchy and others in elections for various academic posts in Paris. He claimed repeatedly that he, Libri, "had solved before Abel the equations related to the division of the lemniscate," expressing hurt that Jacobi had not credited him.

Libri also made the mistake of attacking Dirichlet. It would be the first and last time that anyone had taken it upon himself to question Dirichlet's mathematics (one of Libri's simplistic criticisms, directed to the introduction of **1840**, was based on a confusion of necessary and sufficient conditions). Liouville's friend and collaborator Jacques Sturm was the first to refute Libri's attacks on Dirichlet; Liouville would publish substantiating excerpts of letters received earlier from Dirichlet.

The *Comptes rendus* for 1840 are filled with abbreviated references to some of the claims and counterclaims primarily involving Libri, Liouville, and Arago, who now had reason to regret his having been taken in by Libri's earlier claims.[5]

Libri's decline in Paris took longer than his meteoric rise. The rise appears to be related to his having won the support of François Guizot, Minister of Public Education and noted historian, possibly because of Libri's publicized background in politics and their joint interest in the evaluation of historic documents. What brought Libri's eventual downfall was not theft of intellectual property but of books and manuscripts, something only Liouville appears to have suspected even a year before Dirichlet's visit. Jesper Lützen calls our attention to the punctuated entry in Liouville's notebook dated April 12, 1838:

> I have been told that the <u>*amiable*</u> Libri has been appointed librarian of the *Bibliothèque royale* and Treasurer!!![6]

Eventually, members of the Académie apparently grew tired of Libri's supercilious claims, without, however, exploring the problem he presented, as Liouville had done.

---

[3]Lützen 1990:53; also see Demidov 1983, referred to in Lützen 1990:54.

[4]Lützen 1990:54–55.

[5]Lützen 1990, especially in Chapters 2 and 3, contains a detailed account of Libri matters, including those pertaining to Dirichlet.

[6]Lützen 1990:55.

Another decade would pass before the French government acted on the persistent rumors that books and manuscripts had gone missing after Libri's visits to a site and prepared to issue a warrant for his arrest.[7]

10.3 Return to Berlin

While in Paris controversies continued, Dirichlet was back in Berlin, resuming his normal life. Before the beginning of the winter term 1839/40, Dirichlet went to Göttingen to visit Gauss. There he unexpectedly came together with Jacobi, who had been attending the meeting of the Society of Scientists and Physicians, which in September 1839 had taken place in Bad Pyrmont. Apparently Wilhelm Weber, too, had attended this meeting. According to letters Jacobi wrote to his wife and his brother, all three spent about a week with Gauss.

The Akademie and the University

The year 1840 would mark an initially subtle change in Prussian administrative attitudes toward mathematics and the institutions with which Dirichlet was affiliated. Friedrich Wilhelm III died and was replaced by the more unpredictable Friedrich Wilhelm IV. Altenstein, too, died in 1840; he had retired in 1838 and was replaced as Minister of Cultural Affairs by Friedrich Eichhorn, a patriotic jurist of long service to Prussia. Whereas Altenstein had taken a special interest in raising mathematical and scientific standards in Prussia, Eichhorn would be preoccupied with complex, largely financial, affairs related to the churches and did not see eye to eye with Humboldt, as his predecessor had done. The ministerial purse strings for mathematical affairs tightened. Nevertheless, by the end of the summer term in 1842, Dirichlet's university salary was raised from 600 to 800 Thaler.

In the Akademie, Dirichlet successfully nominated Liouville as external member. In addition, he obtained support for Wilhelm Weber's work in Leipzig by having the Akademie furnish Weber with 300 Thaler to purchase a chain covered with copper wire for measuring the speed of galvanic currents.[8] It appears to have been intended for the possibility of subsequent expansion of the earlier small Gauss–Weber telegraph. In addition, Dirichlet continued to read annual memoirs at meetings of the

---

[7]In the revolutionary year 1848, when Guizot, by this time in the Department of Foreign Affairs, lost his job, Libri was ready. He left for England, shipped some 30,000 books and manuscripts there, and was well received across the Channel, where feelings of nationalism overrode possible doubt, and where even Augustus DeMorgan treated him as an innocent victim of political persecution and French intrigue. In 1850, Libri was convicted of theft in France. He sold off books and manuscripts in England as needed for funds. There were two highly publicized auctions in the 1860s, and the French government negotiated for the return, through purchase, of a certain number of items. Others would continue to surface well into the twentieth century. Libri himself moved back to Italy where he died. For details, see Lützen 1990. In recent years, more news items concerning stolen materials have appeared, along with book-length works on Libri.

[8]Harnack 1900, 1:776.

Akademie until his trip abroad in 1843 and recommended topics for prize submissions at the Akademie and the university.

Dirichlet's lectures increasingly reflected results he had published. In addition, a wider mathematical readership would profit from expanded memoirs such as **1839–40**, one of his most widely read publications. A comprehensive, multi-part work, incorporating several of his preceding smaller memoirs, it was an initial introduction to what would come to be known as analytic number theory.

By the early 1840s, Dirichlet's students included Gotthold Eisenstein, Eduard Heine, Leopold Kronecker, Philipp Seidel, and others whose accomplishments would become noteworthy.

Eisenstein's unusual mathematical aptitude had come to his teachers' attention when he was still in high school, at which time he already had attended lectures by Dirichlet at the university. In 1842, he had accompanied his mother to Great Britain to join his father who was looking for job opportunities there. Although they did not stay long, the trip is significant because he looked up William Rowan Hamilton in Dublin; Hamilton gave him one of his publications to be forwarded to the Akademie in Berlin. Upon his return, Eisenstein presented it to the Akademie, along with a memoir of his own. This led to Crelle's becoming aware of him, and Crelle informed Humboldt of the unusual young man in their midst. Humboldt would support Eisenstein through recommendations, financial assistance, and personal reassurances from that time until Eisenstein's death in 1852.

Eduard Heine, related to Dirichlet through the marriage of his sister Albertine to Dirichlet's brother-in-law Paul Mendelssohn Bartholdy, first matriculated at the university in Berlin in 1838. After his first semester, he went to Göttingen, where he studied with Gauss and Moritz Abraham Stern before returning to Berlin in 1840. Aside from working with Dirichlet, he attended lectures there by Steiner and Encke. He dedicated his doctoral dissertation on differential equations, accepted in April 1842, to Dirichlet. (Since Dirichlet was still "designate," he could not be the official thesis supervisor; nominally these were Dirksen and Martin Ohm.) After a postdoctoral year in Königsberg—Dirichlet having arranged his studying there with Jacobi—Heine began his career in Bonn, where he served as privatdozent from 1844 until 1848 and as extraordinary professor from 1848 to 1856, when he became ordinary professor in Halle.

Leopold Kronecker had become interested in mathematics as a pupil of Kummer, who taught in the Gymnasium of Liegnitz which Kronecker attended. Kronecker matriculated in Berlin in 1841, spent a summer term in Bonn, but for the winter term 1843/44 went to Breslau, where Kummer was now on the faculty. Returning to Berlin while Dirichlet was abroad, he completed his requirements for the doctorate, and his dissertation, "On complex units," was submitted at the end of July 1845, shortly after Dirichlet's return. The official thesis supervisor was Encke, who persuaded the faculty, however, to allow Dirichlet to serve on the committee and take over the mathematical part of the oral examination for Kronecker's doctorate.

Seidel came to the university in Berlin in 1840 and worked with Dirichlet and Encke until Dirichlet sent him, along with Heine, to Jacobi in 1842. Because of the subsequent absence of both Dirichlet and Jacobi, Seidel, on advice from the astronomer Bessel, moved to Munich where he continued to pursue a twofold interest in astronomy and mathematics. His doctoral dissertation, a study of telescopic mirrors, was followed by his *Habilitationsschrift* dealing with the convergence and divergence of continued fractions. He remained in Munich, where, rising through the ranks at the university, he became ordinary professor in 1855.

Family Matters

The end of the year 1842 had been dark. Dirichlet's mother-in-law, Lea, had broken her arm in a fall (the result of being run into by a pedestrian). Despite a long and painful recovery, she maintained her schedule, continued to be a sociable presence, but died early in December. Christmas was gloomy and Dirichlet took Walter to Leipzig where Felix welcomed them and sent back to Berlin a glowing account of his young nephew.

10.4 Jacobi

As noted, in 1842 Dirichlet had arranged with Jacobi that Seidel and Heine should come to Königsberg to attend Jacobi's lectures on mechanics. It was doubly interesting for professor and students. Only four students attended the course because Jacobi had discouraged most of his own students from attending so that he could discuss topics for which they were not prepared. It was the last significant lecture course Jacobi was able to complete in Königsberg. Early in 1843, he became too ill. After a serious cold, he was diagnosed with diabetes.

Although diabetes was still considered a mostly fatal illness, the timing of Jacobi's diagnosis was relatively fortunate. Biot and Apollinaire Bourchardat, a pharmacist, professor of hygiene and diabetes specialist, only a few years before had published influential memoranda on the polarimetric diagnosis and diet, respectively, of the disease. Jacobi read up on their results, and the combination of testing Bouchardat's proposed diet and Franz Neumann's taking polarimetric measurements of Jacobi's glucose content seemed to result in some improvement in his condition.[9] However, in late February 1843 Bessel had warned Gauss of Jacobi's possibly fatal illness, and a month later Dirichlet decided he and Rebecca should go to Königsberg to see Jacobi "one last time."

Königsberg

A two-week visit in April 1843 proved to be entertaining and constructive. When the Dirichlets arrived in Königsberg, they were greeted warmly by Jacobi's colleagues at the university: the astronomer Bessel; the physicist Franz Neumann; and two recent Jacobi students, Friedrich J. Richelot, now extraordinary professor, and the privatdozent Otto Hesse. Also frequently present were two future mathematicians, the ten-year-old Alfred Clebsch and his friend Carl Neumann, son of Franz, both still

[9]Davis and Merzbach 1994:143.

pupils at the renowned Altstädtische Gymnasium. Although Jacobi was weak and depressed when the Dirichlets arrived, he brightened up considerably during their stay.

Wilhelm Cruse, Jacobi's Königsberg physician, wrote to Jacobi's brother Moritz, since 1837 active at the Academy in Saint Petersburg, suggesting that the fading of Jacobi's depression was largely attributable to Dirichlet's visit and that he had learned from Dirichlet that, although Jacobi was too weak for long, sustained activity, his mental capacity was basically sound. By May, Jacobi himself claimed improvement in his condition, based on his increased sense of well-being and confirmed by the decreasing amounts of glucose observed through the polarimeter in use by Neumann in Königsberg.

After Dirichlet's departure, Jacobi felt well enough to express his regrets that they had not seen more of each other. He blamed his own weakness and Dirichlet's almost uninterrupted invitations elsewhere, along with Dirichlet's unwillingness to rise before nine in the morning. Indeed, Dirichlet alternated visits to him with other social engagements; but Dirichlet's late rising probably had less to do with his social activities than with adhering to his by now well-established schedule of withdrawing to work in the evenings and staying up well into the night.

Since Jacobi had improved sufficiently to resume work but not to bring his notes to publishable form, Dirichlet took back sixty folio-sized sheets of Jacobi's incomplete papers on number theory, to see what was needed to polish them for publication. In addition, Dirichlet used the visit to have Jacobi's physician, Dr. Cruse, prepare a write-up of his diagnosis and recommendations to be forwarded to Dr. Schönlein, since 1839 the king's personal physician in Berlin. Schönlein concurred with Cruse's diagnosis and treatment as well as the recommendation that Jacobi spend the winter in a southern climate.[10] This pleased all concerned; Jacobi had attempted unsuccessfully for some years to obtain a transfer from Königsberg, whose climate already previously had been deemed injurious to his health.

Upon his return to Berlin, Dirichlet immediately spoke to Humboldt and then advised Jacobi to send a request to the king via Humboldt. By the end of May, Jacobi had a reply from Humboldt and a personal get-well note from the king confirming that his leave had been granted along with a travel stipend.[11]

10.5 Preparations for a Vacation

Despite Fanny Hensel's misgivings in 1840 as to whether it would be possible ever to move Dirichlet to an Italian journey, Jacobi's circumstances presented the opportunity for which the family had been waiting. Dirichlet's setting in motion the decisions for Jacobi's forthcoming trip, as well as his subsequent actions, demonstrate that he was capable of acting with alacrity when there was concern for those dear to him.

[10] Ahrens, W., ed. 1907:98–100.

[11] Königsberger 1904:308–9.

A week after Jacobi had received the king's positive response, Dirichlet submitted his own request for a paid leave for the winter term of 1843/44. He noted that he had not had relief from teaching in the sixteen years that he had been employed at Prussian universities, and that his wife's health required a winter in a southern climate.[12] This request, too, was granted.

Finally, Steiner, who had been unwell, apparently suffering from kidney stones, a prelude to the increasingly serious kidney ailment that would eventually lead to his death, had been seen by Dr. Schönlein who advised him to visit some spas. Steiner asked for leave and received a travel stipend of 300 Thaler, which he used to join Dirichlet and Jacobi.

Jacobi invited his student Carl Wilhelm Borchardt, who had personal resources, to accompany him, and Steiner suggested they bring along the twenty-nine-year-old Ludwig Schläfli as translator. Schläfli, at that time a high school teacher in Thun, had impressed Steiner with his unusual linguistic talents when they had met in Bern the previous year; he not only knew Sanskrit and other esoteric languages, but was apparently fluent in Italian as well as French, which would be useful to the travelers.

Schläfli had only recently decided to become a mathematician. It subsequently turned out that he benefitted greatly not only from increased contact with Steiner but also from Dirichlet's coaching him in number theory during the six months that he spent with the group. In that time, he was helpful through his linguistic competence as well as his assisting Jacobi by copying out revisions of papers Jacobi had been preparing for publication.

The anticipated journey kept all concerned occupied. Jacobi was treated to cold baths to further increase his strength. Dirichlet presented a paper to the Akademie at its mid-June meeting dealing with problems that involve the determination of an unknown function under the integral.[13] Borchardt received his doctorate in Königsberg under Jacobi's oversight on July 2, having submitted a dissertation on nonlinear differential equations.

Since Dirichlet's summer term did not end until later, Rebecca left ahead of him with the two boys on July 5, accompanied by a servant and a maid. For the rest of the month, Dirichlet and his mother were looked after by Fanny Hensel, while Jacobi stayed with Dirichlet on and off for the three weeks until the end of the summer term and enjoyed meals with Dirichlet's relatives and their mutual acquaintances living at Leipzigerstraße 3.

The thirty-two-year-old Rebecca for the first time was taking a lengthy trip alone, that is, without an adult family member as accompaniment. Her letters to her sister reflect her consciousness of this fact, particularly as she is describing impressions of her trip that followed a route familiar to the Hensels, and her commenting at one time that she is still afraid of family reprimands, although she must admit that she has never received one from Dirichlet. The first part of the scenic trip took her to Heidelberg in an open carriage; from there she was able to proceed to Karlsruhe by train. Originally intending to meet Dirichlet in Badenweiler, she decided to surprise him by taking the

[12] Biermann 1959a:72.

[13] This presentation (**1843b**) is recorded by title only in the *Monatsberichte* for the year 1843.

"Rapid Post" to Freiburg, where she visited the Woringen family, long-time family friends from Berlin. They had just established themselves in Freiburg to which Franz Woringen had been called that spring as ordinary professor of law at the university.

Dirichlet did not arrive when expected the next day, but Jacobi and Borchardt did. After Rebecca had begun to worry whether Dirichlet might have gone straight to Badenweiler, he showed up the next morning. All were staying at the same hotel and with much jubilation they joined the Woringens for breakfast, picked up the children, and continued to Italy.

It was most likely on the next part of the trip that Steiner, proving that even a gruff, single-minded, synthetic geometer will still be a boy, persuaded Schläfli that the impressively uniformed conductor of the Post was the Crown Prince of Sardinia.[14]

10.6 Switzerland and Italy North of Rome

Now that their real vacation had begun, their initial major stops were in Switzerland. First they stayed in Vevey, where they had an unexpected delay because Ernst had a cold, possibly received during an excursion to Montreux. They felt an enchantment on and around Lake Geneva which none of the renowned Italian lakes they were to visit would equal. Dirichlet was left with admiration for the snow-covered mountain tops, sight of which lured him to several excursions on his own, just as it would draw him back years later. By the end of September, they were going on via Martigny, the Col de Balme, the Wallis, the Simplon Pass (which, on the Swiss side, Rebecca likened to a bravura aria), and, after some unnecessary anxiety as to how Dirichlet could pass customs into Italy with his inseparable stash of cigars, they proceeded via Como to Milano. While still enjoying the scenery, Rebecca had some difficulties becoming acclimated to the human environment for the first few days before leaving Como.

Once in Milan, they were reunited with Jacobi and Borchardt, who had left them in Freiburg. First came the obligatory, impressive visit to the Cathedral, parts of which were examined more closely the next day. Then the adults went to the Ambrosian Library, where, as "Membres de l'Académie," they were invited to examine the cupboards filled with rare manuscripts and drawings that were opened for their perusal.

In Pavia, they were impressed by the mosaics in the Certosa di Pavia, the first samples of Florentine mosaics they had encountered. A few days later, the dark blue of the Mediterranean made up for the exhaustion of Genoa, and the way along the Riviera di Levante surpassed any expectations they may have had. Spezia gave Rebecca the opportunity to display a little one-upmanship vis-a-vis the Hensels: "How can you speak of Italy without knowing the Gulf of Spezia. For that alone

[14] Ahrens, W., ed. 1907:106ff. contains various excerpts of Jacobi's letters from Italy, including one telling of this episode.

you must go back one more time." They stopped over in Florence, leaving there by mid-November. Jacobi and Borchardt had attended a meeting of scientists in Lucca, and everyone was reunited in Rome.[15]

10.7 Rome

From the distant approach, continuing with the traditional, welcoming entrance at the Ponte Molle, the city enchanted the expectant travelers. What added to their enjoyment was the warm reception they received from the small community of mathematicians and physicists. Since these were heavily geometrically oriented, it finally gave Steiner a chance to shine and would show some reflections in Dirichlet's later work.

After three weeks, Jacobi could report to his wife that he and Steiner visit art collections daily, tend to meet Borchardt at noon, take tea at Dirichlet's, and that he is being showered by the Abbé Tortolini with mathematical works by a variety of Italian authors such as Fagnano, Ruffini, and Mascheroni which, he explained, he had longed for and now received as gifts. Tortolini was the best-known mathematician Jacobi and Dirichlet met; their other two closest scientific companions were Macedonio Melloni and the Padre Domenico Chelini. Both were particularly hospitable to their foreign colleagues.

Melloni, with whom Dirichlet had occasion to discuss reports of recent results in physics, hoped to see an account of his experiments in press before too long. As head of the Vesuvius Observatory, he was able to take the visitors on some memorable excursions. Padre Chelini, who visited the Dirichlets frequently, was not only a mathematical colleague and subsequent author of books on rational mechanics and geometry, but a calming presence during small disagreements. Jacobi wrote his wife that she and the rest of his family should be nice to him when he returned home because otherwise he would simply return and go back to Rome. There, when he yelled at Padre Chelini, the Padre was not troubled but took his outbursts in stride!

Here, as throughout Italy, Jacobi found to his annoyance that he was continually mistaken for his brother Moritz who had gained special fame after the publicity that followed his discovery of electrotyping ("Galvanoplastik"). This was true even when he and Dirichlet called on Mary Somerville, then living in Italy. The incident amused Dirichlet so much that both he and Jacobi returned from their visit in high good humor. Eventually, Jacobi got into the habit of saying "I'm not me, I'm my brother."[16] Rebecca, on the other hand, felt flattered when someone who had heard her play the piano mistook her for her sister Fanny.

Christmas in Rome was celebrated with a tall laurel tree rather than the accustomed fir tree. It was decorated with roses, yard-long grape branches, oranges, and Roman candied fruits, while surrounded with a wreath of apples, nuts, and laurel leaves.

[15] Königsberger 1904:315.

[16] Königsberger 1904:315.

Rebecca described this, and the numerous presents that were exchanged, in great detail, writing to Fanny in Berlin. Those joining in the celebration were Jacobi, Steiner, Borchardt, and several painters of Wilhelm Hensel's circle, namely Julius Elsasser, A. F. Geyer, August Kaselowsky, and Julius Moser. As would be his habit during their stay in Italy, Kaselowsky made himself especially useful to the Dirichlets, not least in helping Rebecca rebuild the tower of Dirichlet's cigars which kept falling down. He also delighted the family in Berlin by capturing various notable sights of the trip, such as the laurel Christmas tree, in vignettes attached to the letters for Fanny. They missed Wilhelm Hensel's toasts but, when they became a bit morose thinking of their relatives left at home, Rebecca proposed a toast to all the future brides of the bachelors present, which cheered everyone up again.

A pleasant surprise was an audience Dirichlet and Jacobi had with Pope Gregory XVI on December 28, an event that would be mentioned in some newspapers. While Rebecca and Fanny were amused as they attempted to picture the awkward attempts that lanky Dirichlet and hefty Jacobi had made to pay proper respects to the pontiff, Jacobi was impressed by the Pope's astronomical knowledge. As he wrote to Bessel, their audience lasted more than one-half hour, during which His Holiness cheerfully discussed a variety of subjects, while standing, despite his seventy-nine years. He spoke to them of Newton, Kepler, Copernicus, and Laplace, knew of the ratio of the squares of periods of revolution to the cubes of the mean distances, talked with admiration of Copernicus but seemed to think the Copernican system was not yet proved rigorously, and wanted to acknowledge its validity only if someone could discover the parallax of a fixed star. As Jacobi wrote, they both found it interesting that Jacobi could describe to the Pope how a famous countryman of Jacobi's, through long and sagacious labors, had discovered such a parallax and that, according to the unanimous judgment of all astronomers, his discovery stood without doubt. Jacobi was reporting this to the discoverer himself, namely Bessel, who had been studying the parallax of 61 Cygni since the late 1830s.

By spring 1844, the group had the dubious pleasure of finding themselves in the midst of Roman carnival. Dirichlet threw his flowers and candy with disdain, whereas the continually delighted Borchardt joyously joined in the spirit of the celebration, both by viewing the activities from his spacious balcony and by mingling with the crowds. Jacobi retreated from the festivities to visit the Dirichlets and on at least one evening read portions of the Odyssey to those present there.

Toward the end of April, Jacobi and Steiner left for Naples, where they stayed for three weeks before returning to Rome, prior to Jacobi's heading back to Germany. Jacobi had received a report in February that he would not need to return to Königsberg, which he had felt for a long time to be injurious to his health. Initially it had appeared that he would be transferred to Bonn. It turned out, however, that he received an appointment to Berlin, as member of the Akademie, with a salary and with authorization to lecture at the university. He returned to Berlin by June 1844 and received a memorable farewell dinner in Königsberg later in the year.[17]

[17] Königsberger 1904:326–29.

10.8 Illnesses

After Jacobi's departure from Italy, Dirichlet decided to take an excursion to Naples and Sicily to view sites previously missed, and, at least metaphorically, to follow Archimedes' footsteps on the sand. It was not a good idea.

In October 1844, Mama Dirichlet, in Berlin, told Fanny that Dirichlet had fallen victim to "the fever." It seemed to have passed, but soon reappeared. Kaselowsky had accompanied Rebecca back to Florence. By the first week of November, it became apparent that both Dirichlets were sick and Fanny was terrified when she learned not only that their two friends, Kaselowsky, who had assisted the Dirichlets throughout their Italian journey, and Julius Elsasser were seriously ill, but that Rebecca had jaundice. Fanny was almost relieved when the Hensels learned, via the communications that their maids had exchanged, that Rebecca was pregnant, in addition to the other problems. It seemed to be a more normal explanation for the unusual pain and discomfort that Rebecca had reported. Still, everyone who had seen her recently remarked that she looked terrible. Dirichlet was said to be convalescing from his "fever," alternately described as "the Roman fever" (most likely malaria) or the dreaded "nervous fever" (typhus). The recurrences of his spiking temperatures and Rebecca's pregnancy made it clear that they could not return by year's end, and that his leave had to be extended.

By early December 1844, Dirichlet wrote home confirming the state of affairs. Jacobi took over some of his classes in Berlin. The Hensels offered to come to Florence to help out. Rebecca accepted, and Fanny, Wilhelm, and Sebastian left Berlin for Italy on January 2, 1845.

10.9 The Birth of Flora

When the Hensels arrived in Florence on January 19, matters seemed to have improved slightly. Walter met them on the street, Rebecca, assisted by Kaselowsky, greeted them at the door, and they found an apartment across the narrow street so they could converse window to window. The Hensels' landlady provided thoughtful services and everything was comfortably furnished. Both families settled into a routine resembling that in Berlin, except that, instead of going to classes, Dirichlet looked after the children, while Fanny looked after Rebecca who was able to take small walks. Rebecca was not supposed to be due until March, but, since she had always run late during her previous pregnancies, there seemed to be plenty of time for everyone to settle in, rest, and recuperate.

Dirichlet spent his usual studious late evenings in his room; they had breakfast at noon, and then the main meal at 5:00 p.m. Since Wilhelm Hensel could not find the necessary backdrops or model for his painting, everything appeared sufficiently under control for him to leave for Rome by January 25.

Surprise! From the ninth to the twelfth of February, Rebecca was in great pain; the physician came by, but remained non-committal and left. On February 12, Fanny, who had decided to spend the night, returned and realized that she had to call the physician back. She prepared some bedding, and the physician reappeared about 11:00 p.m.; there was no other appropriate help available. As a result, Fanny and Dirichlet found themselves in the unaccustomed roles of serving as midwives while a very healthy Flora was born sometime after 11:30 p.m., during which process Rebecca was supported by the two people to whom she felt closest.

The packet with the necessary linens and other materials had not been expected for a few more days. The "Florentine" was wrapped in an old woolen jacket of Dirichlet's, a cutoff sleeve covering her head. Since it had turned unusually cold and stormy, they could only light a fire in one room where the maid held the bundle during the night; Fanny and Dirichlet had a cup of tea at 2:00 a.m.

It was a joyous occasion for multiple reasons. Fanny and others who had been around Rebecca had expected a miscarriage rather than a live birth. Not only did baby Flora appear strong and healthy but Rebecca began to improve rapidly. Also, despite subsequently published date discrepancies, it is apparent that Flora was born close to, if not on, Dirichlet's birthday, February 13, a date that had continued to be important in Dirichlet's life. Her birthday would be celebrated on February 12.

10.10 Return to Berlin

After Fanny had seen to it that all necessary materials were available, doing some hand-sewing of her own where necessary, she left for Rome with Sebastian on March 15 to meet Wilhelm Hensel. She only then was told that he had been sicker than any of them since he had arrived in Rome, but had not wanted to worry her and the others, so had not written about his illness. His may have been the actual typhus. Because of Hensel's weakened state, they could not do all the things to which she had looked forward, but they had plenty of companionship: His fellow artists and former students tried to meet their needs and provided them with support and diversion.

Dirichlet had to return to Berlin in time to meet his classes for the summer term, which began April 7. The Hensels left Rome, arrived in Florence on May 20, stayed for three weeks, but left there on June 15 with Rebecca and the four children. Except for an accident the last day of their return that involved the turning over of the coach in which Rebecca was riding, they had a pleasant trip with stops to see family members and friends. Finally, they were back in Berlin and reunited with Dirichlet. At last, the Dirichlet family could move into their new abode at Leipzigerplatz 18 that Fanny had prepared with great care for their originally planned return the previous summer.

Chapter 11
Publications: 1839–1845

The publications that Dirichlet offered from late 1839 until his travel to Italy in 1843 had a common theme. It was that of progressing on the new pathways opened as the result of his reflecting on the successful technique he had applied to the proof of the theorem on arithmetic progressions. He wished to explore further the possibilities of linking number-theoretic questions to infinitesimal analysis, in particular the use of his L-series and functions, suggested by Euler's Chapter 15 of the Introduction to Analysis of the Infinite [Euler 1748].

Throughout, Dirichlet continued to follow closely Gauss's terminology and results, with special emphasis on the sequence of Gauss's publications on residues and the latter portions of Section 5, as well as Section 7, of the *D.A.* These intricate parts of Gauss's magnum opus led Dirichlet to the determination of class numbers and of his unit theorem.

At the same time, Dirichlet used his publications to familiarize readers with introductory complex analysis, number theory, and connections between the two. This was necessary for them to be able to follow his recent research results. In Prussia, many lacked the necessary background in number theory—aside from one summer course in 1833 on the introduction to higher arithmetic, his previously announced lectures on that subject would only in the winter term 1837/38 begin to attract a sufficient number of students to be held. In France, readers had to add to the expositions of Lagrange, Legendre, and their precursors the more recent Gaussian terminology and topics.[1]

The publications of this period begin and end with lengthy memoirs of particular significance. The first, **1839–40**, essentially outlines the basics for what would come to be known as Analytic Number Theory. The last, **1842b**, provides Dirichlet's most explicit account of his determination of class numbers. Both, appearing in the *Journal für die reine und angewandte Mathematik*, were elaborations of original results he had presented to the Akademie and would be especially influential and widely read.

[1] Legendre 1830, vol. 2 finally had incorporated significant parts of Section 7 (cyclotomy) of Gauss's *D.A.*, but still lacked some of the necessary proofs.

U. C. Merzbach, *Dirichlet*, https://doi.org/10.1007/978-3-030-01073-7_11

The publication **1839–40** may have been his memoir most frequently acknowledged in the secondary literature; Dirichlet himself would refer to it for background to statements he made in later publications.

11.1 Analytic Number Theory

The long memoir **1839–40** in which Dirichlet explained his new direction had the title "Researches on diverse applications of infinitesimal analysis to the theory of numbers." Divided into three parts appearing in two volumes (19 and 21) of Crelle's *Journal*, he began the memoir with a historical introduction highlighting some past events leading to his work. Once again, he reminded readers that the results of his memoir on arithmetic progressions, **1839a**, had not been proved rigorously prior to his 1837 presentation. Outlining his program of further study, he also called attention to the fact that the memoir **1838b**, which had appeared in volume 18 of Crelle's *Journal*, already had contained references to several of the applications he had in mind.

He now proposed to share his results in all necessary detail, beginning with the problem of determining the number of different quadratic forms whose determinant D is some positive or negative whole number, or, what amounts to the same thing, the number of quadratic divisors which belong to the expression $x^2 - Dy^2$. Without mentioning their names, he here juxtaposed Gauss's terminology of the determinant with that used by Legendre when referring to quadratic divisors. He indicated that the analysis leading to the solution of this problem will facilitate the way to finding "new and very simple demonstrations of several beautiful theorems due to M. Gauss, but which this illustrious geometer has only established by means of very complicated considerations in the second part of the fifth section of his *Disquisitiones Arithmeticae*."[2]

Dirichlet continued by giving the reader a quick synopsis of the *D.A.*'s fifth section, noting that, while all of it deals with second-order-degree forms, there are two distinct parts. The first, through article 223, completes and refines the results given by Euler, Lagrange, and Legendre on the subject, whereas the second part, after an explanatory introduction, almost entirely gives results due to Gauss himself. Dirichlet observed that this second part, noteworthy because of its profound methods and variety of results, was practically unknown to geometers. Acknowledging that it presented the greatest challenge, Dirichlet quoted the statement that Legendre had made more than three decades earlier:

> One would have liked to enrich this Essay with a much larger number of the excellent materials which compose the work of M. Gauss; but the methods of this author are so particular to him that one could not, without very circuitous extensions and without being subjected to the simple role of translator, profit from his other discoveries.[3]

[2] *Werke* 1:413–14.

[3] Legendre 1808: preface.

Dirichlet concluded the introduction to his own memoir by expressing the hope that, aside from his new results, his work

> can still contribute to the advancement of science by establishing on new bases and by bringing nearer some elements of the beautiful and important theories which until now have only been within reach of the small number of geometers capable of the intentness of spirit necessary in order not to lose the thread of ideas in a long sequence of calculations and of very compound arguments.[4]

This major memoir, described as "classic" by Kronecker,[5] is composed of eleven sections. The first five are essentially preparatory to Dirichlet's discussion of the new material he wished to present. In these, he drew heavily on theorems and definitions Gauss had put forth in the fifth section of the *D.A.* This contributed to the unusual length of Dirichlet's memoir; but he found it necessary to present an extensive synopsis of material found in Gauss's *D.A.*, because the *D.A.*, published in Latin in 1801, only had had a French translation in 1807 and was not very widely available. Yet, in order to follow Dirichlet's own presentation, it was important to be familiar with at least the highlights of Gauss's work.[6]

In the first section, Dirichlet addressed the following:

> Let the letters k and ρ denote two positive quantities, of which the first is a constant, the second a variable. Consider the sum of the infinite series
>
> (1) $\quad \frac{1}{k^{1+\rho}} + \frac{1}{(k+1)^{1+\rho}} + \frac{1}{(k+2)^{1+\rho}} + \cdots .$
>
> As this sum increases beyond every finite limit when the variable ρ becomes infinitely small, let us see what is the simplest function of ρ which could serve to measure this increase, or, in other words, the ratio whereby the preceding expression converges to unity as ρ converges to zero.[7]

Using known properties of the Gamma function and taking logarithms, after transforming several expressions resulting from the series (1), Dirichlet proved in detail a theorem of which he would make frequent use:

> Let
>
> (2) $\quad l_1, l_2, l_3 \ldots, l_n, \ldots$
>
> be an infinite number of positive constants different from zero, either unequal or partly equal; let $f(t)$ be a discontinuous function of the positive variable t, which expresses how many terms are included in the sequence (2) whose value does not exceed that of t. Then, if the function $f(t)$ can be put in the form
>
> (3) $\quad f(t) = ct + t^{\gamma}\psi(t),$

---

[4] *Werke* 1:414.

[5] Kronecker 1865:285; see Kronecker *Werke* 4:229.

[6] A reprint of the 1801 edition appeared in volume 1 (1863, reprinted in 1870) of Gauss's collected works. Until 1889, when the first German translation appeared, there were no other editions of the *D.A.* besides those mentioned. For lists of subsequent editions and translations, see Merzbach 1984:1, 3, 44–45, 47, 49–52, or the later Goldstein et al., eds, 2010:xi, which, in addition to a 1989 reprint of the French translation, includes Spanish, Japanese, and Catalan translations published in the 1990s.

[7] *Werke* 1:414–15.

where c and γ denote positive constants the second of which is less than one, and the new function $\psi(t)$, disregarding its sign and the size of the variable t, remains less than the positive constant C, I say that the sum

(4) $\phi(\rho) = \frac{1}{l_1^{(1+\rho)}} + \frac{1}{l_2^{(1+\rho)}} + \frac{1}{l_3^{(1+\rho)}} + \cdots,$

in which ρ denotes a positive variable, will be such that, for an infinitely small value of ρ, one will have the value

(5) $\phi(\rho) = \frac{c}{\rho},$

i.e., the ratio of the sum $\phi(\rho)$ to the fraction $c\rho$ converges to one, when ρ becomes less than every given size.[8]

After completing the proof, he added that this theorem could be extended, but, not needing such a generalization for his intended applications and noting that it presents no special difficulty, he stopped short of providing it. He did, however, call attention to two other lemmas, noting that these belong to infinitesimal analysis.

11.2 Primes in Quadratic Forms

A memoir titled "On a property of quadratic forms," **1840a, b**, and presented to the Akademie's plenum on the fifth of March 1840, was an expansion of the theorem on arithmetic progressions, **1839a**, which, Dirichlet again reminded the readers, had provided the "first strict proof of the theorem that every arithmetic series, whose first term and difference are integers without common divisor, contains infinitely many prime numbers." He now extended the statement to quadratic forms, i.e., expressions like $ax^2 + 2bxy + cy^2$, subject to the restriction that the numbers $a, 2b, c$ have no common factor.

As pointed out in the Akademie's *Bericht* (**1840a**) and a French translation in Crelle's *Journal* (**1840b**), for the most part the proof depends on the same principles as those given in 1837 (**1839a**), but these require some modifications, illustrated by the special case where the determinant is a negative prime number $-p$, which, aside from the sign, has the form $4n + 3$, and is one of the so-called regular determinants as Gauss defined them in the *D.A.*, art. 306, VI. Dirichlet's proof focused on this special case.

He began by letting h denote the number of distinct forms ϕ belonging to $-p$. He reminded the reader that, under the prescribed condition, each of the forms can be obtained by successively joining one of these, say ϕ_1, to all the rest. Next, he divided all positive odd primes except for p into two classes, distinguished by $-p$ being a quadratic residue with respect to each member of the first class. Combining the relationships Gauss had deduced in connection with regular determinants and the previous techniques utilized in **1839a**, letting $h = 2\lambda + 1$, Dirichlet outlined in fewer than three pages his proof that each of the forms

[8] *Werke* 1:415–16.

$$\phi_{-\lambda}, \phi_{-(\lambda-1)}, \ldots, \phi_0, \phi_1, \ldots, \phi_{(\lambda-1)}, \phi_\lambda$$

contains an infinite number of primes.[9]

It is of interest that, by calling particular attention to this case, he brought to the reader's notice Gauss's statements concerning the multiplication of classes and the large number of tabular instances which Gauss had previously discarded before his definition of regular determinants, after which Gauss had commented that the topic of regular and irregular determinants "appears to depend on the most hidden mysteries of higher arithmetic and to make room for the most difficult investigations."[10]

Dirichlet chose not to call attention to the fact that Legendre's statements concerning this extension of the theorem on arithmetic progressions, only published in 1830,[11] had been just as lacking in proof as had Legendre's earlier pronouncements concerning the arithmetic progression theorem.[12]

Dirichlet's own proof of the complete theorem was spelled out in more detail by Heinrich Weber in [Weber, H. 1882] and, based on this, by A. Meyer in [Meyer, A. 1888].[13]

11.3 Extract of a Letter to Liouville: The Unit Theorem for Degree 3

An extract from a letter Dirichlet had written to Liouville on January 22, 1840, was published in both the *Comptes rendus* of the Académie for February 17 and in volume 5 of Liouville's *Journal* (**1840c** and **1840d**). It was Liouville who was responsible for these two publications. Liouville subsequently wrote to Dirichlet, apologizing for publishing the mathematical details without first having obtained Dirichlet's authorization for doing so. He explained that, when he shared some of these details with various associates, they were so insistent that he should insert an excerpt in the *Comptes rendus* that he went ahead and did so.[14]

In the letter that created such interest in Paris, Dirichlet had provided Liouville with a revealing summary of his previous work and his current efforts, including a sketch of a limited proof for degree 3 of the statement that came to be known as his unit theorem.

First, after complimenting Olry Terquem on his translation **1839e** of Dirichlet's memoir on arithmetic progression, he informed Liouville that he had had the idea of extending the same analysis to quadratic forms. Explaining that he had been very much occupied with extending to quadratic forms with complex coefficients

[9]See *Werke* 1:500–502.

[10]Gauss *D.A.*, end of art. 306.VI.

[11]Legendre 1830, 2:102–3.

[12]See the sharply critical assessment of this work by Legendre in Weil 1983:329–30.

[13] See Bachmann 1894/1921:272.

[14]Tannery 1910:2–3; translated in Lützen 1990:61.

and indeterminates, that is, of the form $t + u\sqrt{-1}$, the theorems which hold in the ordinary cases of real integers, he noted that, in particular, if one seeks to obtain the number of different quadratic forms which exist under this hypothesis for a given determinant, one arrives at the rather remarkable result that the number in question depends on the division of the lemniscate; just as in the case of real forms and a positive determinant it is connected to the section of the circle. He added that he had been especially pleased in this work by the portion derived from geometric considerations, particularly from the theory of the perspective properties of figures. "With this help, the question, which to begin with, and considered in a purely analytic manner, seems extremely complicated, becomes nearly as simple as when it is a question of real determinants."[15]

Dirichlet continued by explaining that these investigations had led him to a theorem which he described as remarkable by its simplicity and which, he thought, would be of some importance for the theory of indeterminate equations of degrees higher than the second, a matter still very little cultivated.

> If the equation
>
> (1) $s^n + as^{(n-1)} + \cdots + gs + h = 0,$
>
> with integral coefficients, has no rational divisor, and if among its roots
>
> $$\alpha, \beta, \ldots, \omega$$
>
> it has at least one which is real, I say that the indeterminate equation
>
> (2) $F(x, y, \ldots, z) = \phi(\alpha)\phi(\beta)\ldots\phi(\omega) = 1,$
>
> where, to abbreviate, one has set
>
> $$\phi(\alpha) = x + \alpha y + \cdots + \alpha^{(n-1)}z,$$
>
> always has an infinity of integral solutions."[16]

Dirichlet remarked that, in order to establish this theorem, it is necessary first to show that there exists at least one integer m such that the equation $F(x, y, \ldots, z) = m$ has an infinity of solutions. He noted that in dealing with degree 2 one arrives at the known result by the use of continued fractions.

Dirichlet concluded by stating:

> Among the numerous consequences that one can derive from this theorem, there is one which, so to say, presents itself, and consists in this: The functions that Lagrange first considered in the *Mémoires de Berlin*, later in the Additions to the Algebra of Euler, and which reproduce by multiplication, if they can obtain a certain value are, as a consequence, susceptible of the same value for an infinity of systems of values of indeterminates x, y, ..., z, supposing always that the algebraic equation from which these functions take their origin to satisfy the conditions stated above.[17]

---

[15] *Werke* 1:621.

[16] *Werke* 1:622.

[17] *Werke* 1:623.

11.4 The Theory of Complex Numbers

On May 27, 1841, Dirichlet read to the plenum of the Akademie a memoir on the theory of complex numbers that he described as preliminary to a larger work intended to transplant several questions he had earlier resolved for real integers to the realm of complex numbers, using his previous method.[18]

The initial report **1841b** printed in the Akademie's *Bericht* was reissued in Crelle's *Journal*, **1841d**, followed by the main memoir **1843a** in the Akademie's *Abhandlungen*, and a French translation **1844** by Hervé Faye in Liouville's *Journal* the next year.

1841b, d

In the brief report concerning the memoir, it was merely noted that, since, with regard to both results and method, the memoir follows closely previous results of the author, it seemed appropriate to briefly mention some of the questions treated previously.

Accordingly, consideration was given to **1839a**, the 1837 memoir on arithmetic progressions. Aside from again remarking that this contained the first rigorous proof of the frequently used statement, the report stressed that this proof is noteworthy because, despite the purely arithmetic nature of this statement that is to be established, it rests largely on considering variable continuous quantities. It derives from the formation of infinite series which, like those already treated by Euler in [Euler 1748], are created by multiplication of an unending number of factors. Dirichlet pointed out, however, that these new series differ from Euler's in the characteristic that in the factors, of which each contains a term of the series of primes, there are also powers of roots of unity whose exponents coincide with the so-called indices of the prime numbers; if, along with all the others, these are related to a system of primitive roots.

As soon as one pursues this path, which appears to be easily followed, a difficulty occurs, which could make the whole process illusory. This has to do with the fact that the sums of certain convergent series are different from zero. Initially, it had been suspected that it was impossible to sum these series. It was possible to achieve the proof—despite the fact that it is frequently just as difficult to do so for a finite form of the series used in the argument as for the original—but that was so complicated it appeared desirable to find another approach that was shorter and corresponded more closely "to the nature" of the question.

Dirichlet reported that after numerous unsuccessful attempts, he finally arrived at the unexpected result that "the mentioned series are related to a problem whose solution fills a gap experienced for a long time in one of the most important parts of the theory of numbers. The theory of which we speak is that of quadratic forms, first founded by Lagrange, later arrived at a high degree of development by Legendre and especially by Gauss." Dirichlet was referring to the dependence of the forms on their determinants, reminding the reader that, as Lagrange had shown, to each determinant, whether positive or negative, there corresponds only a finite number of essentially different forms; he added that "the same great geometer" had provided

[18] This larger work would become **1842b**.

the procedure according to which these essentially different forms can be represented for every numerically given determinant.

Dirichlet pointed out that the question concerning the general connection between the number of forms and the determinant will not be settled, however, by knowledge of this procedure that can be carried out only in certain cases. It is this question which receives its answer in the investigations to which he had referred at the outset.

He noted that the results were discussed in detail in **1839–40**. For his present study, he wished to stress merely that the dependence of the number of forms on the determinant is represented with considerable difference depending on whether the determinant is negative or positive. In the first case, the dependence is purely arithmetic; in the second, the expression for the number of forms contains certain relationships of the coefficients that belong to the auxiliary equations which appear in the division of the circle.

The new investigations, of which the first part is contained in **1843**, have the purpose of extending these previous results to the theory of complex numbers. He pointed out that

> we owe the thought of introducing complex integers into higher arithmetic to the famous author of the *Disq.arith.*, who was led to this expansion by his investigations on the theory of biquadratic residues, whose fundamental theorems only appear in their greatest simplicity and entire beauty when one relates them to complex prime numbers. The importance of this expanded concept is not limited to the mentioned application; rather by its introduction one opens up a new area for the investigations of higher arithmetic in which one finds an analog for every property of real numbers, which not rarely equals or even surpasses the former in simplicity and elegance.[19]

11.5 Certain Functions of Degree Three and Above

In October of the same year, 1841, Dirichlet read a memoir to the Physical-Mathematical Class of the Akademie on "Some results of investigations concerning a class of homogeneous functions of degree three and above."

1841c

In the brief notice published by the Akademie, Dirichlet explained that the homogeneous functions with integral coefficients that concerned him are those that contain a number of undetermined integers equal in number to their degree and that can be decomposed into linear factors with irrational coefficients. He pointed out that in the case of degree 2 these coincide with binary quadratic forms.

Dirichlet justified his interest in the more general functions by the fact that just as the theory of the binary quadratic forms constitutes one of the most fruitful parts

[19] *Werke* 1:507.

of arithmetic, so numerous properties of these expressions of higher degree appear to promise considerable expansions not only for number theory but for related disciplines. In this presentation, he wished to deal merely with the problem he described as follows:

> To find all representations of a given number by a given function of the kind just described, or to become convinced that the given number is not capable of such a representation.[20]

To clarify the considerations on which the solution for this problem rests, Dirichlet first took up the case of the second degree, although, as he noted, this solution was already well known. This required examining all solutions of the undetermined equation

$$ax^2 + 2bxy + cy^2 = m \tag{1}$$

where $b^2 - ac = D$ is positive and not a square. The solution he presented hinged on knowing two values satisfying the [Pell's] equation

$$t^2 - Du^2 = 1. \tag{2}$$

Denoting any two such values by T and U, assumed to be positive, and letting (X, Y) be any solution of (1), Dirichlet referred to Euler, who had remarked that one can obtain infinitely many solutions determined by the formula

$$ax + (b\sqrt{D})y = \pm(aX + [b + \sqrt{D}]Y)(T + U\sqrt{D})^n; \tag{3}$$

here n designates an integer, and the rational parts and coefficients of $\sqrt{D}$ on both sides are to be set equal. Dirichlet observed, however, that Euler's remark is not sufficient for establishing the connection between (1) and (2). It does not provide a means to find an initial solution (X, Y), and, as Lagrange had shown, the equation (3) does not necessarily encompass all solutions of (1), even if T and U are chosen as the smallest values satisfying [Pell's] equation.

To establish the missing connection, Dirichlet now called attention to the fact that the solutions contained in (3) form a group which will continue to contain the same solutions if the solution (X, Y) is replaced by any of the others that can be derived from it.[21] From this, it follows that all the solutions of (1) can be distributed into such groups and that a complete solution of the problem will depend merely on knowing one solution from each group, since (3) will then produce the entire group.

Dirichlet concluded the discussion by noting that, although in the second case the equation (7) will admit infinitely many solutions, they cannot all be derived from a single one by raising to powers. Rather, in this case there are two basic solutions

---

[20] *Werke* 1:627.

[21] Dirichlet here used the term "group" in an informal (non-mathematical) sense.

which generate all the rest by multiplication and raising to powers. Without knowing just what these are, it suffices to have two that cannot be transformed into each other and then to apply the preceding procedure.

11.6 A Generalization re Continued Fractions and Number Theory

The following spring, on April 14, 1842, Dirichlet read a historic memoir titled "Generalization of a statement from the doctrine of continued fractions along with some applications to the theory of numbers." It was published in the Akademie's *Bericht* for that year.

1842a

This is the statement to which the title of the memoir refers:

"If α is an irrational value, there are always infinitely many related numbers x and y for which the linear expression $x - \alpha y$ is numerically less than $\frac{1}{y}$."

Dirichlet noted that this statement had been known for a long time from the theory of continued fractions. The generalization, which Dirichlet described as being as simple as it is fruitful, is the following:

> Let $\alpha_1, \alpha_2, \ldots, \alpha_m$, be given positive or negative values such that the linear expression
>
> (1) $\quad x_0 + \alpha_1 x_1 + \alpha_2 x_2 + \ldots + \alpha_m x_m,$
>
> in which
>
> (2) $\quad x_0, x_1, \ldots, x_m$
>
> denote indefinite positive or negative whole numbers, can vanish only if $x_1 = x_2 = \ldots = x_m = 0$, and so also $x_0 = 0$; then there are always infinitely many systems (2) for which it is not true that:
>
> $$x_1 = x_2 = \ldots = x_m = 0,$$
>
> and for which the expression (1) is numerically smaller than $\frac{1}{s^m}$; by s one understands the largest of the numerical [absolute] values of $x_1, x_2 \ldots, x_m$.[22]

He observed that, to prove this theorem, it will suffice to show that a system of the required property can be found for which, additionally, the numerical [absolute] value of (1) is smaller than a previously determined quantity δ. To obtain this, he suggested taking a positive integer n, which satisfies the condition:

$$\frac{1}{(2n)^m} < \delta$$

and in the expression (1) attaching to each of the numbers $x_1, x_2, \ldots, x_m$ all the values contained in the series $-n, -(n-1), \ldots, -1, 0, 1, \ldots, n-1, n$.

If, for each of these $(2n+1)^m$ systems one determines x_0 in such a fashion that the expression (1) obtains a nonnegative value less than 1, this will result in $(2n+1)^m$

---

[22] *Werke* 1:635.

proper fractions, of which at least two must lie in the same of the $(2n)^m$ intervals bounded by the values $0, \frac{1}{(2n)^m}, \frac{2}{(2n)^m}, \ldots, \frac{(2n)^{m-1}}{(2n)^m}, 1$.[23]

Subtracting from one another two expressions corresponding to such values results in a new expression of the form (1), in which the numbers $x_1, x_2, \ldots, x_m$ do not all vanish simultaneously, and, aside from the sign, none of these numbers exceed $2n$; additionally, their numerical [absolute] value will be less than $\frac{1}{(2n)^m} < \delta$, and hence also less than $\frac{1}{s^m}$.

As a consequence, there exist infinitely many systems (2), to which this statement applies. Since (1) does not vanish for any of these, by the preceding it must be possible to find a new one, distinct from the given ones. For this, one merely needs to select the smallest numeric value of (1) for δ which corresponds to one of the known systems.

Dirichlet remarked that there are analogous statements for two or more simultaneous expressions of the form (1), which can be proved by the same simple considerations. After noting two examples, he also remarked that the preceding statements and proofs given can be extended to complex numbers with minor modifications.

At the end, Dirichlet added that, thanks to these results, the lemma which is the basis of the generalization for Fermat's [Pell's] equation $t^2 - Du^2 = 1$ can be proved by elementary means.[24]

The short report assumed considerable historic significance because in the proof Dirichlet used the so-called *Schubfachprinzip*, or pigeonhole principle, his well-known example of applying an obvious concept to validate statements that may otherwise require rather complicated proofs. In later years, this particular publication caught Minkowski's eye. The first chapter of Minkowski's *Diophantische Approximationen* of 1907, titled "Application of an elementary principle," opens with the following introduction:

> The considerations of this chapter will be based on a simple principle of which Dirichlet in his time made several deep applications. It states: "If $n + 1$ things are divided among n pigeonholes there must be at least one pigeonhole that will contain more than one thing."[25]

Throughout the twentieth century, especially among students of the geometry of numbers and of diophantine approximations, this opening to Minkowski's last major publication would focus attention on Dirichlet's box, or pigeonhole, principle and the methodology it represents.

---

[23]Note that the intervals serve as *Schubfächer*, and this is a clear application of the "pigeonhole principle." See Chapter 1 of Minkowski 1907.

[24]Also see **1840d**.

[25]Minkowski 1907:1.

11.7 Complex Quadratic Forms and Class Numbers

In an eighty-page-long memoir, **1842b**, published in volume 24 of Crelle's *Journal* with the title "Investigations on quadratic forms with complex coefficients and indeterminates," Dirichlet reviewed the basic concepts and definitions that formed part of his methodology that had appeared in the preceding years.

The memoir not only summarized results concerning complex quadratic forms, called attention to the relationship of these to Pell's equation, but also established Dirichlet's class number formula for binary quadratic forms, the result for which it is primarily remembered.

11.8 Comments

With his expansion of complex number theory, Dirichlet supported a foundation for complex function theory. Also, his frequent use of geometric analogies and examples brought to the attention of his readers and listeners the close relationship between numbers and geometry.

By summarizing the approach spelled out during the previous decade, Dirichlet was able to formulate these reviews and later results in a smoother form than had been possible with the lengthy procedures of the earlier memoirs, particularly those preceding his use of L-series and the associated L-functions. This lent comparative elegance to his final achievement of this period: The application of analytic number theory to the class number determination for complex binary forms.

Chapter 12
A Darkling Decade

The initial period after the Dirichlets' return from Italy was pleasant and hopeful. On his way back to Berlin, Dirichlet had stopped to visit Felix Mendelssohn Bartholdy and other family members and friends in Frankfurt. Felix had written to his sisters, who were still in Italy, that Dirichlet was lively and well, sported an enormous beard, and could hardly be recognized, not only because of the beard, but because he appeared younger, heavier, and generally healthier than he had been when they last saw him.

The appearance was deceiving.[1] Whereas Dirichlet apparently could cover a lack of energy, there were other changes. Dirichlet seemed particularly absent-minded. While in Italy, he had been focused on working out the proof for dimension greater than three of the unit theorem that would cap his recent number-theoretic discoveries. According to Kummer, he had succeeded while listening to the Easter music in the Sistine Chapel. But the family saw little evidence of such intense mental concentration. Instead, they noticed that Felix who had been looking for a letter that his sister Fanny was to have sent had it handed to him in Frankfurt from Dirichlet's pocket.

In addition, Dirichlet had become less attentive to his mode of dress. Rebecca, who wrote him weekly until they were reunited in Berlin, asked him, varying a family expression, whether he remembered to wash, bathe, and comb his hair; and Felix found it necessary to rearrange his brother-in-law's necktie before they went to dinner with Madame Jeanrenaud, Felix's mother-in-law. Such incidents may simply be attributable to Dirichlet not having Rebecca nearby to pay attention to sartorial details. Noticeable, however, are verbal and pictorial depictions of him during the following decade, within and outside the family circle, that contrast with his previous manner and appearance. We shall note examples of these below.

[1]The theme of this chapter and the ending of Chap. 14 refer to Matthew Arnold's last stanza of "Dover Beach."

U. C. Merzbach, *Dirichlet*, https://doi.org/10.1007/978-3-030-01073-7_12

12.1 The University

Once arrived in Berlin, it was time for Dirichlet to resume university business and exchange news of their most recent research results with Jacobi who had assumed his new position as resident reading member of the Akademie in Berlin upon his return in 1844.

Jacobi wrote his brother during the summer of 1845 that, having been unable to accomplish much since 1839, he was now intensely involved with his mathematical investigations. Jacobi's improvement may have been the result at least partly of his receiving continuing diabetic counseling and polarimetric measurements. After the measurements taken in Königsberg by Franz Neumann, in Berlin the polarimetric tests would be continued by the chemists Eilhard Mitscherlich and Heinrich Rose. On the first of June, Jacobi wrote to Bessel that he and Dirichlet saw each other daily. We surmise that Dirichlet could discuss with him his recent work on potential theory, his expansion of number-theoretic studies to the determination of mean values, and his proposal of the division problem.

Dirichlet's classes during the decade from summer 1845 to summer 1855 predominantly dealt with applications of the theory of the integral calculus. These ranged from applications to number theory, probability, the study of series, and problems of attraction, to partial differential equations.

12.2 The Heidelberg Offer

By the end of 1846, Fanny Hensel wrote in her diary that she had been about to mark the year as a red letter period, when she was stopped by disturbing reports. Dirichlet had been offered a professorship in Heidelberg, and there was uncertainty as to whether or not he would accept.

Dirichlet's mathematical colleagues in Berlin who held regular appointments were junior to him, if not in age or length of service, then certainly in achievement and international recognition. Although the number of students in his classes continued to grow, he sensed that only few of them would continue to grow in achievement. In addition, he was becoming tired of the repetitive, elementary lectures at the Kriegsschule that he had given for nearly twenty years, interrupted only by his Italian journey. The recent denial of a salary raise in Berlin appeared to favor Heidelberg's chances that he would be willing to leave Prussia.

The offer had come in the form of an inquiry from the university in Heidelberg to Gustav Magnus. Immediate action was taken by those desperate to have him stay. Jacobi wrote to the king on December 12 notifying him of the inquiry from Heidelberg which had contained the comment that they wished to approach Dirichlet before turning to any others with lesser names; Jacobi also informed the king of Eichhorn's explanation to Dirichlet that because of limited funds he should not expect a salary increase for a number of years. In addition, Magnus had alerted Humboldt, and August Boeckh informed the faculty.

In the ministry, there was a long and animated discussion as to what means should be taken to prevent his leaving. Sebastian Hensel would claim that a member suggested offering Dirichlet the title of "Geheimrat," whereupon Johannes Schulze stated that he would not wish to be the one to transmit word of that honor to Dirichlet: that the individual bearing the message would be tossed down the stairs by Dirichlet and that Dirichlet lived on a second floor with very steep steps. At any rate, by the end of January 1847 Dirichlet's salary was raised to 1500 Thaler and the matter appeared settled.

Fanny Hensel continued to marvel at the mild winter and beautiful spring.

12.3 Growing Tensions at the Akademie

While the weather continued to be warmer and more pleasant than usual, exchanges in the Akademie grew heated and distinctly unpleasant. The beginning of the tense atmosphere there may be linked to the "Affair Raumer." In January 1847, it was the turn of the historian Friedrich Raumer, since 1829 Secretary of the Historical-Philosophical Class of the Akademie, to give the official annual address honoring Frederick the Great (Friedrich II).

Raumer, a constitutional monarchist of long standing, chose to focus on Friedrich II's statement of tolerance ("in my realm everyone has to be able to attain salvation in his own fashion"). His timing in choosing the theme was deliberate. Recently, this statement of Friedrich II's had been attacked by the orthodox with strong arguments designed to convince Friedrich Wilhelm IV as well as other conservatives in Prussia. Raumer extended his theme to draw lessons in his address for future kings, if they did not follow, and oppose attacks on, the philosophy of Frederick the Great.

The problem was that Raumer's address was attended by Friedrich Wilhelm IV and his princes. The king, complex, frequently vacillating, but enjoying attendance at Akademie functions and often speaking of individual freedoms, was deeply insulted when members of the Akademie audience, sitting a few rows behind him, began to titter during Raumer's unmistakeable analogies. He left, threatening never to attend another Akademie meeting. This caused an understandable reaction. The Akademie depended on the royal administration's beneficence and had enjoyed this king's goodwill.

On February 1, Encke, a member of the Akademie's Secretariat by virtue of being Secretary of the Physical-Mathematical Class, wrote to the Akademie's membership attacking Raumer and urging his removal from the Akademie. The king asked the Minister of Cultural Affairs to take a hand; a committee was appointed, the king indicated that he did not wish to distance himself from the entire Akademie but only from the offender (Raumer) and to gain assurance against future occurrences of the kind that had taken place. The majority of the essentially conservative members of the Akademie, though disapproving of the perceived discourtesy by Raumer toward their ruler, had felt that Encke's letter was so unseemly that they refused to have it printed. After another two months of committee appointments, negotiations, half-hearted apologies, and an unfortunate decision by Eichhorn, the Minister of Cultural Affairs, to offer some of the conciliatory exchanges between the king and the Akademie to the press, the Akademie accepted Raumer's resignation from the Secretariat. It wished nevertheless to retain him as a member. This offer was left open until the end of the year. Raumer declined it, however, and left the Akademie altogether on March 22.[2]

Dirichlet took no overt role in these discussions, which would drag on past March, throughout most of 1847. Aside from alerting the Akademie to Kummer's applying his ideal complex numbers to Fermat's Last Theorem (**1847**), he was kept busy with his courses for the summer term, responding to Humboldt's queries concerning the algebra of the Hindus for Humboldt's next volume of his *Kosmos*, and communicating with Schläfli, Borchardt (who was visiting Paris and Oxford), and others about mathematical issues, many of which had arisen during the Italian visit. Nevertheless, the affair, which contributed to a growing rift among members of the Akademie, did not leave Dirichlet untouched.

12.4 Family Tragedies

For the rest of the year 1847, ideological issues took a backseat to two major, unexpected family tragedies. The first, the death of Fanny in May, brought a major change to the conduct of the Dirichlet household, aside from affecting its members emotionally. It would be followed by the equally unexpected death of Felix in November.

In April 1847, Rebecca's birthday was celebrated with a performance of Fanny Hensel's trio for piano, violin, and cello. For some time, Fanny had suffered from nosebleeds of varying intensity. She had also worried about occasional numbness in her arms affecting her piano playing as early as July 1843; for this she had sought and received medical attention. Yet she had repeatedly stated during 1846 that she felt better and happier than she had for a long period of time. While rehearsing for the performance of her trio, however, she felt a numbness in her hand. Less than a month later, on May 14, while practicing for a Sunday household concert, she had another nosebleed, somewhat more severe than the ones from which she had suffered periodically. In addition, the numbness in her hand recurred, so that she interrupted her playing and remarked "it must be a stroke, just like mother." Helped to bed, she was dead by eleven that evening. Buried next to her parents, surrounded by flowers, according to her aunt "Hinni" Mendelssohn "the nightingales sang her a lullaby."

Wilhelm Hensel after some time gave up the Hensel household, thereby displacing his sister Minna who had still lived with him and Fanny; she now moved to their sister Luise. Luise not only had to modify her own household but took on the not inconsiderable burden of sorting and clearing out her brother's abandoned apartment. Wilhelm, whose creative drive had begun to wane since his return from Italy, now seemed totally lost in his widowed state. He had given up his painting palette,

[2]For a detailed account see Harnack 1900:929–45.

although he continued to carry his sketch pad, to produce pencil drawings, and to be entertaining in social settings.

Rebecca Dirichlet took over the care and feeding of Sebastian, as well as of Wilhelm, whenever the latter was present. Without the older sister who had been her confidante and best female friend, Rebecca would attempt for the rest of her life to continue feeding Fanny's son Sebastian physically, emotionally, and intellectually.

Less than half a year after Fanny's death, Felix, established in Leipzig, had to call off several concerts because he was feeling unwell. In late October, he suffered three strokes during a period lasting just over a week. By the evening of the fourth of November, he, too, was dead. Ed Devrient, who had been living in Dresden since 1844, took the train to Leipzig the morning after Felix's death; he recorded the events in his diary and noted the reactions of those who were near: Clara Schumann was crying; Felix's brother Paul wanted "to have all the pain and the importance of the occurrence for his own." It was Cécile, Felix's widow, who gained Devrient's respectful admiration: She had him come to see her, was restrained in her grief, and wanted to express her sympathy to Devrient for having lost such a loyal friend.

Devrient recorded that Leipzig turned out in grand style to honor Felix Mendelssohn Bartholdy's memory. As the coffin passed through the streets, richly covered with silver embroidered velvet and "a forest of giant palm leaves," the population hung out of windows and filled the streets. The procession that followed the coffin contained the city's notables, officials, clergy, and musicians playing suitable compositions.

12.5 Political Turmoil

The dark clouds hanging over the family were soon subsumed by mightier events. In 1832, Lea Mendelssohn Bartholdy, sitting on their common back veranda, had read the morning newspaper accounts of the political changes in Paris to the Dirichlets and Hensels. Now revolutionary movements came closer to home.

1848

After the Paris uprising of February 1848 that marked the onset of "the year of revolution," protest meetings spread over most of the German states. In Prussia, the strongest opposition to the ruling powers initially occurred in the Rhineland. Here continuing industrialization had resulted in substantial displacement of workers; in Cologne the year began with a third of the population being on welfare. Berlin was better off economically; but for some time there had been moderate protest meetings and signs of unrest, such as a "Potato War" in which people at the market were pelted with produce. Disturbances grew after the city council rejected a popular petition, and the council requested formation of a non-military security force to curtail the angry public demonstrations. Before this could be organized, in March a growing crowd surrounded the palace and became dispersed by the cavalry with considerable bloodshed throughout the center of the city and some neighborhoods.

Among those killed was the brother of the mathematician Eduard Heine on March 18. Their sister was the Dirichlets' sister-in-law, Albertine Mendelssohn Bartholdy. Imprisoned for about twenty-four hours was Gotthold Eisenstein. He may have been simply among a crowd of protesters but was found in a house that contained weapons used by the revolutionaries.

The situation affected most members of Dirichlet's extended family and friends. The Dirichlets locked Sebastian Hensel in his room to avoid his rushing out into the turmoil and getting himself killed. Jacobi, on the other hand, acceded to his oldest son Leonard's request and joined him in defending part of the city from both the mob and the cavalry. Wilhelm Hensel armed himself as the head of a group of artists intent on safeguarding various cultural sites, including the king's palace, and received plaudits for leading his men in a successful peaceful defense of the buildings they were protecting. At the palace, they were joined by a group of students that included Bernhard Riemann who had been in Berlin for several semesters to supplement the more meager mathematical offerings then available in Göttingen.

Dirichlet decided to protect the palais of the Prince of Prussia, the later Emperor Wilhelm I, who had hightailed it to England. The building, next to the Royal Library, had been marked as "National Property." While pacing up and down with his weapon near a gunpowder-holding structure, puffing on one of his usual cigars, Dirichlet was stopped by one of the guards, who pointed out to the professor that smoking was not permitted in the vicinity of firearms and gunpowder. The guard happened to be one of his students. Dirichlet gave up his cigar.

In Berlin, the tumults subsided gradually. The victims of March 18–19, known as the March Heroes, were carried past the palace, followed by a considerable procession that included, among many others, Alexander von Humboldt and Jacobi, who estimated the number of participants at 300,000. The king stood in deference.

In Frankfurt, a convention known as the Frankfurt Diet would meet for a lengthy period, engaging in thoughtful and complex negotiations with the intent of drafting a document that would lead to a unification of the German states, with a constitution and a ruling monarch. The group is said to have contained some of the most learned members called together to draft a country's rules of governance. After numerous complex negotiations involving Austria and several of the German principalities, however, Friedrich Wilhelm IV, who was to serve as ruler under the plan, refused to sign the document, stating that the other German princes should have had a part in its construction, but actually unwilling to go along with the curbs on absolute power presented by the proposed constitution; it would be dead before the end of 1850.

Many constitutional monarchists, who included "democrats" of the period, had had their hopes raised that the proposals of the Frankfurt Diet would lead to a peaceful compromise. Instead, there were renewed short, bloody revolts in parts of Germany such as Baden and Dresden. In Berlin, as elsewhere in the German states, those whose hopes had been shattered saw a bleak future ahead. Among those most distressed by Friedrich Wilhelm IV's decision were the intellectuals of Berlin, some of whom had received personal favors from the king but all of whom remembered an earlier statement of Friedrich Wilhelm III, later supported in a modified form by

Friedrich Wilhelm IV, that had appeared to endorse some version of a constitutional government.

Aside from his protecting the palais of the Prince in 1848, Dirichlet took no overt, public actions during this difficult period. True to his lifelong custom of being firm in his opinions without engaging in public discussions, priority disputes, or lengthy private arguments, he refrained from demonstrations, attendance at any of the constitutional or democratic clubs, or participation in political controversies. Among like-minded friends and family members, he shared his opinions and judgments freely. We know from Varnhagen's diaries that they would analyze social and political issues while walking down Unter den Linden on Dirichlet's way home after class in the afternoon. Rebecca wrote to her nephew in January 1849 that he should take Dirichlet as a model for not responding to Sebastian's conservative father Wilhelm Hensel, who daily expressed ideas that must affect Dirichlet like a red cloth waved in front of a rooster. Another time she wrote that she and Dirichlet only discuss politics in his room, which she enters for that purpose.

Dirichlet's views as a constitutional monarchist were no secret, however. Occasionally, he joined small groups of those supporting constitutional government at the Cafe Kranzler. When Jacobi sent his brother an excerpt from a March 1849 supplement of Berlin's leading tabloid newspaper that announced the course catalog, it included a list of "the red contingent on the faculty," containing the names of seventeen men, largely scientists such as Rudolph Virchow and Paul Erman, along with Jacobi, Dirichlet, and others; an additional comment noted that [the musicologist] Adolph Marx accompanies them with "democratic music." Jacobi explained to his brother that this newspaper was one that "knows everything" especially how often the Deputies of the Left go see girls.[3]

Dirichlet did take open action when it came to something he considered a defamation of his character or a face-to-face lie. We noted earlier his response in connection with the Breslau appointment, when his actions had been called duplicitous. Now, in the winter of 1848/49, he confronted a faculty member. The university had sent a subservient address to the king in November 1848. The faculty member in question had told Dirichlet that he thought no man of honor could sign the statement. Fifty professors, including Dirichlet, Magnus and Dirksen, had not signed it. Apparently, the man in question did not think of himself as honorable: for his name appeared as one of the published signatories. Dirichlet confronted him earnestly and audibly; according to Varnhagen, the man was deeply ashamed and silent. Varnhagen made a note of this incident in his diary as an example of people signing the opposite of what they think when an official entity is involved. He considered the faculty statement a true blemish on the university, whose rector and senate had not long before denied any expression for freedom under the pretext that it was not appropriate for the institution to enter into political matters.[4]

---

[3]Ahrens, W., ed. 1907:219.

[4]Varnhagen, K., *Tagebücher* 5 (1862):349.

Apparently in the same connection, Rebecca wrote at the end of November 1848 that Dirichlet had so thoroughly "flattened" all professors in a meeting of the Akademie that they sat there like dripping dogs.

The Kinkels

In 1850, Rebecca would be a participant in a historic episode that received international attention, without anyone but those immediately concerned being aware of her role in a spectacular drama.

The continuing revolts in parts of the Rhineland and in Baden had involved Gottfried Kinkel, whose wife Johanna was the daughter of one of Dirichlet's former teachers at the Gymnasium in Bonn and had been an occasional visitor and participant in Fanny Hensel's Sunday concerts at Leipzigerstraße 3 in the 1830s. Gottfried Kinkel, in 1848 an extraordinary professor of art history in Bonn, had become very active on behalf of republicanism. In March 1848, it was he who had led a group in Bonn, carrying a large flag with the black–red–gold colors, and addressing a sizeable crowd.

After participation in an unsuccessful, unarmed, but well-publicized storming by Bonn democrats of an armory in Siegburg in May 1849, Kinkel attached himself to uprisings in Baden where he was wounded and captured at the end of June. A military tribunal sentenced him to life imprisonment. When a higher court in Berlin quashed the sentence because there had been strong arguments for a death penalty, the king, supposedly influenced by Bettina von Arnim, confirmed the life sentence with the proviso that the sentence be carried out in a civilian prison, rather than the fortress previously specified. As a result, Kinkel spent the next six months in a Pomeranian prison, said to have been occupied with the usual work program of "spinning wool." By April 1850, he was taken back to Cologne where a special inquest concerning the Siegburg affair was being held. He managed to be absolved in that case, helped by an unusual display of oratorical skill on his own behalf. He was returned to prison, this time to the well-known, large, and secure structure in Spandau.

Carl Schurz and Rebecca Dirichlet

Johanna Kinkel, who had been making a variety of attempts to ameliorate her husband's fate, wrote to Carl Schurz not long before the Siegburg inquest, asking for his advice and assistance in an escape plan for Gottfried Kinkel.

Schurz, at this time himself a fugitive, had attended one of Kinkel's lectures at Bonn in the winter term of 1847/48. Kinkel, rather than confining himself to talking about art and cultural history, introduced his students to the fine points of rhetoric and took notice of Schurz who had declaimed Marc Antony's famous "Friends, Romans, countrymen..." oration as a class exercise. Kinkel invited Schurz to his home and they soon became friends, participating in one of the numerous democratic clubs and running a related Bonn newspaper. Schurz also had participated in the Siegburg affair, gone to Baden, escaped to Straßburg, and from there to Switzerland.

Johanna Kinkel initially had thought the time of her husband's appearance before the inquest might be opportune for an escape. But Schurz read between the lines that she hoped he himself would undertake an attempt to free Kinkel. Although subsequently ambivalent about the unexpected fame the successful outcome of this enterprise had brought him, Schurz proceeded to ask a like-minded cousin to obtain a travel passport for him in the cousin's name. Feeling safe with his new passport as long as he went unrecognized, he left Zurich for Germany, with a stopover in Paris. In August 1850, he returned to Cologne where he met with Johanna Kinkel. She explained to Schurz that, in their censored correspondence, she had informed her husband that a plan for his flight was underway by writing him in great detail about her musical studies, in which she made frequent reference to a "Fuge." Since they were both well-versed in Italian, Kinkel understood what the censors did not, that she was referring to a flight (*fuga*), not a musical composition. She told Schurz that the money she had been collecting for her husband's escape had grown considerably. They agreed it would be sufficient to execute their plan and that she would send it to "a trusted person" in Berlin from whom he could get what he needed.

The person whom Johanna Kinkel knew in Berlin and trusted was Rebecca Dirichlet. When she had participated in some of the concerts at Leipzigerstraße 3 in the 1830s, Johanna had occasionally joined Rebecca in four-handed piano playing. In her memoirs, she later wrote of her surprise at Rebecca's musical talent and of her commenting on this to Rebecca. Rebecca had responded only half-facetiously "my older siblings stole my artistic reputation. In any other family I would have been highly praised as a musician and might even have conducted a circle. But next to Felix and Fanny I could not achieve any recognition."[5] Now, in April 1850, Johanna Kinkel tried unsuccessfully to visit her husband in Spandau. Before leaving Berlin, she stopped by to see Rebecca who was surprised and deeply moved by the visit.[6]

Schurz spent the next two months in preparations. He went to see Rebecca Dirichlet, the "trusted person." To protect her, he would never mention her name, however; even in his posthumously published twentieth-century autobiography the closest identification is to "a lady connected with the Mendelssohn family" and elsewhere "a relative of the famous Felix Mendelssohn Bartholdy." His description of their meeting, too, was guarded.[7]

Schurz's complex preparations included bribing several accomplices, including some of the guards, finding means of transportation for Kinkel and himself, and, of course, getting Kinkel out of Spandau. The guards were willing to strengthen Kinkel with extra food. All but one, however, were unwilling to be accessories to an escape.

When everything seemed ready, the escape was planned for November 5. Unexpectedly, it turned out that an essential set of keys was unavailable because an absent-minded inspector, unaware of any plan, had pocketed these keys that day, instead of putting them in the regularly assigned place. The following evening everything went according to plan. Schurz waited below at midnight and watched Kinkel slide down a rope from his high prison window, avoiding being hit or heard because of falling

[5]Quoted in Feilchenfeld 1979:56–57.

[6]Hensel, S. 1904:115, quoting from a letter of Rebecca, dated April 20, 1850.

[7]Schurz 1906:302ff., quoted in Feilchenfeldt 1979:58; also see Wersich, ed. 1979:50–71, which contains pertinent contemporary illustrations and bilingual excerpts taken from Schurz's *Lebenserinnerungen*.

brick fragments. They had arranged for an accomplice to take them to the border of Mecklenburg where they would be safer from the Prussian police. By the time they reached Mecklenburg the horses were tired, the alarm had spread, the newspapers were full of the unbelievable escape, with contradictory accounts of sightings; yet, with the help of a relay team of friends and paid volunteers that had been set up along the entire route, they arrived at their destination Warnemünde, from where they sailed, arriving in Edinburgh after approximately two weeks.

Johanna Kinkel had gone to Paris in the meantime, where she and her husband would be reunited. They decided to make a home in England; there Kinkel taught German literature, and she gave music lessons to support them and their children. It was not an easy life. In 1851, Kinkel went on a brief lecture tour in the USA, but decided it was preferable to stay in England. Help from home was limited because the Prussian authorities were still checking on possible accomplices to the escape from their assumedly impenetrable prison.

Dirichlet, too, had exercised caution to avoid incriminating family or friends in the affair. When Johanna Kinkel asked him for some references after the Kinkels had landed in England, he demurred, pointing out that it would do her more good to call attention to the fact that Felix, who was well known in England by this time, had admired her musical accomplishments. As it turns out, Varnhagen von Ense helped strengthen useful contacts for the Kinkels in England.[8] The timely entries in his diaries suggest that he may have been privy to some of the escape plan.[9]

12.6 Return to Surface Normalcy

In Berlin, everyday life had given an appearance of return to normalcy, although there were continuing retributions against those who had been involved in the uprisings of 1848 and 1849. Among these, we single out several mathematicians, notably Georg Rosenhain, Jacobi, and Eisenstein, all three of whom had come to public attention in 1848.

In addition, attitudes in institutions such as the Kriegsschule, the university, and the Akademie continued to be affected by prior divisions.

Samples of Retributions Against Mathematicians

Rosenhain, a student of Jacobi's, had been a privatdozent in Breslau since 1844. He was expelled from there in 1848 because of political activities that year. He had

[8] Letter from Varnhagen to Rebecca quoted in Feilchenfeld 1979:67.

[9] After Johanna's death in 1858, Kinkel remarried and, though by then settled in England, in 1866 accepted a call to the ETH in Zurich as professor of archeology and art history. Schurz had turned to the USA in 1852, where he would gain fame as a leading German-American liberal thinker and writer, coming to notice not only as an early supporter of Abraham Lincoln and fighting in some of the major battles of the Civil War, but subsequently serving as Senator from Missouri, Secretary of the Interior, and continuing to write and to speak with the oratorical skills that Kinkel had helped him develop.

won the 1846 Paris Académie's gold medal for a memoir that would be published in 1851[10]; it provided the solution of the inversion problem for hyperelliptic integrals of order $p = 2$. He had edited several of Jacobi's lectures when still a student. He had Humboldt, Dirichlet, Jacobi, and others, including even Johannes Schulze, recommend him for another position because of his competence and his having no other proper means of support. Yet it took six years before he was able to obtain an appointment as privatdozent at the university in Vienna. This outcome was the result of Dirichlet's having turned to the appropriate authority in Austria. Rosenhain only returned to Prussia in 1857, when he gained a position as extraordinary professor in his native Königsberg, which he retained until 1886, a year before his death.

Jacobi shared honors with Dirichlet in their being considered Prussia's foremost mathematicians. Non-specialists saw that Jacobi's publications were far more prolific and diverse than Dirichlet's, perhaps not noticing that Dirichlet seemed to strive for his model Gauss's motto "pauca sed matura." Jacobi, however, had been in attendance at one of the "constitutional clubs" in Berlin (he would claim he had gone there on advice of his doctor who had recommended the excitement of such a meeting might be better for his health than the quinine he was taking).[11] Dove, as committee member at the club, had asked various members to give testimonials. When he mentioned it to Jacobi, Jacobi responded with such a convincing discourse that he was applauded for oratory described as equal to that of the ancients. Jacobi himself stated afterward that he could not recall the points made in this impromptu presentation, which he considered to have been impartial; the audience seemed divided in opinion as to the degree of his impartiality. The result of the ensuing tremendous publicity was financial loss.

For some time, Jacobi had had to face financial difficulties. Unexpectedly, he had been left with a far smaller inheritance than had been assumed. His assignment to the Berlin Akademie carried a stipend supplemented by income from Königsberg, but he was still hoping to receive an appointment as ordinary professor at the university in Berlin. When he requested this, the ministry asked the faculty for an opinion. The response was negative on two grounds: The faculty pointed out that with three ordinary professors—Dirichlet, Dirksen, and Martin Ohm—mathematics was already amply represented. In addition, it was felt that, aside from his not having been part of the university previously, Jacobi's recent public appearances were not in the best interest of the institution. This opinion was not signed by the three mathematicians; but Encke, as well as Magnus and Mitscherlich, concurred with the majority.[12] Jacobi's request was denied. He now had to choose between returning to Königsberg or losing any income beyond the stipend from the Akademie. He chose to stay in a room in the capital, his wife and seven children settling in the cheaper town of Gotha. Jacobi himself spent much time with Dirichlet during this period.

[10]Rosenhain 1851.

[11]Königsberger 1904:448.

[12]Königsberger 1904:454.

In the spring of 1849, Jacobi reported to his brother that he and Dirichlet were busy reading works of Euler recently edited by P. H. Fuss in St. Petersburg.[13] Fuss had sent a box with the two volumes to them and Jacobi noted that among the previously unpublished works by Euler they came upon a number of theorems that Euler had found by induction and which belong to Gauss's most famous discoveries. In thanking Fuss for the present of the volumes, Jacobi added that Dirichlet, although the *schreibfaulste* man on earth, would probably write Fuss as well, since they speak often and at great length about the treasures newly found in those volumes.[14]

Eisenstein's case is more complex. While Dirichlet was in Italy, Humboldt had decided to inform Gauss of Eisenstein, which led to Gauss adding his support by writing an introduction to a collection of Eisenstein's papers. In 1844, Crelle's *Journal* carried twenty-five publications by Eisenstein, which brought him widespread attention. From then until 1848, he received numerous modest financial tokens of support from the court and the Akademie. Following a recommendation from Kummer, he was awarded an honorary doctorate in Breslau. Meanwhile his health deteriorated, as did his psychological state concerning which he himself, while still in school, had called attention, referring to hypochondrial tendencies.

As previously noted, Eisenstein was incarcerated in Spandau in March 1848 for being in a house in Berlin where shots were fired from weapons determined to have been used by the revolutionaries. He disclaimed having taken part in any political activity. However, his constitution had been weak since childhood; he was the only survivor of a number of siblings most of whom are said to have died of meningitis. The arrest and beatings in 1848 further affected his health. In addition, he felt that he had no support in Berlin. Jacobi, whom he had followed with great admiration, had publicly accused him of plagiarism in 1846, a charge which, Jacobi later explained, had been leveled largely as a pedagogical device to teach Eisenstein the importance of crediting his sources. This questionable bit of pedagogy could not have had a positive effect on someone as debilitated as Eisenstein. Jacobi's motivation may have been twofold: for he had published, as long as a decade before, some of the results Eisenstein now claimed, and presumably did not wish his own work to be overlooked.

Eisenstein had received his habilitation in 1847 and began his lecture courses in Berlin that summer. His lectures were well-received although he gave some of them from his bedroom when too ill to go to the university. After the March 1848 arrest, his modest annual salary of 500 Thaler was cut to 300, and, despite very positive reports from Dirichlet and others, regular visits from Dirichlet, and election to the Akademie in March of 1852, his condition continued to deteriorate physically and emotionally until he died in 1852.

[13]Euler 1849.

[14]Ahrendt, W., ed. 1907:205.

Eisenstein had had short-lived, pleasant, mathematical contacts with Joachimsthal, Riemann, Heine, and others. But the only lasting mathematical friendship existed with Stern in Göttingen, whom he had met while visiting Gauss.[15]

The Kriegsschule

The Kriegsschule which had been closed during the disturbances, from March 17, 1848, to October 15, 1850, and again from October 17, 1850, to February 15, 1851, reopened. During this time, Dirichlet and the other civilian instructors at the school had received no remuneration. The winter term 1851/52 introduced a change in the opening and closing dates from October 15 to October 1, and from July 15 to July 1. The mathematics curriculum had improved; calculus was now admitted as part of the regular instruction cycle. The orientation of the student body had changed, however. After 1848, the young future officers had become far more conservative and nationalistic than their predecessors of the late 1820s and 1830s who had admired the military expertise of Napoleon's officers whom their elders had fought in the Wars of Liberation.

The University

The mathematics faculty at the university had taken on a somewhat unusual composition. Dirksen, who had been unwell, had taken a leave of absence beginning with the 1847/48 term and, with his wife, had gone to Paris, where he died in July 1850.

This left Dirichlet and Martin Ohm as the only ordinary professors. Dirichlet, however, could not enjoy the full privileges of that position until 1852, after he had held his long-deferred public Latin lecture that had been the stipulated requirement for his full admittance to the faculty in Berlin, agreed upon twenty-four years before. Martin Ohm, on the other hand, had had a long career teaching both at the secondary (Gymnasien and Kriegsschule) and university level. But he had a low reputation among students. They saw him as an outmoded, elderly vestige of another era, wearing Biedermeier clothes and teaching elementary subjects at a very deliberate pace, while constantly referring to his own books. They were quite unaware that his teaching style was part of a methodology that had once helped transform mathematics education in Prussia to a new, higher level. Neither the students nor some of his research-oriented colleagues would have guessed that a century later some historians of mathematics would consider the sequence of his books as precursors to Bourbaki for having provided a rigorous, highly structured approach to the fundamental concepts he was teaching.

There were two extraordinary professors on the faculty. One was Steiner, who, having taught synthetic geometry in that position since 1834, would continue to hold his lectures until the fall of 1862, a few months before his death. The other

---

[15]Contemporary rumors concerning a dissolute lifestyle, which led to at least one official query, were subsequently discounted by the reliable twentieth-century historian Kurt-R. Biermann. On the other hand, Rebecca Dirichlet, not given to repeating groundless gossip, appears to have had undocumented information that led her to express regrets in writing to Sebastian at the time of Eisenstein's death that he had been a tragic, hopelessly wasted genius who had followed his worst instincts since youth.

was Johann Philip Gruson, who, until his death at age 90 in 1857, offered classes primarily designed for teachers of mathematics. It appears that he did not actually lecture after the 1840s, however. By that time, his early publications, including a German translation of Lagrange's Theory of Analytic Functions, appear to have been largely forgotten.

Jacobi taught at the university as a "reading member" of the Berlin Akademie, but was not a member of the regular faculty.

Aside from these men, mathematics was taught by a number of promising privatdozenten: Eisenstein, who died in 1852; Joachimsthal, who left for Halle in 1853; and Borchardt, first appointed in 1848, and then who, like Jacobi, taught as a reading member of the Akademie, beginning with his membership there in 1855, the year he took over the editorial duties for the *Journal für die reine und angewandte Mathematik.*

After their habilitation in 1853, recommended by Dirichlet, two more privatdozenten would join the group: They were Friedrich Arndt and Reinhold Hoppe. Arndt had received his doctorate in Greifswald under J. A. Grunert with a dissertation on continued fractions. He taught in Stralsund's secondary schools for eight years and gained a reputation for successfully guiding his Berlin students into higher mathematical studies. He served as extraordinary professor for four years before his death in 1866. Hoppe, a solitary figure who had received a doctorate in Halle in 1850, was less successful as a lecturer but became a prolific publisher. Primarily geometrically oriented, he eventually (1872) succeeded his one-time teacher Grunert as editor of the *Archiv der Mathematik und Physik.*

During the short period after 1852 that Dirichlet was entitled to serve as official doctoral adviser, he acted in that capacity, jointly with Martin Ohm, for five Berlin doctorands who received their degrees in 1853 and 1854. They were L. L. Wituski, Hermann Suhle, O. K. F. Janisch, Rudolph Lipschitz, and Johann Franz Stader.[16] The best known of these, who would always consider himself a student of Dirichlet's, was Rudolph Lipschitz. Born near Königsberg he studied there with Franz Neumann before going to Berlin and working under Dirichlet's tutelage. His doctoral dissertation, titled "Determinatio status magnetici viribus induentibus commoti in ellipsoide" reflected the influence of both men.

Visitors

There were occasional highlights in the routine. Some of these were provided by visitors from abroad, two of whom left us interesting descriptions of Dirichlet at this time. They are the Russian Pafnuty Chebyshev and the Englishman Thomas Archer Hirst.

Chebyshev, returning from Paris, where he had received guidance from Liouville, spent several days in Göttingen. He described his visit in October 1852 in an official account of his trip to Western Europe:

> It interested me very much to make the acquaintance of the celebrated geometer Lejeune-Dirichlet. Among the investigations made by this savant in Analysis, the first place belongs

[16]For their dissertation topics, see Biermann 1988:350.

to his principles of the application of the infinitesimal calculus to research of the properties of numbers.

But only a certain part of his researches of this question has been published until now; as for the rest of his work, we know nothing except that some definite results remain without demonstration. The investigations of M. Lejeune-Dirichlet interest me especially, because I have occupied myself with similar questions; in my article presented to the Imperial Academy of Sciences of St. Petersburg under the title "On the function which determines the totality of numbers less than a given limit," I have demonstrated that the formula found, by analogy, by Legendre, for determining the quantity of prime numbers less than a given limit, must be replaced by another one; this result has been so unexpected that M. Lejeune-Dirichlet, speaking of his researches touching on this question, says nothing of the inexactitude of Legendre's formula.

During my stay in Berlin I found an occasion each day to speak with this geometer on the aforementioned researches as well as on other points of pure and applied Analysis.

I attended with particular pleasure one of his lectures on Theoretical Mechanics.[17]

Chebyshev's remark about Dirichlet not commenting on the problem with Legendre's formula is interesting but not indicative of Dirichlet's unawareness of the issue. Most likely it was simply an example of his reticence to comment, in a casual conversation with someone he had just met, on Legendre's flaw, particularly if the younger man had just made a notable discovery of his own on the topic, without apparently being aware of Dirichlet's related publications.[18] There also appears to have been a linguistic problem. We know that Chebyshev was used to French; most likely he and Dirichlet conversed in French. It is possible that German was more difficult for him. For that reason, it is likely that Chebyshev was not familiar with Dirichlet's Akademie publications but probably knew from conversations with Liouville of Dirichlet's major results.

There was another visitor from abroad the same month that Chebyshev was in Berlin. This was Thomas Hirst, an English mathematician who had just received his Ph.D. from the university at Marburg and was visiting Berlin during the fall term of 1852/53. He called on Dirichlet one morning and attended some of his lectures. After his morning visit, on October 13, 1852, Hirst wrote in his diary:

[Dirichlet] is a rather tall, lanky-looking man, with moustache and beard about to turn grey with a somewhat harsh voice and rather deaf: it was early, he was unwashed, and unshaven (what of him required shaving), with his "Schlafrock", slippers, cup of coffee and cigar... I thought as we sat each at the end of the sofa, and the smoke of our cigars carried question and answer to and fro, and intermingled it in graceful curves before it rose to the ceiling and mixed with the common atmospheric air. "If all be well, we will smoke our friendly cigar together many a time yet, good-natured Lejeune-Dirichlet."

By the end of that month, Hirst commented on Dirichlet's lecture style he had been observing:

Dirichlet cannot be surpassed for richness of material and clear insight into it: as a speaker he has no advantages–there is nothing like fluency about him, and yet an eye and understanding make it indispensable: without an effort you would not notice his hesitating speech. What is

[17] Chebyshev *Oeuvres* 2.

[18] Chebyshev 1851 and 1852.

> peculiar in him, he never sees his audience–when he does not use the black-board at which time his back is turned to us, he sits at the highest desk facing us, puts his spectacles up on his forehead, leans his head on both hands, and keeps his eyes, when not covered with his hands, mostly shut. He uses no notes, inside his hands he sees an imaginary calculation, and reads it out to us—that we understand it as well as if we saw it too. I like that kind of lecturing.[19]

Hirst's description conforms to several of Dirichlet's attributes that are familiar to us: His manner of lecturing, by his lack of fluency allowing time for a student to follow the argument presented; the nearsightedness that had saved him from the draft thirty years earlier and probably made it impossible for him to make eye contact with listeners in a lecture hall; writing on the blackboard from notes without a constant need to take his glasses off and putting them on again; and his continuing capacity to draw on his memory for even complex arguments.

His morning appearance at home, too, is not altogether surprising on the basis of what we know of his evening working routine and late rising in the morning. What is new to us are the references to the harshness of his voice, his hearing difficulty, and a description, included in another diary entry (February 20, 1853), of his rushing out as soon as class was over–these characteristics may have been manifestations of his physical deterioration.

Adding up the political and family-related turmoil and disappointments that would greet Dirichlet not long after his return to Berlin in 1845, and his more frequent references to lack of mental endurance during the ensuing years, it is difficult to avoid the conclusion that recurring spikes of his "Roman fever," acquired on his excursion to Sicily, were leaving an increasing effect on his physical condition, noticeable by the 1850s, and a foretaste of his weakening cardiological state.

12.7 Göttingen 1849 and 1852

Ironically, one positive outcome of the unsuccessful attempts at revolution in 1848 had been the revocation of the Hanoverian punishments against the "Göttingen 7," resulting in Wilhelm Weber reclaiming his professorship in Göttingen in 1849.

1849

The year 1849 also marked a celebration in Göttingen: the 50th anniversary of Gauss's doctoral dissertation. Dirichlet and Jacobi attended the event. If they had had hopes that they could have some interesting mathematical discussions with Gauss, they were disappointed. Jacobi complained that it was quite impossible to have a decent conversation with Gauss. Dirichlet at least was pleased that he had been able to salvage part of Gauss's original manuscript of the *D.A.*—a chapter of the *Analysis Residuorum*—that Gauss was using to light his pipe.[20]

---

[19] 31 Oct 1852. For a biographical sketch and additional excerpts from Hirst's diary see Gardner and Wilson 1993.

[20] See Merzbach 1981 for further details concerning this manuscript portion.

The impression Gauss was leaving suggests some truth in the warning Poggendorff in Berlin had given to his good friend Wilhelm Weber when he learned that Weber was planning to resume his work in Göttingen. Not only was he afraid Weber might be diverted from his own research if persuaded to do something in which Gauss was more interested. He also suggested that Gauss was getting old and might be losing his mental agility. This impression is validated in a letter (1845/46) from Gauss's daughter Therese in which she mentions the slowing of his thought processes, and in a sequence of letters that Gauss himself wrote to Humboldt in the late forties and early fifties.[21] Gauss celebrated the doctoral anniversary by reading to the society on July 16 and then letting them publish his fourth proof of the Fundamental Theorem of Algebra[22]; the first proof had constituted his doctoral dissertation.[23] [Gauss 1849] was his last substantial mathematical publication. He spent considerable time during his remaining years trying to polish his Russian and find suitable books in that language, which he had undertaken to learn in 1840.

1852

In 1852, Dirichlet would again visit Göttingen during the fall vacation. At this time, Riemann was busily working on his *Habilitationsschrift*. As he would write his father in a letter quoted by Dedekind, Riemann called on Dirichlet at his hotel and subsequently joined him at a luncheon given by geologist Sartorius von Waltershausen that included Dove and Listing, among others. Riemann asked Dirichlet, whom, according to Dedekind, Riemann considered the greatest then living mathematician aside from Gauss, for some advice. The next morning Dirichlet came to see Riemann, spent two hours with him, and provided him with notes that he needed for his *Habilitationsschrift* and which, Riemann noted, spared him hours of searching in the library. Dirichlet also discussed his dissertation with him and was so friendly that Riemann expressed his astonishment, not having expected it "because of the great distance" between them: "I hope he also won't forget me later."[24]

Riemann, who did not easily "mix and mingle," would have a chance to see more of Dirichlet during this visit, in company of Wilhelm Weber, at the latter's home as well as during a group excursion to the Hohen Hagen. They would, of course, communicate frequently after Dirichlet moved to Göttingen three years later. But the 1852 visit was an important event in the life of Riemann, who not only absorbed the mathematical pointers he received from Dirichlet, but was also bolstered by the encouragement Dirichlet gave him. For Dirichlet, the visit provided a respite from the tragic events that had been accumulating around him.

---

[21] Biermann 1977.

[22] Gauss 1849.

[23] Gauss 1799.

[24] Dedekind 1876:546.

12.8 The Death of Jacobi

The preceding year, 1851, had brought a heavy loss and new responsibilities for Dirichlet. In January, Jacobi had caught the flu. At first, he seemed to recover and resumed his research activities, but by February 11 he again became ill. It turned out that this time he had caught smallpox. Given his already weakened state he had no chance of survival. Within a week, he was dead.

On April 9, Dirichlet asked the Minister of Culture to grant him the guardianship of Jacobi's children.[25] By the end of July, the Akademie successfully requested that the Royal Cabinet grant 320 Thaler for the education of those seven children. After a series of further stepwise fragmentary contributions for the upkeep and education of Jacobi's family, ranging from a pension from Königsberg to some interest from the remaining Jacobi family estate, Jacobi's widow received 800 Thaler for the next five years. This was supplemented further after requests from a number of members of the Akademie, including Encke, who previously had opposed Jacobi in all matters except his scholarship. Finally, in 1868, Jacobi's widow was granted an annual support of 500 Thaler.

Having received Jacobi's papers from his widow during the summer of 1851, Dirichlet, with the help of Jacobi's former student Borchardt and of Ferdinand Joachimsthal, would begin the massive task of organizing these, editing the second volume of Jacobi's *Opuscula*, and by 1852 holding a widely acclaimed memorial address for Jacobi before the Akademie in Berlin.

As Dirichlet reported in Crelle's *Journal* in August of 1851 (**1851**), the managerial task of organizing the Jacobi *Nachlass* was formidable, and the threesome was assisted briefly in this work by Kummer and Rosenhain, who fortuitously were spending time in Berlin. The initial plan outlined by Dirichlet called for a carefully considered division of labor. First, the sorted material would be reviewed by an additional number of friends who would seek out those materials most suitable for publication. After this stage, they would publish these edited papers in subsequent issues of Crelle's *Journal*; only after completion of this task would the papers appear in a set of collected works.

Dirichlet, commenting on the overarching influence of Jacobi's teaching activities, explained that several of Jacobi's friends had offered to publish Jacobi's most important lectures. Since Jacobi (like Dirichlet himself) had left no lecture notes of his own, these would be based on the exact transcripts several of his best students had produced, which Jacobi himself (unlike Dirichlet) had collected. Those already decided upon included Jacobi's lectures on elliptic functions, on cyclotomy and its application to number theory, on analytical mechanics, and on the general theory of curves and surfaces.

Several events conspired to delay this work longer than Dirichlet may have anticipated. Joachimsthal, since 1845 a privatdozent in Berlin, in 1853 left for an ordinary professorship in Halle, where he had earned his doctorate after studies in Berlin and Königsberg. That meant the work of dealing with Jacobi's papers now was

[25] Biermann 1959:73.

divided between Dirichlet and Borchardt. As previously noted, after Crelle's death in 1855, Borchardt succeeded him as editor of the *Journal für die reine und angewandte Mathematik*; Dirichlet had left Berlin earlier in 1855. In 1859, after Dirichlet's death, Borchardt would take over the guardianship of Jacobi's children as well as Jacobi's papers and sought with considerable success to gather additional scattered correspondence and notes for publication.

Jacobi's papers from the *Nachlass* began to appear in the *Journal* in 1857; the editors included Borchardt, Clebsch, Sigismund Cohn, Heine, Oswald Hermes, E. Luther, and Richelot. In addition, the collected works would include papers not printed in the journal, edited by Franz Mertens, Borchardt, E. Lottner, Hermann Kortum, Albert Wangerin, and August Bruns. Borchardt published Jacobi's lectures on elliptic functions based on theta series in the first volume of the collected works.

Jacobi himself had edited a volume of his *Opuscula* in 1846; Dirichlet edited the second volume in 1851. In 1871, Borchardt edited the third volume. The collected works that superseded these three *Opuscula* consisted of seven volumes, appearing between 1878 and 1891, and edited by Weierstrass, with Georg Hettner taking over work on the last two of these because of Weierstrass's failing health.

Dirichlet's memorial address of Jacobi, held in 1852, not only records Jacobi's life, working habits, opinions and achievements, but also documents significant aspects of Dirichlet's character and thought processes as well. In discussing Jacobi's mathematical contributions, he presented us with an interesting portrayal. Of particular interest in this respect is his treatment of Jacobi's work in its relationship to his predecessors and contemporaries. He cited Jacobi's explanation to his uncle why, after early years of indecision, he chose mathematics over philology: Explaining the struggle it had cost him to give up the study of the ancients, Jacobi had stated:

> I must now give them up entirely. The immense colossus which was brought to the fore by the work of Euler, Lagrange, Laplace, requires the most immense strength and endurance of thought if one wants to enter its inner nature instead of just scratching the surface. To master it, without having to fear at each moment to be suffocated by it, one is driven by a force which does not allow rest or pause until one stands above it and has an overview of the entire work. Only then, when one has comprehended its spirit, is it possible to work calmly to complete its individual parts and to bring the entire great mass as far forward as one's powers allow.[26]

Dirichlet did not allude to the occasional, independent overlap in topics between his own work and that of Jacobi. He chose an unusual course in discussing Jacobi and Abel's preeminent contributions to the founding of the theory of elliptic functions, however. He first provided an introduction to the short history of elliptic functions, briefly covering the high points of the contributions of Fagnano, Euler, and Legendre, praising the last-named for having recognized the seeds of an important branch of analysis in the discoveries of his two predecessors. When it came to the work of Abel and of Jacobi himself, he treated both equally. Reading this eulogy for Jacobi, one is left with the impression that Dirichlet sought to rectify the lack of proper homage paid to the young Abel at the time of his death more than two decades before.

[26]Quoted in **1853a**; see *Werke* 2:229.

In reading Dirichlet's tribute to Jacobi, one is struck also by the difference in style between that of this talk and that of his memoirs and of the few letters he wrote. The informal, yet impressive, approach he used here is found in almost none of his other writings. The close approximation to Dirichlet's style in the lectures G. Arendt published posthumously (see Chap. 16), suggests that this easygoing manner of communicating which Dirichlet apparently used in his lectures is closest to that which he used in his tribute to Jacobi. This would account for the many positive comments his students made about the presentations they attended and the widespread interest this eulogy attracted over the years. It represents a marked contrast to the formal, main portions of his memoirs and to his known letters, especially to his rather stilted official correspondence.

12.9 Family Deaths: 1848–1853

The years since the beginning of the 1848 revolutions had brought other deaths, particularly to Rebecca's family. Though none of these carried the feeling of loss for Dirichlet that he experienced with the passing of Jacobi, they reinforced the sense of diminution that this period represented for him and Rebecca.

The first to die, of the group we mention, was the seventy-eight-year-old Joseph Mendelssohn, who passed away on November 26, 1848. He had been the oldest son of Moses Mendelssohn and it was this son concerning whom his father had expressed deep regrets for having to take him away from scholarship to make him a servitor of Mammon. "He does not care for medicine; and as a Jew he must become a physician, a merchant, or a beggar." As it turned out, he would become sufficiently successful in the banking business to turn Mammon to the service of others, Alexander von Humboldt being only one of his beneficiaries.

Joseph Mendelssohn had not converted and had not turned away from intellectual pleasures such as reading his favorite Dante, or maintaining a small salon where he entertained the likes of Hegel, Börne, Ranke, and Humboldt. His welcoming vineyard property Horchheim, near Koblenz, had been a favorite summer gathering place for the family, especially convenient in earlier years, when Felix was active in Düsseldorf and Dirichlet visited his parents and friends in the Rhineland. As Rebecca noted in a later letter to Sebastian, although her brother Paul, who had been closest to the deceased, had probably felt his death the most, they all had become considerably diminished once again by the departure from life of this quiet man.

Within four years, Joseph was followed in death by his youngest brother, Nathan, who had become a scientific instrument maker. He was the last of Moses Mendelssohn's six children to die. Among other relatives whose unexpected deaths would be significant for the Dirichlets was that of Felix's seven-year-old son, little Felix, the same week that they mourned Jacobi, and of Cécile, who survived her young son by two years, dying only two weeks short of her thirty-sixth birthday.

12.10 The Death of Gauss

Winter 1854–55 brought head colds, fever, flu attacks, and a general feeling of being unwell to the Dirichlet household. Flora's tenth birthday was celebrated with fifteen "little crumblings," but Dirichlet's went by quietly because of everyone's colds. The month of February did not pass without excitement, however. It was the news, not totally unexpected, yet sudden, and clearly foreshadowing a major decision for Dirichlet, of Gauss's death.

On Friday, February 23, 1855, Wilhelm Weber wrote to Dirichlet as follows:

> Dearest friend,
>
> This morning at 2 o'clock Gauss died. Yesterday evening until 9 o'clock I, with my niece and Ewalds, was in the room next to his where he slept and did not reawaken. His daughter Therese, Ewalds, Baum, and Listing remained with him overnight. I learned of the death this morning and with Sartorius immediately went over, where we found him sitting in his armchair as though he had fallen asleep, without any traces of pain. He suffered more during the previous days, partly from the asthmatic hardships with which his illness had begun already last Easter, and partly from the dropsy, which toward the end had spread over his entire body and in the last days hindered him from lying in bed, so that he could rest at night only by sitting in the easy chair.
>
> I only write these lines rapidly from the observatory.[27] Sartorius is busy reporting the same news to Humboldt. A few weeks ago the king had sent the sculptor Hesemann down from Hannover, who did a beautiful medallion of Gauss. This morning we have invited the same by telegraph to come down today to take the death mask for a bust.
>
> I shall soon write you in more detail.
>
> Yours WW[28]

12.11 The Call to Göttingen

Weber followed this letter with a barrage of further communications, to Dirichlet as well as to the pertinent officials in Göttingen and Hanover, attempting to persuade everyone involved to have Dirichlet come to Göttingen as Gauss's successor. Meanwhile, Ernst had the mumps, and Rebecca wrote to Sebastian Hensel that she feared it was in the air that they should go to Göttingen, although there had been no official word and she hoped perhaps they would acquire an astronomer and "leave us where we are."

By April, the official word came, or, in Rebecca's words, "der dei ist gekastet." In a letter informing Sebastian of the forthcoming upheaval, she reflected on the move

[27] [where Gauss had lived].

[28] Berlin. Staatsbibliothek. Handschriftenabteilung. Dirichlet Nachlass. Correspondence. Weber.

further loosening the familial bonds that had held her close to her Berlin. They had become tenuous for some time: the deaths of her parents, the move to Leipzigerplatz 18, the deaths of Fanny and Felix, brother Paul's living in the Jaegerstraße, the sale of Leipzigerstraße 3, Walter and Sebastian no longer living in Berlin—was the bimonthly roast beef dinner with the remnants of this once close-knit family better than nothing? She wrote Sebastian that she had suggested to Weber he should establish a chair for agriculture in Göttingen but that Weber had informed her such a chair already existed and was occupied![29]

Reactions, Final Prussian Obligations, and Departure

As soon as he had accepted the official offer from Hanover concerning the professorship in Göttingen, Dirichlet went to see Varnhagen and others to let them know of his decision. According to Varnhagen, there was widespread outrage among faculty colleagues and Berlin's liberals, directed primarily against the inaction of Karl Raumer (the orthodox opposite of his namesake, the historian Friedrich Raumer), who had succeeded Eichhorn in 1850 as Minister of Education. Varnhagen referred to him as the "Minister of Ignorance."

In a well-meaning but misguided attempt to keep Dirichlet in Berlin, Friedrich Wilhelm, son of the Crown Prince and future Emperor Friedrich III, blamed the minister and suggested to his uncle, the king, that he ask George V, the King of Hanover, to release Dirichlet from his recent commitment. This almost cost Dirichlet the goodwill of the king. Friedrich Wilhelm IV was furious with Raumer. Now Humboldt, who shared the general upset but also knew that Dirichlet would not go back on his word to Hanover, with his usual diplomacy had to keep the king's anger from spilling over on Dirichlet. After delicate intervention, Humboldt could inform Dirichlet that the court wished him well, though seeing him leave with heavy heart.

On August 15, Dirichlet received a royal confirmation: the medal "pour le mérite." This, too, had been assured by Humboldt, who had communicated to a sufficient number of those voting for the award the services Dirichlet had rendered. Varnhagen wrote in his diary in early August that when Dirichlet made his farewell visit to him, he told Varnhagen that he had learned from Humboldt to expect the medal, as nine votes were already on hand. Dirichlet was able to amuse Varnhagen by also relating that upon his leaving Humboldt's house, Humboldt's servant had told him confidentially "Herr Professor, you'll get the *Orden*. We already have nine votes for you." The servant had added, "You are right to leave. Why didn't the minister know how to appreciate your value!"

Dirichlet's departure from Berlin was a blow to both Varnhagen and Humboldt. Varnhagen had noted in his diary that he had wept, when by himself, after hearing from Dirichlet of his decision to leave. He also observed that in recent years the Dirichlets had become his best acquaintances; that he did not see them often but that when he did, it was good. Humboldt, who had not been feeling well, had commented when Dirichlet came by to bid him farewell, "you will not see me again." Both of the old gentlemen had not only been aware of their age and felt a personal loss, but also

---

[29]Letter of April 20, 1855, quoted in Hensel, S. 1904:186–87.

had seen a symbolism in Dirichlet's departure; it foretold the continuing disruption of the results of intellectual and political endeavors they had championed over many years.

Dirichlet gave his last Berlin lecture course in the summer term 1855 on Number Theory and the Applications of Integral Calculus. He attended to miscellaneous duties such as writing a recommendation for Cayley in England, presenting an excerpt of a paper by Borchardt on symmetric functions to the Akademie, and suggesting Weierstrass as an addition to the Berlin faculty.

Dirichlet's mother had returned to Berlin with a teenaged grandniece and took over supervision of children and household, enabling Rebecca to travel to Göttingen where she rented appropriate living quarters (Gotmarstraße 1) within three days. Rebecca had previously written to Sebastian that, despite positive assurances, she imagined Göttingen as a horrible "Kuhschnappel" but was prepared to be surprised.[30]

The moving vans left Berlin on September 7, 1855.

[30]Kuhschnappel is the fictional site of one of Jean Paul's most popular novels (...*Siebenkaes*), especially widely read by the generation of Sebastian Hensel. I was recently informed by Meredith McClain of the existence of a real (not horrible) Kuhschnappel in Saxony and its local historian, Andreas Barth.

Chapter 13
Publications: 1846–1855

During the decade spanning his return from Italy and his departure for Göttingen, twenty-one publications by Dirichlet appeared. Discounting translations and excerpts, these represent thirteen separate contributions. They differ from his previous work in several respects.

Only one publication, **1846b**, presented a significant new result. That was his memoir on the unit theorem. Many were short, not exceeding five pages. Most fill in details of theorems and proofs that relate to subject matter covered in his university lectures or to suggestions and proofs originating with Gauss in the *D.A.* Specifically, there are numerous explanations of terms and statements familiar from the *D.A.*, which Dirichlet here reviewed in some detail wishing to show how his own methodology simplifies Gauss's arguments; he usually did so without explicit references to the relevant articles of the *D.A.*

Several of the memoirs in this decade reflect Dirichlet's study of Lagrange's *Mécanique Analytique* and conversations on mathematical physics that he had had in Italy.

Some of these publications would be of particular significance for serving as the basis of future work by others. Among these, three areas stand out. They are the sketch of a proof for the unit theorem, which he had merely outlined for the three-dimensional case in 1840; a note on the stability of equilibrium, with additions, a translation, and a reprint included in [Lagrange 1853], in which he firmed up a basic proof by Lagrange; and four notices dealing with several fundamental definitions and properties of the potential.

As mentioned previously, aside from the Latin *Habilitationsschrift* for Breslau, the reports and memoirs issued by the Berlin Akademie, and the account in the *Repertorium der Physik* that were written in German, prior to 1846, all of Dirichlet's other publications had been written in French. After his return from Italy, however, he produced the Latin memoir on the composition of binary quadratic forms that would end his "designate" status in the Berlin faculty; also, all but two of his publications in Crelle's *Journal* were now written in German. The exceptions were two French memoirs on potential theory, **1846c** and **1857c**; the second of these had previously appeared in Liouville's *Journal* as **1856b**.

U. C. Merzbach, *Dirichlet*, https://doi.org/10.1007/978-3-030-01073-7_13

Thanks to Liouville, French-language readers nevertheless later had at their disposal many of Dirichlet's memoirs: Six would be translated for Liouville's *Journal* as **1847b**, **1856e**, **1856f**, **1857d**, **1857e**, and **1857g**, joining three that had previously been published there and another, **1856c**, drawn from his correspondence with Liouville. Liouville published five additional translations, **1859a–e**, the year of Dirichlet's death.

13.1 Stability of Equilibrium

The stability of equilibrium is the first topic Dirichlet treated on January 22, 1846, in resuming his talks at the Akademie after his return from Italy the previous summer. Dirichlet's talk was reported in the *Bericht* of the Akademie, **1846a**, as having the title "On the conditions of the stability of equilibrium," but came to be more generally known in the same year through a slightly modified version of this report in Crelle's *Journal*, **1846e**, and a translation into French, **1847b**, by Kopp in Liouville's *Journal*.

It should be noted that Dirichlet's memoir, narrowly focused on Lagrange's proof concerning the stability of equilibrium, should not be confused with subsequent conversational opinions Dirichlet may have uttered concerning stability of the planetary system. The only known written item by Dirichlet on the latter topic is a piece of blotting paper uncovered by Kronecker in the Dirichlet *Nachlass* on which Dirichlet had scribbled "Exposition d'une nouvelle methode de calculer les perturbations planetaires."[1] Kronecker's discussion of results erroneously attributed to Dirichlet was prompted by an introductory comment in *Acta Mathematica* made by the editor, Mittag-Leffler. As Kronecker showed, the misunderstanding concerning Dirichlet's work may have resulted from a misinterpretation of a comment in Kummer's eulogy of Dirichlet. Despite Kronecker's detailed explanation in [Kronecker 1888b], a number of subsequent authors continued to merge the two stability topics, attributing to Dirichlet a published contribution that never existed.

1846a

The object of Dirichlet's presentation on the stability of equilibrium was to provide a rigorous proof for the basic theorem that, as he observed, goes back to Daniel Bernoulli's statement on the integral of *vis viva*. For the purpose at hand, Dirichlet considered the equation

$$\sum mv^2 = \phi(\lambda, \mu, \nu, \ldots) + C,$$

where m denotes the mass of each point in the system under consideration, v the velocity, and C an arbitrary constant. The significance of the function ϕ is that it depends merely on the nature of the forces acting on the system; the condition that, for certain values of the variables, the system is in equilibrium coincides with the

[1] Kronecker 1888b; see his *Math. Werke* 5:476.

requirement that the differential of ϕ vanishes, or that, generally, the function will be a maximum or minimum for every state of equilibrium of the system. If a maximum occurs, the system is stable. Dirichlet considered this statement "undeniably one of the most important of all of mechanics."[2]

For the proof of the statement, Dirichlet referred to Lagrange's treatment in the *Mécanique Analytique*, which, as far as he knew, had been reproduced by all subsequent authors. Dirichlet, too, was probably led to the topic by his study of the *Mécanique Analytique*, which Lagrange had divided into two parts: statics and dynamics, equating statics with the theory of equilibrium and dynamics with the theory of motion.

Dirichlet criticized Lagrange's proof based on the fact that in this proof expansion of the function is limited to second-order terms, which Dirichlet questioned on two grounds: First of all, that it assumes the higher orders remain small enough to be dropped; secondly, that it disregards conditions that must be placed on the lowest order for which not all terms vanish. Dirichlet then went on to provide a proof independent of these restrictions but stipulating two conditions: The first one was the one previously used, that the position of equilibrium corresponds to the vanishing values of $\lambda, \mu, \nu, \ldots$; the second one supposes that $\phi(0, 0, 0, \ldots)$ is likewise zero, which, he pointed out, is legitimate because of the arbitrary nature of C.

Dirichlet remarked that, whereas Poisson had given an addition to Lagrange's proof, this, in fact, did not add anything to the validity of the proof, as it was based on the assumption that every second-order term would exceed the totality of all higher-order members.

1846e

In the slightly modified version, **1846e**, of his presentation published in Crelle's *Journal*, Dirichlet added an additional criticism of various authors, among whom, in a footnote, he again included Poisson. It deals with a supposition concerning multiple, successive states of equilibrium, which Dirichlet found surprising as it could be refuted by observing the motion of an ordinary pendulum. That concluding sentence illustrates why Kronecker in his Kummer *Festschrift* would refer to this particular memoir by Dirichlet as showing the advantage of using the simplest means, not only those serving to describe mechanical processes but those that are sources of understanding, something Kirchhoff had mandated.[3]

1853c

The preceding critique by Dirichlet became even better known to subsequent generations because in 1853 Joseph Bertrand included it in his third edition of Lagrange's *Mécanique Analytique*, along with an addition by Louis Poinsot to Lagrange's proof.

---

[2]See *Werke* 2:6.

[3]Kronecker 1882; see Kronecker *Werke* 2:354. Also see the opening of Kirchhoff 1876, the *Mechanik*.

Bertrand noted that Dirichlet had simplified Lagrange's demonstration while also rendering it more rigorous.[4]

13.2 The Unit Theorem

On March 30, 1846, Dirichlet communicated the verification of his unit theorem, **1846b**, to the Physical-Mathematical Class of the Akademie. He had first suggested the theorem in his letter to Liouville published in 1840 (**1840c,d**). Having restricted himself to the case $n = 3$ in that letter, he now sketched a general proof.

1846b

The Report on the session in which this memoir was presented was prefaced with the remark that "Mr. Lejeune Dirichlet made some communication concerning an investigation carried out by him which has the theory of complex units as topic and is to be made known in another place in the future."[5] Because of the significance of this sketch and the lack of a more extended publication by him on the topic "in another place," we outline this short presentation here.

Dirichlet began by introducing the function

$$F(\omega) = \omega^n + p_1\omega^{n-1} + p_2\omega^{n-2} + \cdots + p_n = 0, \tag{1}$$

where the p_i are coefficients having no rational factor and whose roots will be designated by $\alpha, \beta, \ldots, \rho$.

Upon forming expressions of the form

$$\phi(\alpha) = t + u\alpha + \cdots + z\alpha^{n-1},$$

and

$$\phi(\beta) = t + u\beta + \cdots + z\beta^{n-1}, \ldots,$$

where $t, u, \ldots, z$ are undetermined integers, the product

$$\phi(\alpha)\phi(\beta)\cdots\phi(\rho)$$

will be a homogeneous function with integral coefficients of $t, u, \ldots, z$ which has the noteworthy property that it reproduces itself by multiplication, and hence by raising to powers.

---

[4]For Poinsot's addition and for Dirichlet's **1853c**, see Lagrange 1853:389–98 and 399–401, respectively; for Bertrand's note on Dirichlet's rigorous simplification of Lagrange's proof, see Lagrange 1853:64.

[5]*Werke* 1:641fn.

Dirichlet noted that this was first observed by Lagrange. As being of primary importance for the theory of functions formed in this fashion, he singled out the answer to the question for what system of values $t, u, \ldots, z$ they will equal unity, that is, the complete solution of the equation

$$\phi(\alpha)\phi(\beta)\cdots\phi(\rho) = 1. \tag{2}$$

This, he commented, is to be regarded as a fundamental problem of this theory.

Dirichlet first observed that one can disregard certain special solutions of the equation for which the factors $\phi(\alpha), \phi(\beta), \ldots, \phi(\rho)$ are roots of unity; these can be easily determined. For the rest, every given solution when raised to an undetermined positive or negative power will generate an infinite number of new solutions. Likewise, he considered it evident that given two or more solutions one can combine their undetermined powers by multiplication for the same purpose. In the special case where $F(\omega) = \omega^2 - D$, the equation is transformed into the well-known Pell's equation for which all solutions are obtained from one fundamental solution by raising to powers and multiplying by ± 1. This brings up the question whether there is a similar property that holds for the general equation, and whether there, too, fundamental solutions exist, from which all solutions can be formed by raising to powers and by multiplying. Dirichlet answered the question by the theorem, "noteworthy because of its great generality," that came to be known as the unit theorem:

> If h denotes the total number of real and pairwise imaginary conjugate roots of the equation (1), then there are always $h - 1$ fundamental solutions such that if one raises these to powers and multiplies them with one another and adjoins to the general product thus formed each of the previously mentioned special solutions as factor, each of the solutions of (2) will be represented once and only once.[6]

Dirichlet added that, for the degree immediately following the second, this could be proved without significant difficulties, which is what he had expressed in dealing with the third degree in **1841c** several years previously. But proving the theorem via induction in the full generality in which it subsequently appeared presented the greatest difficulties, which, he remarked, he could only overcome after many fruitless attempts. He reported, however, that continued occupation with this topic finally simplified the proof so much that he could now describe it in understandable terms with few words.

He set about doing so by stating first that the essential "nerve" of the proof consists of finding $h - 1$ independent solutions, whereby he considered those which when raised to arbitrary powers and multiplied with one another will never give the obvious solution $t = 1$ and $u = 0, \ldots, z = 0$, unless all the exponents of raised powers equal zero. For if $h - 1$ such solutions are known, the method used in **1841c** can be applied to either solve or show the impossibility of solving the equation $\phi(\alpha)\phi(\beta)\cdots\phi(\rho) = r$. If $r = 1$, the procedure gives the solution of equation (2); with a few transformations, it can assume the form expressed in the theorem.

---

[6] *Werke* 1:642.

To get to the proof that $h - 1$ solutions independent of one another always exist, Dirichlet referred the reader to his **1842a**, containing generalizations of properties of continued fractions. The theorems given there can be used to show that there is always a solution of the equation (2) for which the numerical value of each of the expressions $\phi(\alpha), \phi(\beta), \ldots, \phi(\rho)$ corresponding to real roots, as well as every product of any two belonging to conjugate imaginary roots, lies arbitrarily below or above unity, provided one excludes the impossible combinations where all are either larger or all are smaller than unity.

This now permitted him to increase the number of independent solutions one by one until he reached $h - 1$ of them.

He concluded by observing that his results can be generalized both with regard to the number of fundamental equations and also with regard to replacing the numbers considered as coefficients or variables of the homogeneous function by complex numbers of an arbitrary form. His characteristic finish:

> The same principles remain applicable to all these generalizations, which provides the most favorable testimony that these principles are taken from the true nature of the topic.[7]

13.3 Potential Theory

Dirichlet's contributions to potential theory, **1846c**, **1846d**, **1850c**, and **1852a**, all appear to have been prompted by the publication a few years previously of Gauss's memoir [Gauss 1840] on the subject.

1846c

The first of Dirichlet's four memoirs, published in Crelle's *Journal*, was titled "On a general means of verifying the expression of a potential relative to any mass, homogeneous or inhomogeneous." He introduced it as follows:

> Given any mass, bounded in every sense whether forming a continuum or composed of several separate pieces, and having a finite density in each of its points, if one sums all the elements of this mass, each divided by its distance to some point m, one will have what geometers have agreed to call the potential of the mass with respect to this point. One knows that the triple integral thus defined, which is always a determined function of the rectangular coordinates x, y, z of the point m, enjoys a great number of important properties.

Dirichlet wished to limit himself to recalling only the following properties on which rests the remark that forms the object of this note.

(1) The potential v and its first-order differential coefficients

$$\frac{\partial v}{\partial x}, \ \frac{\partial v}{\partial y}, \ \frac{\partial v}{\partial z},$$

which express the components of the attraction exerted by the mass at the point m, are finite and continuous functions of x, y, z in every extent of space.

[7] *Werke* 1:644.

(2) There always exist determined limits that the products

$$xv, \quad yv, \quad zv, \quad x^2\frac{\partial v}{\partial x}, \quad y^2\frac{\partial v}{\partial y}, \quad z^2\frac{\partial v}{\partial z}$$

should not exceed at any point of space.

(3) If one excepts special points about which the density varies abruptly, each of the three derivatives

$$\frac{\partial^2 v}{\partial x^2}, \quad \frac{\partial^2 v}{\partial y^2}, \quad \frac{\partial^2 v}{\partial z^2}$$

will always be finite and single-valued, and these simultaneous values will be such that

$$\frac{\partial^2 v}{\partial x^2} + \frac{\partial^2 v}{\partial y^2} + \frac{\partial^2 v}{\partial z^2} = -4\pi\rho,$$

where ρ designates the density at the point (x, y, z) which must be considered equal to zero when this point is outside the mass.

After a brief consideration of some examples of the excepted points, Dirichlet remarked that in those cases where these assume two values the specifics are irrelevant as long as the values are finite.

What he wished to prove concerned the stated properties and was something that as far as he knew had not been demonstrated before. It was the statement that these properties completely characterize the potential v and that there exists no other function v' which has all of these properties. He proceeded to give a proof by contradiction.

Having accomplished this, he noted that this result furnishes a general means of verifying the expression of a potential when the potential is given throughout the extent of the space considered. He repeated that for this to hold it is sufficient that all the enumerated conditions are satisfied.

He pointed out that his procedure can be applied to the celebrated case of a homogeneous ellipsoid, but cautioned that it would only be useful for purpose of verification, not for finding an unknown potential. He worked out some examples, also noting the applicability for an interior point as well as the necessary condition for a point exterior to the ellipsoid.

Dirichlet concluded this short note by pointing out that the argument given requires only minor modification if one wishes to obtain a similar result when the mass is distributed over one or more surfaces, rather than being considered in three dimensions.

1846d

As projected at the end of the preceding publication, it was followed with a brief report in the Akademie on the characteristic properties of the potential on a mass distributed over a single or multiple surfaces. Dirichlet enumerated the following four properties as completely determining the expression of the potential in those cases:

(1) The potential v is a finite, continuous function of the rectangular coordinates x, y, z throughout the entire space.

(2) The potential's three partial derivatives with respect to the coordinates are also continuous everywhere outside the surfaces; however, at a point on these, there is a discontinuity. This consists of the following: Let the potential at such a point be designated by v, and at points lying on the normals extended in both sides at a distance of ϵ, by v' and v''; then the quotient $\frac{v'+v''-2v}{\epsilon}$, for ϵ positive and infinitely small, has the limit $-4\pi\rho$, where ρ designates the density at the point of the surface.

(3) The equation $\frac{\partial^2 v}{\partial x^2} + \frac{\partial^2 v}{\partial y^2} + \frac{\partial^2 v}{\partial z^2} = 0$ holds everywhere outside the surfaces.

(4) At an infinite distance from the surfaces, the same conditions hold as those for a mass extended in three dimensions.

Dirichlet remarked that, aside from the usefulness of proving that the properties listed completely characterize the potential because it furnishes a way of verifying an expression for a potential obtained by some other means, the characterization is of even more essential interest. He stressed the identical nature of Gauss's important assertion, according to which a mass can be distributed over given surfaces in such a way that the potential corresponding to such a distribution receives finite, continuously changing values, and the statement that in a homogeneous mass occupying the entire space, if at the outset there are vanishing temperatures at an infinite distance, then, under the influence of constant heat sources distributed over the surfaces, after an infinite period of time there will be a constant temperature everywhere. He commented that the identity of the two statements, and the new connection established thereby between two branches of mathematical physics already presenting such a major relationship, becomes immediately evident, since the known conditions that determine the permanent state of heat coincide with those whereby the potential corresponding to the desired distribution of the mass is characterized.

It clearly pleased Dirichlet to provide another example of the close mathematical relationship between two different areas of study—in this case within physics.

1850c, 1852a, 1857d

On November 28, 1850, Dirichlet read a paper to the Akademie's plenum titled "On a new expression of mass distribution on a spherical surface when the potential is to attain an arbitrarily given value at each point of the surface."

The Monthly Report **1850c** of the Akademie only provided that title for the lecture. But in 1852, the full memoir **1852a** appeared in the Akademie's Mathematical Memoirs for 1850 with the title "On a new expression for determination of the density of an infinitely thin spherical bowl, when the value of its potential is given at each point of its surface."

It would be translated into French by Jules Hoüel for Liouville's *Journal* where it appeared almost five years later as **1857d**, with yet another title modified from

1852a, "On a new formula for determining the density of an infinitely thin sheet when the value of the potential of this sheet is given at each point of its surface."[8]

Dirichlet introduced the memoir by referring to Gauss's theorem (Gauss 1840; see Gauss *Werke* 5:195) that every surface can be covered by an infinitely thin layer of mass whose potential obtains an arbitrarily given value at each point of the surface, provided that this potential varies continuously along the entire extent of the surface. Dirichlet added, however, that, given the present state of science, the determination of such a distribution is possible so far for only certain special surfaces, among which, as Gauss had noted, is the surface of an entire sphere.

Dirichlet continued with the following statement: Let R be the spherical radius, r the distance of every point in the space from the center, θ the angle between r and a fixed straight line, and ϕ the angle between the surface and r; then one can expand the potential V, given in terms of θ and ϕ, according to the known spherical functions. Next, let $V = \sum X_n$ be such an expansion, where the sum ranges from $n = 0$ to $n = \infty$, and let $\rho = \sum Y_n$ be the similar expansion of the density ρ to be determined. Referring to [Laplace's] *Mécanique céleste*, he noted that by using the expansion for the density one can represent the potential v for every point in the space. There will be two expressions, one valid in the interior of the space and one on the outside, each adequate for the purpose. The first of these is

$$v = 4\pi R \sum \frac{1}{2n+1} \left(\frac{r}{R}\right)^n Y_n.$$

Since this expression is valid up to $r = R$, where $v = V$, and the same function can only be expanded by spherical functions uniquely, a comparison with the first series results in $Y_n = \frac{2n+1}{4\pi R} X_n$ and therefore

$$\rho = \frac{1}{4\pi R} \sum (2n+1) X_n.$$

Referring to **1837e**, he recalled that he had shown that every arbitrarily given function, defined for the entire spherical surface, can be expanded in a convergent sum of spherical functions as long as it nowhere becomes infinite. Therefore, he stated, the convergence of the series for V holds without doubt. This is not so for the series $\sum Y_n$. The density can become infinite at certain points or lines; it remains uncertain whether in those cases the series remains convergent and actually represents the density at all points where the density remains finite.

Dirichlet explained that he had found it not without interest to examine this question further and thought that his earlier memoir would provide the necessary means; he believed this would also be worthwhile because it could lead to a simpler form of the expression for ρ. That expression involves a fourfold infinite operation, a double summation, and a double integration. The last cannot be removed because the density at each point must depend on all the potential values to be assumed continuously

[8] Liouville also had published a French translation of Gauss's memoir [Gauss 1840] in his *Journal* in 1842.

throughout the entire extent of the surface. On the other hand, the density can always be represented without series by a double integral, which many times can even be reduced to a single one.

Dirichlet divided his discussion into six sections, which outlined his results with minimal space devoted to the arguments. In the first section, basing himself on his earlier memoir, he derived an expression for the density ρ when the series is convergent. In the second section, he considered the case when the series is not convergent so that the derivation for ρ is no longer valid. He followed this in the third section by establishing a rule for determining the density ρ by transferring the result established when $\theta = 0$ to an arbitrary point, adding appropriate cautionary statements concerning the nature of the integral. In the fourth section, he illustrated with an example the case that the series for ρ is divergent. In section five, he considered means of determining the coefficient A_n, previously determined only for the special case where $k = \frac{1}{2}$. Having found a finite expression for the density, Dirichlet applied this to a special case in the sixth section of the memoir. This involved his relatively rare use of elliptic integrals, referring to Legendre and Jacobi.

Dirichlet concluded the memoir by considering the possibility that the surface differs slightly from a sphere; he adapted this case to that of the sphere without having to undergo the series expansion.

13.4 Reduction of Ternary Quadratic Forms

On July 31, 1848, Dirichlet presented an account "Concerning the reduction of positive quadratic forms with three indeterminate whole numbers" to the Mathematical-Physical Class. Again, the Akademie's *Bericht* **1848** provided a summary; the full memoir **1850a** would be published in Crelle's *Journal* two years later, and a translation into French, **1859a**, in Liouville's *Journal* did not follow until 1859, the year of Dirichlet's death. As we shall observe, this memoir would have substantial influence.

1848

The Akademie's *Bericht* was introduced with a short historical review. Dirichlet reminded the reader that Lagrange was the first to show that every binary quadratic form can be reduced by transforming it into another form satisfying certain conditions of inequality. He noted that Lagrange also proved that in every class of positive forms there exists only a single form of this kind so that in this case the various reduced forms corresponding to a given determinant can serve as representatives of the respective classes. Dirichlet observed that after Gauss had considered ternary forms from a general point of view in the *D.A.*, it became necessary to extend Lagrange's result for binary quadratic forms to such forms in order to expand the further development of the theory, or to find the conditions of inequality among its coefficients that would be satisfied by one and only one form in each class.

In 1831, Ludwig August Seeber of Karlsruhe, at the time professor of physics in Freiburg/Breisgau, had produced the required extension in a work on positive ternary quadratic forms. According to a later remark by Encke, who had been a fellow student with Seeber in Göttingen, Seeber had been interested in explaining, through the laws of mechanics, the manner in which solids are formed from their smallest components. Because of the close relationship of positive ternary quadratic forms to the inner structure of solids, Seeber felt that it was important to expand Gauss's treatment of these forms.[9]

Seeber's work was 248 pages long and complicated. Gauss, to whom the work was dedicated with great respect, wrote a diplomatic review, noting that although it was regrettable that the extent of Seeber's work may be considered intimidating, justice required that one bear in mind the following:

> When a difficult problem or theorem is to be solved or proved, the first step, to be recognized with appropriate gratitude, consists of recognizing that a solution or proof is found, and the question whether this could not have been done more easily or simply is an idle one so long as the possibility is not decided along with the deed. Therefore, we consider it untimely to linger with this question.[10]

Dirichlet only alluded to this statement by Gauss in **1848** but quoted it in the extended **1850a**.

Again, a statement by Gauss here presented a challenge Dirichlet could not resist. He remarked that the exceedingly complicated nature of Seeber's method had long ago tempted him to try to establish the theory of reduced ternary forms in a simpler way. For sake of brevity and transparency, he felt he must retain the geometric form in which the study had been conducted, using as basis "the noteworthy relationships between quadratic forms of two or three elements and certain spatial structures."

Gauss had used parallelopipeds in his review of Seeber. Pursuing this same approach, Dirichlet now divided Seeber's results into two main statements:

> 1. Every system of points ordered in parallelopipeds can be split so that for the corresponding elementary parallelopiped the sides of its surfaces are no larger than its diagonals nor are the edges larger than the diagonals. 2. For a given system, such an ordering is usually unique.

By suitable manipulations of the parallelopipeds, Dirichlet proved the first of the two statements and sketched a path for the proof of the second.[11]

In the later full publications of his memoir, **1850b** and **1859a**, Dirichlet added a theorem Seeber had based on an induction and Gauss had derived in his announcement of Seeber's publication. According to this, the product of the three first coefficients in a reduced form does not exceed twice the absolute value of the determinant. Dirichlet provided a short proof of "this pretty statement" derived from his procedure.

The historical impact of Dirichlet's memoir is significant. The first, obvious, reaction was astonishment that he had succeeded in reducing Seeber's 248 pages to a

---

[9]Encke 1862. In this comment, Encke used the opportunity not only to mention to members of the Akademie his own relationship to Gauss but to explain that Gauss's interest in Seeber's work was consistent with Gauss's [Shakespearean] motto "Thou Nature art my goddess, to thy law/My services are bound."

[10]Gauss 1831; see Gauss *Werke* 2:191.

[11]For details, see *Werke* 2:24–26.

manageable, transparent, yet rigorous twenty pages. It seems appropriate to suggest that it was Gauss's analysis of the problem found in his review of Seeber's work—actually a rather lengthy analysis—that enabled Dirichlet to wade successfully through Seeber's complex opus. The more lasting effect of Dirichlet's memoir was the one it had on the subsequent development of the geometry of numbers. Minkowski, in referring to **1850b**, stressed that Dirichlet's "great advance consisted of his not operating with the cumbersome, computational expression of inequalities whereby Seeber had defined reduced forms, but with the well-recognized inner significance, to have the reduced form depend on certain least distances in the associated point system."[12]

Minkowski, who throughout the years emphasized the influence on himself of Dirichlet and Hermite, did not fail to remark on the fact that **1850b** appeared in the same volume of Crelle's *Journal* as Hermite's letters to Jacobi. Keeping in mind various assessments of Dirichlet's work by Minkowski in addition to the one just cited, fanciers of conceptual genealogies may ponder whether Dirichlet and Gauss qualify as grandfathers of the geometry of numbers, with Hermite (who disliked geometry) serving as a benign but more distant uncle to Minkowski, the generally acknowledged father of the field. What is indisputable is Minkowski's repeated adaptation not only of Dirichlet's technique but of some of his favorite phrases, such as referring to the "inner significance" of certain concepts or procedures.[13]

13.5 Mean Values in Number Theory

On Thursday, August 9, 1849, Dirichlet read to the plenum a memoir on the determination of mean values in number theory. The *Monatsberichte* for that year only provided a reference by title (**1849**). The full account was published in the *Mathematische Abhandlungen* for 1849 that, with the usual two-year delay, appeared in 1851 (**1851b**); in 1856, a translation into French by Hoüel (**1856e**) was published in Liouville's *Journal*.

1851b

Dirichlet introduced the topic of his short address by noting that although number-theoretic functions can rarely be represented by analytic expressions and seem to proceed without rules, nevertheless in their mean values there appears a greater

---

[12]Minkowski 1891; see his *Ges. Abh.* 1:244–45.

[13]In addition to references we have cited that Minkowski made to Dirichlet on other occasions, it is of interest to note how much emphasis he placed on Dirichlet's work in the paper [Minkowski 1896] presented at the 1893 Chicago Congress. Dirichlet's influence on Minkowski was to be of considerable importance for another direction of twentieth-century number theory. Although a more extended discussion of the impact of **1850b** would lead us beyond the chronological boundary of our volume, it is worth noting that it was this memoir, along with **1839a** and Kurt Hensel's p-adic number theory, which, after 1920, would guide Helmut Hasse to the Hasse–Minkowski theorem and his "global-local principle."

conformity to a rule the farther one pursues their series; that is, there are definite simple expressions which represent the mean values, increasing in accuracy the further one progresses with the series. He likened this to the behavior of a curve that approaches another curve, whose asymptote it is, more and more closely. He called attention to several noteworthy expressions of this kind that Gauss had listed toward the end of the *D.A.*'s fifth section, and which pertain to the theory of quadratic forms.[14]

Dirichlet continued by explaining that, since neither of these interesting results, which there are communicated only in passing and without verification, had been proved so far, nor are methods for treating similar questions known generally, he occupied himself already several years previously with seeking suitable means for this. However, aside from some new results, he made public nothing of his work at that time [1838] since he saw a possibility by continued efforts to considerably simplify the treatment of such problems and, in particular, to make them independent of the integral calculus. Subsequently other researches drew him away from this topic, but when he again resumed the study of this question, he became convinced that in many cases one arrives, by quite elementary considerations, based on a very simple transformation of series, at an asymptotic expression of the mean values.

For the present, he wished to restrict himself to a series of problems for which the indicated procedure suffices; but he stated that he planned to follow this with a memoir where he planned to treat more difficult problems the solution of which requires connecting the transformation with other auxiliary procedures.

Dirichlet divided his short memoir into seven sections. He noted that the problems he wished to treat all depend on a certain transformation. In typical fashion, he suggested that the true nature of this transformation would be best made clear by starting with one of the simplest problems and carrying out a procedure for its solution as far as possible before drawing on the needed transformation itself.

The problem he chose concerned the number of divisors of an integer n. Denoting this by $f(n)$, the question consists of finding the sum

$$F(n) = f(1) + f(2) + \cdots + f(n).$$

He first obtained an approximation whereby $F(n) = n \log n$, noting that although one can obtain a bound for the error, his procedure does not help decide whether the asymptotic expression for F(n) contains a member of order n with constant coefficients, meaning whether with increasing values of n, $\frac{1}{n}F(n) - \log n$ will approach a fixed limit.

Dirichlet next turned to cases where that function is expressed by an equation which includes a series containing the function to be determined in its general term, so that the only thing known is the recurrent relationship between successive values of the function. As a simplest example of this, he noted the function $\phi(n)$, which expresses the number of primes relative to n in the sequence $1, 2, \ldots, n$, that is so significant in the theory of numbers.

---

[14] See the *D.A.*, art. 304, and the last item (Sect. 13.13) in this chapter.

As his last example, in Section 7 Dirichlet considered the function $\phi(n)$ that designates all the decompositions of n into two factors without common divisor. Letting ρ denote the number of different primes contained in n, $\phi(n) = 2^\rho$. From this, he found that the asymptotic expression for $\psi(n)$ will have the form $\alpha n \log n + \beta n$. This led him to recognize the mean value as being

$$\frac{6}{\pi^2}\left(\log n + \frac{12C'}{\pi^2} + 2C\right).$$

Dirichlet now called attention to the fact that this relates to the mean value of genera corresponding to a negative determinant $-n$ given in the *D.A.*'s art. 301.

13.6 Three-Squares Decomposition

In June 1850, Crelle published a memoir by Dirichlet in his *Journal* having the title "On the decomposition of numbers into three squares."

1850b

Dirichlet introduced this short memoir by noting that the theory of decomposition of an integer n that is not of the form $4k$ or $8k + 7$ into three squares without a common divisor is one of the more complicated of higher arithmetic if one wants to develop this theory fully. This means not only proving that such a decomposition exists, but determining the number of all possible decompositions. He observed that this had first been shown by Gauss, in the *D.A.*'s art. 229, utilizing results for binary forms with determinant $-n$, but could also be handled independently of this argument, as he had done.[15]

Since there are cases where one need only assume the existence of a decomposition, it seemed useful to Dirichlet now to supply a simple proof for that decomposition. He referred to the proofs by Cauchy and Legendre of Fermat's theorems concerning polygonal numbers as examples of arguments based on the fact that every number except for some previously excluded can be decomposed into three squares. Dirichlet outlined the steps of his argument as follows:

The first thing needed was the theorem he described as well-known that every positive ternary form of determinant -1 is equivalent to the form $x^2 + y^2 + z^2$.

Consider the expression

$$ax^2 + by^2 + cz^2 + 2a'yz + 2b'xz + 2c'xy \qquad (1)$$

with integral coefficients that satisfy

[15] See Dirichlet's concluding remarks of **1839–1840**. Legendre's initial attempts at a proof concerning the sum of three squares are found in [Legendre 1788], which, as André Weil noted in Weil 1983:331–32, may have interested Gauss in the problem.

$$(2) \qquad aa'^2 + bb'^2 + cc'^2 - abc - 2a'b'c' = -1$$

and for which $a, b, c, bc - a'^2, ac - b'^2, ab - c'^2$ are positive. To prove the statement it suffices that for determinant -1, $x^2 + y^2 + z^2$ is the only reduced positive form.

He now turned to **1850a**, published immediately preceding the present memoir in Crelle's *Journal*.

Having given a procedure for the case that the determinant is -1, he added an example when the determinant is -3. He concluded with the observation that his previous arguments demonstrate that every number not divisible by three can be represented by a form $x^2 + y^2 + 3z^2$ with x, y, z having no common divisor and that similar considerations can be applied to numbers divisible by three to which one must add an easily obtained condition, as well as to numbers that can be expressed in the form $x^2 + 2y^2 + 2z^2 + 2yz$.

13.7 Composition of Binary Quadratic Forms

As remarked in our preceding chapter, in May 1851 Dirichlet finally delivered his Latin habilitation discourse, required to change his status in Berlin from being an "Ordinary Professor Designate" to holding a regular ordinary (full) professorship. Originally printed as a separate pamphlet **1851a**, it was published in Crelle's *Journal* three years later, **1854d**; a French translation, **1859d**, appeared in Liouville's *Journal* after Dirichlet's death.

1854d and 1859d

Dirichlet's topic, *De formarum binarium secundi gradus compositione*, dealt with the composition of second-degree binary forms. In his introduction, he explained that this study grew out of his interest, dating back a number of years, in determining the number of classes of binary forms of the second degree which correspond to the same determinant. His wish to extend the results to the theory of complex numbers forced him to reestablish the entire theory of forms which up to then had only been expounded for real integers. He pointed out that he had been successful in presenting the elements of this theory by previously unused considerations that apply to complex as well as to real integers. In this previous study, **1842b**, he had limited himself to the equivalence of forms, their transformation, and the representation of number, as these were sufficient for an ability to understand his memoir. He had not treated the question of the composition of forms which Gauss had developed in full generality in section five of the *D.A.* Gauss, he observed, had used calculations that are so prolix that very few geometers had been able to fully comprehend the nature of composition, especially since "the grand geometer" had himself stated that for greater brevity he had provided synthetic arguments for the most difficult theorems, thereby suppressing the analysis which had given rise to them. For that reason, Dirichlet explained, he hoped that a new and quite elementary presentation of the subject "would please those who cultivate analysis."

Dirichlet divided his exposition into three parts. In the first part, he presented necessary preliminaries consisting of known results, or those easily derivable from known theorems.

The chief statement proved in the first part is the following:

> The values $\xi, \xi', \xi'', \ldots$ which satisfy the congruence $u^2 \equiv D$, according to which the moduli $m, m', m'', \ldots$ are concordant, if one can find a root Z of the same congruence for the modulus $mm'm'' \ldots$, so that one has $\xi \equiv Z \pmod{m}, \xi' \equiv Z \pmod{m'}, \xi'' \equiv Z \pmod{m''}, \ldots$

He noted that it suffices to consider the case where the moduli are odd and prime to D.

Finally, having largely summarized and simplified Gauss's discussion of the *D.A.*'s articles in part III, he derived his chief proposition:

> Given two forms ϕ and ϕ' of the same determinant D, let m and m' be any two odd numbers prime to D, which can be represented respectively by the forms ϕ and ϕ' so that the roots (m, ξ) and (m', ξ') to which these representations belong are concordant. The representations of the numbers mm' which belong to the root $(m, \xi)(m', \xi')$ are always effected by the same form or forms belonging to the same class, however m and m' vary.

Dirichlet considered the theorem fundamental in this doctrine and proceeded with a careful explication.

13.8 The Division Problem: 1851c, 1854c, 1856f

Dirichlet made his first presentation of the year 1851 to the Physical-Mathematical Class of the Akademie on January 20 with a memoir having the title "On a problem concerning the theory of division." The corresponding four-page report appeared in the Akademie's *Bericht* (**1851c**) and three years later was reprinted in Crelle's *Journal* (**1854c**) under the shorter, more informal title "On a problem concerning division." In 1856, readers of Liouville's *Journal* could also find a French translation, **1856f**.

13.9 A Resting Solid in a Moving Fluid

In 1852, the Akademie's *Bericht* contained a brief summary, **1852b**, of a memoir on the movement of a solid in an incompressible fluid medium. Dirichlet pointed out that there was no example of a purely theoretical determination of the changes brought about in a moving fluid by a stationary solid resting in that fluid. He observed that this issue is equivalent to the lack of a study derived from Euler's general hydrodynamic equations of the resistance experienced by a moving solid in a stationary fluid. He commented that Navier, "a mathematician greatly experienced in this type of investigation," had thought that the known methods of integration were not adequate to deal with this problem, even in its simplest form. Dirichlet stated that the problem can be solved, however, if the solid has the shape of a sphere or an ellipsoid. In the report he now presented, he wished to treat only the simpler case of a sphere.

He proceeded as follows: Let c be the radius of a stationary sphere, and let its center be the origin of the rectangular axes of x, y, and z. Consider a homogeneous fluid, initially at rest, having density ρ; let an increasing force σ act on this fluid in such a fashion that at a time t it will have the same intensity and direction but with changing time can vary arbitrarily; in other words, its components α, β, γ will be given functions of t.

13.10 Derivation of Two Arithmetical Statements

On May 23, 1853, Dirichlet read a memoir titled "On a new derivation of two arithmetical statements from a common source" to the Physical-Mathematical Class. It was listed in the Akademie's monthly reports by title only **1853b**, no abstract or copy apparently having been printed. As this was the period in which Dirichlet was still dealing with the management of the Jacobi *Nachlass* and related matters, it may be that he simply did not wish to take time for writing out this May memoir.

13.11 Gauss's First Proof of Quadratic Reciprocity

In 1854, Dirichlet contributed to Crelle's *Journal* a study, **1854b**, of Gauss's first proof of the law of quadratic reciprocity, or, as he chose to call it, following Gauss, "the Fundamental Theorem of the doctrine of quadratic congruences."[16]

Gauss's proof is that published in articles 135–144 of the *D.A.* Dirichlet began his discussion by observing that this first demonstration of Gauss had always seemed to him particularly noteworthy because of the simple thought underlying it and also because it is the one proof that does not stray outside the area of second-degree congruences.[17]

Dirichlet stressed that the undue length and complication of Gauss's demonstration is not due to the methodology used but to the fact that Gauss did not use a notation suitable for the computations involved; this forced Gauss to divide the proof into eight cases with numerous additional subcases. Commenting that this makes it particularly difficult for beginning mathematicians to understand the demonstration, Dirichlet now provided a proof using Jacobi's generalization of the Legendre symbol

[16] As noted previously, Gauss later in the *D.A.* and in subsequent work simply referred to it as the Fundamental Theorem.

[17] This is explained by the fact that this proof, included in part I of the *D.A.*, belongs to the original version of the *D.A.* intended to be a work on congruences. Gauss's change in orientation occurred only with completion of his work on Part V of the *D.A.* See Merzbach 1981.

but following Gauss's basic thought process and pointing to the parts of the proof that must have presented particular difficulties for the young Gauss.

We previously noted that it was the use of Legendre's notation that enabled Dirichlet to provide more streamlined proofs of a number of Gauss's theorems. This memoir, however, is the only instance in which Dirichlet so explicitly and directly addressed the advantage to be derived from simplified notations such as Jacobi's modification of Legendre's symbol when lengthy computations are involved. This may have been partly done in homage to the recently deceased Jacobi, whose mention in this connection would not have been the anathema for Gauss as was Legendre's name in these later years.

13.12 Continued Fractions; Quadratic Forms with Positive Determinant

On July 13, 1854, Dirichlet read a memoir announced in the Akademie's *Bericht* as titled "On a property of continued fractions and their use in simplifying the theory of quadratic forms with positive determinant," **1854a**. The text of the memoir that appeared in the Akademie's *Abhandlungen*, **1855a**, had the shorter title "Simplification of the theory of binary quadratic forms with positive determinant." The title referred to the portion of the *D.A.*'s fifth section that in articles 183–212 contained a discussion, without the use of continued fractions, of forms with positive determinants; articles 183–205 dealt with non-quadratic determinants and articles 206–212 with quadratic ones. Dirichlet rearranged Gauss's discussion, replaced the arithmetic examples with special cases of his own, and simplified the arguments.

A French version, **1857g**, again translated by Jules Hoüel, was published in Liouville's *Journal* in 1857 along with two additions. The first addition had been provided by Dirichlet in August 1857 while visiting Liouville in Toul; the second was an extract of a letter from Dirichlet to Liouville, clarifying some of the memoir's content.

We provide only a brief outline of the original **1855a** but recommend that the reader interested in details of the proofs follow Hoüel's translated version, **1857g**, more closely, as it presented these in more readable fashion. We also include a short explanation concerning the additions.

1855a

Dirichlet's brief introductory paragraph stated that the more higher arithmetic has grown, thanks to the "epoch making work [the *D.A.*] of Gauss" and later contributions of his, the more desirable it is to facilitate the entrance to this branch of analysis by simplifying its elementary part as much as possible. Dirichlet pointed out that he had already previously provided new bases for known statements necessary for this purpose. In this memoir, he proposed to do the same for the study of the theory of quadratic forms with positive determinant, noting that some of the very detailed considerations previously considered necessary can be removed.

Dirichlet divided the memoir into seven sections. The first two summarized some known properties of continued fractions primarily given by Lagrange, although he needed to modify Lagrange's argument concerning the removal of negative members.

In the third section, he considered quadratic forms $ax^2 + 2bxy + cy^2 = (a, b, c)$, first noting that all of these will have the same positive determinant $D = b^2 - ac$. The only restriction placed on this determinant must be that it is not a square.

He introduced two forms

$$ax^2 + 2bxy + cy^2, \ AX^2 + 2BXY + CY^2 \tag{1}$$

and a substitution

$$x = \alpha X + \beta Y, y = \gamma X + \delta Y \begin{pmatrix} \alpha, & \beta \\ \gamma, & \delta \end{pmatrix} \tag{2}$$

that satisfies the condition

$$\alpha\delta - \beta\gamma = 1, \tag{3}$$

and then proved three statements:

> I. Between the equally named roots ω and Ω belonging to the equivalent forms (1) and the coefficients of the substitutions (2)
>
> $$\omega = \frac{\gamma + \delta\Omega}{\alpha + \beta\Omega} \tag{4}$$
>
> always holds....
>
> II. If Equation (4) holds for a pair of equally named roots ω and Ω belonging to the forms (1), and if the integers $\alpha, \beta, \gamma, \delta$ at the same time satisfy the condition (3), then the forms are equivalent and the first is transformed to the second by the substitution (2).
>
> III. Contiguous forms, (a, b, a') and (a', b', a''), where the last coefficient of the first is identical to the first of the second, and the middle coefficients b, b' satisfy the condition that $b + b' \equiv 0 \bmod a'$, are always equivalent.

Section 4 explained the concept of a reduced form and its neighbors to the left and right. We note in passing that this contains an early instance where Dirichlet used the term "absolute value," although he quickly reverted to discussing "numerically" different values and alternated use of the two expressions.

In the following section, Dirichlet obtained neighboring forms in both directions from a reduced form and continued this process until he had obtained an infinite series of equivalent forms. Considering the finite number of reduced forms associated with a given determinant allowed him to divide these forms into periods. This procedure, combined with his detailed analysis of the periods, enabled him to obtain the demonstration of the statement in the next section that, thanks to the formulation of these preliminaries, gave the impression of being considerably shorter than Gauss's proof in the *D.A.* In fact, he had continued to follow the articles from the *D.A.* noted in our introduction to this memoir; but, whereas it is obvious that he used Gauss's definitions, the similarity of his procedures is less apparent by his omitting Gauss's numerous arithmetic illustrations and again referring to the continued fraction discussed at the beginning of his memoir.

In Section 6, he addressed the question of whether two given forms are equivalent or not. This boils down to the question of whether forms from different periods can be equivalent, because given any form, one can derive a reduced form equivalent to it, while forms belonging to the same period are equivalent already. He provided a one-page-long demonstration that forms of different periods cannot be equivalent.

Dirichlet opened the last, seventh, section with the remark that, since now the most difficult statement of the theory of quadratic forms with positive determinant had been proved in simple fashion, he next would merely indicate how the rest of the theory should be modified to conform to the justification of that proof. In doing so, he engaged in a more detailed discussion of the equivalence and transformation of forms.[18]

1857g

Hoüel's translation of the first part of **1855a** followed Dirichlet's memoir rather closely. Minor modifications included a subtle French change in the introduction from referring to the *D.A.* as "epoch-making" to referring to the "memorable epoch of [its] publication." By the third section, the translation constituted a considerably easier-to-follow version of Dirichlet's original.

Dirichlet visited Liouville in Toul in August 1857. At that time, his host was reviewing Hoüel's translation of **1855a** that he was about to publish in his *Journal*. Liouville apparently commented on the difficulty his readers might have in going back to the references Dirichlet had cited for details of the proofs. Dirichlet jotted down the requisite formulas with pertinent explanations and this became the first addendum to the memoir. A second addendum consisted of an extract from a later letter Dirichlet had written to Liouville, in which he had simplified part of the argument in accordance with a procedure he had used in his course. Dirichlet asked, if there was still time, to have this replace his original version; to avoid a delay, Liouville chose to add this version as a second addition while leaving intact the original, which he had considered satisfactory.[19]

13.13 Quadratic Forms with Positive Determinant

Dirichlet's last presentation to the Physical-Mathematical Class of the Akademie was made on July 16, 1855. Given the title "On a property of quadratic forms with positive determinant," this short paper is another example of Dirichlet's taking bait thrown out by Gauss. In this instance, it is the following statement concerning positive determinants:

> It would be a fine task, not unworthy of the efforts of geometers, to determine the law according to which the number of determinants having only one class in each genus is continually diminished; until now we can neither decide theoretically, nor assume by observation with sufficient certainty, whether these eventually break off altogether (which, however, appears

---

[18] Dirichlet gave a reference to his **1842b** and to art. 193 of the *D.A.* To avoid repetition, note our Chapter 16, where we outline the relevant discussion as found in the editions of Dirichlet–Dedekind's number theory lectures, especially paragraphs 82, etc.

[19] *Werke* 2:180–81.

less likely) or at least become infinitely scarcer, or whether their frequency continuously approaches a definite limit.[20]

1855b

As summarized in the Akademie's *Bericht*, Dirichlet began the task that Gauss had suggested by discussing several properties of the equation

$$t^2 - Du^2 = 1$$

because, he explained, the new theorem which he was about to reveal is closely related to that equation. We note that Gauss had called attention to such a relationship in the concluding remarks of the *D.A.*'s art. 304.

Dirichlet assumed that D is not a square and chose to consider only positive integers for the solutions t and u. Letting T and U be the smallest values satisfying [Pell's] equation, Dirichlet recalled that all solutions of the kind just mentioned will be obtained by the formula

$$(T + U\sqrt{D})^n = t_n + u_n\sqrt{D}$$

if one attributes to n all successive positive values. He then demonstrated that among these solutions there are infinitely many for which u_n is divisible by an arbitrary positive number S; the exponents n for which this holds are the successive multiples of N, where N designates the smallest of the indices n.

In his short demonstration, he let $D' = DS^2$, and set $t'^2 - D'u'^2 = 1$; then, if $t = t'$ and $u = Su'$,

$$(T + U\sqrt{D})^N = T' + U'\sqrt{D'}$$

where T' and U' are the smallest values of t' and u'.

Adding appropriate additional conditions, he arrived at the result that, while letting α_i, the powers of the primes contained in S increase indefinitely, the quotient $\frac{S}{N}$ will assume a fixed value, independent of these α_i.

He added that this characteristic has an interesting consequence for the theory of quadratic forms with positive determinant. If to the previous assumptions one adds that D contains no square divisor, denotes by h the number of classes into which the forms belonging to D are distributed, and lets h' take on the similar significance for the determinant $D' = DS^2$, then, as had been proved in [Section 8 of his] (**1839–40**),[21] one obtains the equation

$$h' = \frac{\log(T + U\sqrt{D})}{\log(T' + U'\sqrt{D})} SR,$$

[20]Gauss, *D.A.*, art. 304.

[21]Werke 1:472.

where, he remarked, the factor R depends on the prime numbers $p_1, p_2, \ldots$, but not on the exponents $\alpha_1, \alpha_2, \ldots$.

Commenting that this equation can be put in the form $h' = h\frac{S}{N}R$, it follows that from every determinant D one can derive infinitely many other determinants DS^2, all of which correspond to the same class number. By appropriate choice of D and the primes $p_1, p_2, \ldots$ one can arrange it so that this invariant number of classes will coincide with that of the genera and in this way prove the validity of Gauss's conjecture that the series of positive determinants, which has only one class in each genus, will not break off, whereas it appears that the negative determinants having the same characteristic appear to be only finite in number [*D.A.* art. 304].

1856a and 1857b

The following February, by which time Dirichlet had moved to Göttingen, Liouville's *Journal* contained a clearer version of the preceding summary, "freely translated by the author." In this, **1856a**, Dirichlet had inserted a paragraph refining the proof, and added a closing, additional remark on the unusual nature of Gauss's conjecture. Another year later, this summary was retranslated into German in Borchardt's *Journal* as **1857b**; this version included the additions made in the French translation **1856a**.

In these final versions, **1856a** and **1857b**, Dirichlet commented that the proof of Gauss's suggestion presents an analytic phenomenon all the more remarkable since the negative determinants enjoying the same property seem to be only finite in number and quite limited. He noted that if one goes back to the extensive induction first carried out by Euler, "at an epoch when the theory of quadratic forms founded by Lagrange did not yet exist," for the purpose of finding very large primes, these *numeri idonei* were only 65 in number, and the largest only had the value 1848, abstraction made of the sign the absolute value.[22] It was Gauss who linked Euler's 65 numbers to his study of determinants. In art. 303 of the *D.A.*, he had noted that 65 determinants are the only ones such that all the classes of the forms belonging to them are *anceps* and that all forms contained in the same genus are properly and improperly equivalent. Once again, Gauss had illustrated these possibilities with a substantial number of numerical examples.

Gauss had added to his cited conjecture of the *D.A.*, art. 304, some comments concerning the growth of the mean class numbers in comparison to the number of genera and the square root of the determinants, barely larger with regard to the first and much slower with regard to the second. He commented that these considerations lead to a certain analogy between cases involving positive and negative determinants. We note also that additionally Gauss had indicated he might at some future occasion treat the question of the principles according to which certain mean values of quantities do not progress according to some analytic law but only approximate such a law asymptotically.

[22] For Euler's related efforts, see Weil 1983:224–26.

Dirichlet's addenda reflect his growing curiosity concerning the nature of Gauss's conjectures; his comment concerning the surprising phenomenon of Gauss's conjecture for negative determinants was futuristic.[23]

13.14 Summarizing Comments

As noted, the only publication of Dirichlet's last decade in Berlin that involved a major original result was that containing the outline for his proof of the unit theorem, which had occupied him before the journey of 1843–45 and apparently had been thought through while he was in Italy. Most of the last publications were intended to make rigorously accessible a number of Gauss's results and queries from the *D.A.* and from the memoir on potential theory [Gauss 1840]. They also demonstrated Dirichlet's systematic study of Lagrange's relevant work and his increased awareness of some of Euler's prescient discoveries, something that may have been brought to his special attention while he and Jacobi studied Fuss's 1849 edition of selected number-theoretic work by Euler.

During the decade, his analyses of earlier work took on a more critical tone than previous writings; but mostly these continued to mirror earlier comments made by Gauss. Dirichlet's growing difficulty, related to his increasing physical weakness, in putting new results into publishable form, and in carrying out or even in publishing the details of more involved investigations, will become more apparent in the next chapter.

[23] Attempts to settle the questions Gauss raised in articles 303 and 304 of the *D.A.* have occupied mathematicians to our times. For example, in 2007, H. M. Stark, who resolved several of these issues, in the introduction to his careful discussion of "The Gauss Class-Number Problems" called attention to the conjectures of Gauss's articles 303 and 304; to the impact of Heilbronn 1934 ("[Gauss's] addendum [in art. 304], caused much heartache when in 1934 Heilbronn finally proved that $k(d)$ approaches ∞ as d approaches $-\infty$ ineffectively."); to expanded interpretations; and to related open questions. See Stark 2007:247–50.

Chapter 14
Göttingen

In September 1855, a month before the beginning of fall classes, the "Göttingische Gelehrten Anzeigen" made public what was already common knowledge in the small university town: the appointment of the mathematician from Berlin to the professorship of mathematics at the Georgia-Augusta University.

14.1 The Societät der Wissenschaften

The Societät der Wissenschaften (Göttingen's Royal Society) changed Dirichlet's status from that of external member to ordinary member and the university's curator confirmed that change on September 17. The official notification to Dirichlet came from the secretary of the society, the seventy-three-year-old mineralogist Friedrich Hausmann whose personal and professional life had centered in Göttingen and the Harz since the first decade of the century. The letter welcoming Dirichlet to "the narrower circle" of the society contained an unmistakable reminder of the expectations that accompanied his arrival and the responsibility he carried in being touted as Gauss's successor. His presence was noted as assuaging the pain over the extraordinary loss they had suffered, and Hausmann remarked on his own intense emotion at transmitting to Dirichlet the key to the society's chest (the *Societätskasten*) that "our incomparable and unforgettable Gauss" had used since 1807. Lest the implicit hint was not sufficiently strong, Hausmann added the remark that he shared the society's hope that Dirichlet, who had been so close to the blessed one, would be equally active scientifically in the society.

14.2 The University

If the ghost of Gauss hung somewhat heavily over the society, whose members also were conscious of their seniority vis-a-vis the newcomer, there was little question of Dirichlet's position at the university and, in particular, his rank as the star

U. C. Merzbach, *Dirichlet*, https://doi.org/10.1007/978-3-030-01073-7_14

mathematician. Gauss, though a member of the Göttingen faculty since 1807, had held his appointment as Director of the Observatory and Professor of Astronomy. Dirichlet was appointed Director of the Observatory and Professor of Mathematics, however. This allowed him to turn over the daily work at the observatory to Gauss's former student and assistant, Wilhelm Klinkerfues, who had just earned his doctorate and promotion to observer in March of 1855, having written a dissertation on a new method of computing the paths of double stars.

The Mathematics Faculty

At the time of Dirichlet's arrival, the mathematics faculty consisted of two professors and two privatdozenten. The only full professor was Georg Karl Justus Ulrich, who had taught at the university since his habilitation in 1817; in 1821 he had become extraordinary professor, in 1831 ordinary professor. The extraordinary professor, Moritz Abraham Stern, had spent his academic career in Göttingen since earning his doctorate and habilitation there in 1829. His, in fact, had been the first doctoral examination in mathematics supervised by Gauss; previous Ph.D. students had obtained their degrees in astronomy. Stern had served as privatdozent from 1830 until 1848. The privatdozenten in 1855 were Riemann and Dedekind, both of whom had earned their doctoral degrees under Gauss's auspices, in December 1851 and March 1852, respectively. They had completed their habilitation requirements in 1854 so that both had had little teaching experience, beginning their lectures in the winter term 1854/55.

For twenty-five years, Stern and Ulrich had shared the mathematical teaching burden, sometimes supplemented by Gauss and his two former students assisting him at the observatory: Klinkerfues and Benjamin Goldschmidt, who had died in 1851. Stern had assumed primary responsibility for pure, Ulrich for applied mathematics. They alternated in teaching analytical geometry, calculus, and mechanics, Ulrich stressing applications to mathematical geography and practical problems in all areas. In every other year, Stern taught calculus of variations, continued fractions, and theory of (numerical) equations, a course that Gauss had handled on occasions; Ulrich taught applications to agricultural machinery. After its establishment in 1850 (at the instigation of Stern), they collaborated with the two physicists Wilhelm Weber and J. B. Listing in conducting the physical-mathematical seminar designed to supplement the standard curriculum for the benefit of teachers of mathematics in secondary and collegiate departments of mathematics and physics. Here they came closer to their own research interests, Ulrich conducting work in perspective and other areas of geometry, Stern in number theory and potential theory.

Clearly, Stern had the most to lose by Dirichlet's presence. But by leaving the existing course division unchanged and not involving Dirichlet in the operation of the seminar, Dirichlet and Weber managed to avoid unpleasant changes in the established routine. Dirichlet's courses were simply added to the standard offering, as were those of the two young privatdozenten, Riemann and Dedekind.

The Mathematical Curriculum

Dirichlet's courses in Göttingen were largely the same ones he had taught in Berlin. The first two courses Dirichlet presented, in the winter term of 1855/56, were (1) number theory and (2) integration of partial differential equations and their application to physical problems. He continued to teach the number theory course each winter term, after 1855 adding the description "with emphasis on the doctrine of quadratic forms." In the summer term 1856, he expanded the differential equations course to two courses: (1) potential theory with applications to electricity and magnetism and (2) theory of spherical harmonic functions with applications to physical investigations. The partial differential equations course was repeated in the summer term of 1857, and the potential theory course was given again in 1857/58. His old course on definite single and multiple integrals was given twice, in each of the winter terms 1855/56 and 1856/57.

At the same time, Stern taught theory of numerical equations and analytic geometry in the winter term 1855/56, calculus and mechanics in the summer of 1856, as well as an introduction to continued fractions. The following year Stern again taught calculus, calculus of variations, and analytic geometry in the winter term; by summer 1857, when Dirichlet expanded his old partial differential equations course, Stern went back to a course on numerical equations in addition to his calculus course. He continued to introduce students to these areas throughout the remainder of Dirichlet's presence in Göttingen in addition to his work in the seminar, where he conducted the mathematical exercises and presented special topics such as continued fractions and properties of Eulerian integrals.[1]

Ulrich continued to teach applied topics: "practical geometry" and field work, as well as applications of stability theory to newer machines. He generally handled the more geometrically oriented topics such as trigonometry, stereometry, and applications to mathematical geography.

It was Riemann who expanded the offerings in advanced complex function theory. In the winter term 1855/56, he introduced students to complex functions with special reference to elliptic and Abelian functions. This was followed by theory of elasticity in the summer of 1856. In the winter term 1856/57, he again offered complex function theory, this time with emphasis on the hypergeometric and related transcendental series; this would be repeated in 1858/59. By the summer term of 1857, he returned to elliptic and Abelian function theory, which he repeated the following winter. For the summer of 1858, he turned to mathematical physics, lecturing on the mathematical theory of electricity and magnetism, but by the winter term 1858/59, in addition to the hypergeometric function course, he added a course on "Higher Mechanics."

Dedekind, having taught eight terms, would leave for Zurich after the summer term of 1858. He primarily offered algebraic subjects: In the winter term 1855/56, he read on the algebraic solution of equations (Galois theory). In the summer of 1856, he lectured on Least Squares—Gauss's old course. The following year he spent two hours "on the general theory of curved lines and planes" in addition to his major course on "The theory of algebraically solvable equations, especially those of

[1] Küssner 1982.

divisions of the circle." In the summer 1857, he repeated this course but also turned to "theory of probability and method of least squares." This resembled the probability course Dirichlet had taught in Berlin more than the Gaussian course devoted strictly to methods of least squares. Dedekind's algebra course became his staple offer, to which he added a novel course on analytic geometry with special attention to recent methods. This he repeated his next, and last, term in Göttingen, along with a course on the theory of determinants and their principal applications.

Miscellaneous University-Related Activities

Aside from his no longer carrying the extra burden of the Kriegsschule, Dirichlet's lecture routine did not vary greatly from that during the last years at Berlin: the differential equations courses met Mondays through Thursdays at one o'clock, followed by the number theory course on Mondays through Wednesdays. The content of the lectures did not vary substantially either from the ones he had given in Berlin. Dirichlet's status and consequent involvement at the university did. He supervised few students, but served as Dean during 1856–57; was consulted on numerous appointments, some in foreign languages as well as mathematics; provided recommendations for former and present students and colleagues, including Riemann and Dedekind; and continued to spend extra time discussing mathematics, physics, and teaching policies with Weber, a few of their better students, and his two younger colleagues, Riemann and Dedekind.

Bjerknes, Riemann, and Dedekind

With the expanded offering of subjects, several faculty members attended one another's courses, as did visitors from other universities. Both Riemann and Dedekind proceeded to attend Dirichlet's lectures, as did the Norwegian Carl Anton Bjerknes, among others.

Bjerknes, who had previously studied mineralogy and general mathematics, had obtained work at a Norwegian silver mine where he continued his mathematical studies. In 1855, he received a stipend for further study in Paris and Göttingen. While in Göttingen, he attended Dirichlet's course on partial differential equations. What captured Bjerknes' attention was Dirichlet's argument that a sphere carried along in a uniformly moving fluid will remain stationary. According to his son, C. A. Bjerknes had previously become aware of Euler's critique of Newtonian action at a distance. After hearing Dirichlet, he thought this position could be strengthened further by studying the simultaneous movement of several spheres in a fluid.

Bjerknes himself noted in a publication of 1871 that Dirichlet in **1852b** was the first to treat the problem of the movement of solids in an indefinite, incompressible fluid, restricting his study to a spherical solid in that publication. Bjerknes who by 1871 would become professor of mathematics at the University of Christiania (Oslo) devoted the rest of his life to the study of multiple spheres in a fluid and, while short of a complete solution, like Heinrich Hertz, the younger Bjerknes's doctoral adviser, could show that the individual forces activated do not obey the law of action and reaction.[2]

---

[2]Bjerknes 1915:212–223 (Anmerkungen).

Riemann, though not socially inclined, had no difficulty reestablishing contact with Dirichlet. Riemann had entered the Georgia-Augusta University in Göttingen in 1846. Until then he had been taught at home and in various schools in the Hanover area, while also studying Euler and Legendre on his own. Once at the university, he studied numerical solutions of equations, followed by definite integrals, in lectures by Stern; geomagnetism under Goldschmidt; and the method of least squares with Gauss. As previously mentioned, he spent several terms, from the summer of 1847 through the summer of 1849, in Berlin. Upon his return to Göttingen, he worked largely with Wilhelm Weber, who just had been re-instated on the faculty. During the period of Weber's expulsion, Weber had been replaced by J. B. Listing. With Weber's return, it was decided that there would be two ordinary professorships in physics rather than one. This was justified by having Weber hold the position for experimental physics, Listing one for mathematical physics, and accounts for several contributions by Riemann to experimental physics.

Dedekind lost no time paying his respects to Dirichlet, calling on him the week classes began in October. He was probably discouraged by his modest progress in teaching and research up to that time—no students had signed up for his course at that point. After that first meeting with Dirichlet, Dedekind had written to his sister Mathilde in mid-October that he had visited Dirichlet, had been received to his full satisfaction, and that he would rarely get together with people in the coming winter; that, in fact, music would stop—at least as long as he could bear it. This remark is noteworthy because he was a skilled cellist and pianist who had devoted time to music since boyhood. During the following January, he reported to his sister Julie that he saw Dirichlet almost daily, that Dirichlet had visited him, and that Mrs. Dirichlet had invited him to drop in to tea in the evening. In July, he observed that while he had pleasant relationships with a great many people, he found the most useful for him was his almost daily seeing Dirichlet from whom, he felt, he was only beginning to learn. The summer term still brought him only two students, and he did not find it stimulating.

New stimulation would come that fall.

By October 1856, both Riemann and Dedekind made the acquaintance of Borchardt, who was visiting Dirichlet from Berlin where he had succeed Crelle as editor of the *Journal für die reine und angewandte Mathematik*. The result of this new connection was significant for Riemann as well as for Dedekind. Riemann was pulled back from his largely experimental work with Weber to more mathematical research after Borchardt's visit, and both Riemann and Dedekind would submit contributions for the year 1857 to what was now Borchardt's *Journal*.

It was the first appearance of memoirs by Riemann in the *Journal*. Riemann previously had published four memoirs. One was his doctoral dissertation dealing with a general theory of functions of a complex variable. The next two dealt with problems of experimental physics (distribution of electricity and Nobili's rings). The fourth, which also appeared in 1857, was published in the *Abhandlungen* of the

Göttingen Society; it dealt with functions that can be represented by a hypergeometric series and was an expansion of a piece of his doctoral dissertation, growing out of his course in 1856/57.

Riemann's 1857 memoir in Borchardt's *Journal* was his historic work on the theory of Abelian functions. It is frequently referenced as the source for the subsequent discussions concerning the so-called Dirichlet Principle, usually couched in the context of the calculus of variations and potential theory. For that reason, it is worth noting Riemann's actual statement (here translated):

> As foundation for the investigation of a transcendental it is above all things necessary to establish a system of mutually independent conditions sufficient for its determination. In many cases, namely those involving integrals of algebraic functions and their inverse functions, a principle can serve, which —perhaps stimulated by a similar thought of Gauss's—for some years Dirichlet is in the habit of presenting in his lectures on the forces acting in proportion to the inverse square of the distance: this serves to solve the problem for a function satisfying Laplace's partial differential equation in three variables. For the application to the theory of transcendentals there is one case of particular importance for which the principle in the simplest form stated cannot be used and so can be disregarded as being of secondary importance. That is the case where the function, in the region to be determined, at certain points has prescribed discontinuities.[3]

Riemann continued by presenting the principle in a form necessary for his intended application. In doing so, treating the two-dimensional case, he referred back to specifics in his doctoral dissertation of 1851 and showed how, by adaptation of the principle, he could reach several results stated there.[4]

Dedekind wrote to his family in Braunschweig that the meeting with Borchardt inspired him to some new research that kept him home almost all day. He took a break to accept an invitation for the noon meal at Weber's, but he continued to be busy. His intensive work resulted in two memoirs, finished before the end of October, which appeared in that same 1857 volume of Borchardt's *Journal*. The first of these dealt with higher congruences with regard to a real prime modulus; the second was a proof for the irreducibility of cyclotomic equations. As Øystein Ore would note subsequently, the first memoir laid the foundation to Dedekind's later paper on the connection of his theory of ideals to the notion of higher congruences.[5] While busy with these, his first deeper studies involving Gauss and Dirichlet, he indicated that he might have to give up his project of resuming his cello playing. He wrote home that Mr. Dirichlet had said *no* to the idea of spending more time on music; Mrs. Dirichlet had said *yes*. As it turns out, Dedekind did not have to worry about music stopping altogether because of his intended strict immersion in mathematics. Mrs. Dirichlet prevailed.

---

[3]Riemann 1857; see Riemann 1953:97.

[4]For Riemann's and subsequent treatments of "Dirichlet's Principle," see Chaps. 16 and 17, as well as Monna 1975.

[5]Dedekind 1930–32 (1969), 1:67

14.3 Music

The Göttingen faculty included a number of amateur musicians. When the Dirichlets arrived in town, the home of Jakob Henle, the noted anatomist who had come to Göttingen from Heidelberg in 1852, was one of the centers of musical activity. His colleague on the medical faculty, the gynecologist Eduard Siebold, who had come to Göttingen from Berlin in 1833, was a musical enthusiast who sang, played several instruments, but was noticed particularly for the verve with which he pounded the drums in orchestral performances. He was known to Rebecca because their parents had moved in the same social and musical circles in Berlin. Others in Göttingen either took part in musical activities or encouraged their children to sing, play instruments, or marry musicians. With the arrival of Julius Otto Grimm as professor of music in 1855, the standard of music education was raised considerably, as was the number of appearances by well-known musicians. The presence of the Ritmueller piano factory added to the quality of the instruments in use.

Clara Schumann gave the first of several concerts in Göttingen in the fall of 1855. Her long program included not only works by Beethoven, Brahms, Chopin, Weber, and her hospitalized husband, Robert, but several songs by Felix and Fanny.[6] The following year, Brahms and Joseph Joachim first appeared in Göttingen as well. Brahms, then twenty-two years old, was not yet well-known, but Joachim, whose considerable reputation as a violinist had grown since his days as a child wonder, whose career Felix had furthered in Leipzig as well as in London, and who had recently moved to Hannover, was particularly welcomed by Rebecca. When Joachim wrote to his friend J. O. Grimm in September 1855, shortly after Grimm's arrival in Göttingen, asking Grimm to tell him about his life in Göttingen, he added that he thought Dirichlets would be a welcome stimulus and sent his regards to them.[7] In Joachim's frequent correspondence with Grimm he rarely failed to send greetings to the Dirichlets and he was especially pleased by the chance to visit and play at the Dirichlets' during his several stays in Göttingen. It was at their house, in June 1857, that he first heard Agathe Siebold sing and likened her pure soprano voice to an Amati violin. She, of course, was to become better known as Brahms' close friend, commonly referred to as his "Jugendliebe."

Dedekind gradually felt free to tear himself away from mathematics long enough to attend concerts and by 1857 resumed his own playing, at least on the piano. At home and at Weber's he and his housemate Schlesinger played four-handed piano; at the Dirichlets, he even accompanied the dancers during a large ball on at least one occasion.

---

[6] Michelmann 1929:110.

[7] Joachim 1911–13 1:294–95.

14.4 Adaptation and Social Life

It soon became clear that the mantle of the Mendelssohn Bartholdy hostess had descended from her mother and sister at Leipzigerstraße 3 in Berlin to Rebecca, even before Easter 1856, when the Dirichlets purchased their own home at Mühlenstraße 468 (now 1) in Göttingen. Social exchanges included visits with neighbors and other new acquaintances, musical afternoons and evenings, and a daily open house for Flora's adolescent friends who included Agathe Siebold as well as Wilhelm Georg Baum, son of the professor of medicine, the senior Wilhelm Baum, who lived next door.

The first Christmas in Göttingen was festive and filled with memories of Berlin. Wilhelm Hensel had arrived on the twenty-first, intending to stay two days before spending the holiday with his sister Luise. But the days and meals passed swiftly, aided by the presence of the light-hearted Siebold family, and much supplementary pre-holiday cheer in the form of champagne, Bavarian beer, and liters of wine from Cyprus. Hensel stayed until the twenty-sixth. On Christmas eve, Dirichlet received a silver goblet from his grateful students in Berlin, which Hensel inaugurated for him with one of his memorable toasts. Christmas day was spent at the Siebolds where Hensel continued to demonstrate uncommon cheerfulness by embracing all members of the three generations of ladies present.

According to Dedekind, when the Dirichlets first arrived in Göttingen, Rebecca appeared more aloof than Dirichlet. This assessment, if shared by others, soon changed. She endeared herself to music lovers not only by her presence and the aura of her siblings' names, but by support of musical activities. She provided an additional music center in the Dirichlet home for the local performers. She procured needed sheet music by ordering what was required through her relatives in Berlin—previously Eduard Siebold had spent many nights hand-copying what the local group needed for an upcoming performance. Dedekind would not go without music.

Like Henle a few years previously, Rebecca had marveled at the rustic environment found in Göttingen where "in the mornings the herdsman goes through town blowing his horn, and then cows emerge from all houses to go grazing ... and in the evening they return home in leisurely fashion."[8] But she appeared to adapt quickly and ruminated that sooner or later she, too, would grow her own garden, dig for potatoes, and keep a cow.

Despite the outward adjustment, Rebecca was conscious of her new environment lacking something of the unusual quality of her previous life in Berlin. She noted early in January 1856 that, in this quiet town of small lives, events which might have been lost in the noise of Berlin's streets demanded participation because of the close quarters in which one lived. She would visit Berlin about once a year and comment on the irony of having to stay in a hotel, but the spirit of the city and her life in it remained with her.[9]

[8]Michelmann 1929:40.

[9]Her fondness for life in Berlin stood in marked contrast to the feelings of her brother Felix who disliked the city.

In this way, both Joachim and Varnhagen von Ense might have found an explanation for the warmth Rebecca showed them in their correspondence and on their visits to Göttingen. Joachim wrote to Gisela von Arnim in 1857, "Today Mrs. Dirichlet sent me a few endearing lines accompanying a copper plate engraving of Raphael's violinist. The woman appears so clear and sharp that I cannot explain the warm concern she bestows on me, and which in her case stems far less from self-interest than it does with other Mendelssohn relatives."[10] Similarly, Varnhagen was touched by her delight when she greeted him during a surprise visit. He noted the occasion in his diary: He and his niece Ludmilla Assing had arrived unannounced at the Dirichlets', now comfortably established in their own home. On the staircase they ran into Ernst, who recognized them despite the dark and immediately led them to his parents who were still having tea with old Mama and young Flora. "Frau Rebecca jumped up with a cry of jubilation and wide open arms, embraced Ludmilla then myself, kissed us and called out repeatedly that she had no longer expected such pleasure that day."[11] She also used one of Varnhagen's visits to Göttingen as the occasion for giving a sizeable dinner, Berlin-style, that included friends and dignitaries from various parts of the university. Among them were Wilhelm Baum, Gauss's reinstated son-in-law; Heinrich Ewald Sartorius von Waltershausen; Wilhelm Weber; and a number of other personages such as the Provost and members of the legal, medical, and philosophical faculties.[12]

The few remaining ties to those who had shared her life and that of those no longer in this world, as well as her connections to the fewer left who had remained loyal musically and politically, were a strong binding force for Rebecca and her feelings for the city on the Spree. Those feelings could not be ignored when she received news in those years that both Sebastian and even Walter, her first-born, were engaged and had chosen to turn to agricultural pursuits and a life in East Prussia. It caused her considerable distress and brought a recurrence of the painful neuralgia that kept her from attending either wedding. Nevertheless, when Rebecca confessed to Sebastian that it had taken some time to get used to life in the small town, she added that once music had again begun to surround her she felt at home.

The adaptation was somewhat easier for Dirichlet. Thanks to Rebecca, he was not obligated to attend all the musical and other social events surrounding them, nor did he have to take part in the many visits, management of domestic affairs, correspondence with family, and the like. Instead he participated in meetings of a small *Verein* that included Weber, Listing, Wöhler, Henle, and Dedekind, among others. Here they discussed polarization and double refraction, observed celestial phenomena, looked through a spectroscope recently acquired for the physics laboratory, took short locomotive rides on the continually growing railroad network surrounding them, commented on fluctuations in the price of cigars, and shared similar gentlemanly concerns. Weber, Listing, and Wöhler were Dirichlet's frequent companions during walks on "The Wall." Wöhler also kept Dirichlet informed

[10]Joachim 1911–13 1:418.

[11]Varnhagen 1861–70, 13:39.

[12]Varnhagen 1861–70, 13:43.

of advances in his own field. For example, he wrote to his friend and long-time collaborator Justus von Liebig that he had spent a carriage ride through a large wheat field intent on persuading Dirichlet that Liebig's recently published theories laid out in his *Chemische Briefe* (*Familiar Letters on Chemistry*) were the only sound explanations on the subject of organic chemistry.

As noted, after his classes Dirichlet found time to have tea at home with Rebecca and his Mama that he had not had in Berlin because of the far greater distance between home and classroom, and afternoon time spent with students and colleagues there. In the evenings, he now followed his old routine of withdrawing to his study unless called on to participate in a special event such as a return engagement for dinner with some of his former students from the Kriegsschule, now Prussian officers. Meanwhile, Rebecca, Mama, Ernst, and occasionally Dedekind and his housemate, the privatdozent Schlesinger, would play cards, which tended to be won by Mama. After a while they took up Whist, having been used to the traditional Piquet and 66 until then; Rebecca wrote that at that point, both her mother-in-law and Ernst scolded her for playing badly, and the two privatdozenten cheerfully joined them in the criticism. The Dirichlets were at home.

14.5 Continuing Mathematical Contacts

Dirichlet left most personal correspondence to his wife, just as he had formerly relied on her, and previously his mother, to take care of it. He did, however, respond to his mathematical associates, especially when there was something substantive they had reported or when there was a question about a prior publication or conversation. In addition, there were informal visits and meetings with Kummer, Liouville, Borchardt, Kronecker, and others.

One of his most faithful correspondents was Liouville, who combined personal news with issues concerning the *Journal de mathématiques*, and with invitations for visits in Toul, which Dirichlet frequently followed. In addition, Liouville kept Dirichlet abreast of the health and various feuds of members of the Paris mathematical community.

From Berlin, Borchardt, too, sent a mixture of personal news items and mathematical tidbits concerning his editorship of the *Journal für die reine und angewandte Mathematik*. Whereas Liouville's letters reflected his increasing disappointments in the state of affairs in Paris, Borchardt's remained cheerful and positive, reminiscent of the delights he had taken in minor events on their joint trip to Italy that had amused Rebecca and Jacobi. It was he who was quick to congratulate the Dirichlets on Walter's and Sebastian's engagements, noting that he had had the pleasure of seeing their future daughter-in-law briefly during a visit to Königsberg two years earlier. In a characteristic letter thanking Dirichlet for his nomination to the Akademie, he reported to Rebecca that "the clique lives in similar loose fashion as before, sees one another somewhat less frequently but then fights the more passionately about

'Zukunftsmusik' [the music of the future] where, in contrast to my usual habit, I represent the far right."[13]

The exchanges with Kronecker were probably the most substantive mathematically. Also, it was Kronecker who came to see Dirichlet in Göttingen during his fatal last illness, and it was Kronecker who attended the funeral for Dirichlet. What little is known of the correspondence between Dirichlet and Kronecker edited by Ernst Schering would eventually be published in both Dirichlet's and Kronecker's collected works.[14]

It was Eduard Kummer with whom Dirichlet not only corresponded on professional topics but whom he made aware that all was not well with his health. On a personal level, they were "per Du." He had drunk "Brüderschaft" with Kummer after Kummer's marriage in 1840 to Ottilie Mendelssohn, daughter of Rebecca's uncle Nathan. They may have become even closer after the death of Ottilie in 1848. Now he felt free to share with Kummer his concerns at lacking his former strength; as an example, he mentioned not being able to complete the proofs for the planned memoir on hydrodynamics, only a preliminary announcement of which he had published in the Göttinger *Nachrichten.*

This rare comment of Dirichlet's, acknowledging his lack of strength, ties in to the noticeable change in his personal appearance and the nature of his publications after his return from Italy that we noted at the beginning of Chap. 12 and in Chap. 13. There appears to be little doubt that the recurring spiking temperatures that followed his attack of the "Roman fever" (malaria) he had suffered in southern Italy are responsible for his increased weakness, change in appearance and endurance, and probably were at the basis of the heart ailment only discovered in 1858.

14.6 Publications

The nature and number of publications that emerged from Dirichlet's pen after his move to Göttingen support the mention of his waning powers. Aside from translations into French that appeared in Liouville's *Journal* of previously published memoirs, and reissues of other previously published memoirs, only four new results by Dirichlet appeared while he was in Göttingen. Three of these were published by Liouville in his journal; of these, two (**1856b** and **1856c**) had been contained in letters to Liouville and one (**1857f**) had been written down by Dirichlet while visiting Liouville in Toul in August 1857. During the same visit, he also penned an addendum of a page and a half to the translation (**1857g**) of **1855a** that Liouville was about to publish; this was intended to facilitate understanding of references in the original memoir. The other publication (**1857a**) appeared in Göttingen. It was Dirichlet's summary of the memoir he had read to the Göttingen Society on hydrodynamics and would be the only work he published in Göttingen.

---

[13] Berlin. Dirichlet Nachlass. Correspondence.

[14] *Werke* 2:388–411; Kronecker 1930 (1968): 407–32; originally, Kronecker 1885b.

1856b and 1857c

A three-page contribution, appearing in Liouville's and Borchardt's *Journals* for 1856 and 1857, respectively, is a new proof of the theorem that Dirichlet had published at the beginning of his multi-part work of 1839 and 1840 (**1839/40**) and that he had used as a lemma in obtaining many of his following results. It was titled "On a theorem pertaining to series." As he pointed out, the proof he had given originally assumed a condition that although met in most applications can be avoided. The article **1857c** in Borchardt's *Journal* is an only slightly modified version of the original in Liouville's *Journal* of the preceding year; it was still in French, the only such publication in the Berlin journal after 1846.

1856c

In 1856, Liouville also published a note titled "On the equation $t^2 + u^2 + v^2 + w^2 = 4m$. Extract of a letter from M. Lejeune Dirichlet to M. Liouville" in his *Journal*. It was the result of their having had a conversation about Jacobi's theorem concerning the decomposition of a whole number into the sum of four squares. Jacobi, as Dirichlet recalled, had initially derived his theorem from elliptic series but then provided a purely arithmetic proof. Dirichlet observed that he as well as others had found Jacobi's proof too difficult to reproduce without either having Jacobi's argument in front of him or without having recourse to dealing with series, which essentially was the technique at the basis of even Jacobi's "arithmetic" proof. For that reason, he had sought for some time to find a new proof of the theorem. He now wished to present a proof which simplified Jacobi's, above all facilitating its retention by making clear the arithmetic, or rather the algebraic, fact which forms its principal foundation.[15]

Dirichlet added the following remark concerning the difficulty of absorbing Jacobi's proof:

> Cette circumstance m'a fait chercher il y a déjà longtemps á fonder sur d'autres principes une nouvelle démonstration du théorème dont il s'agit, mais je n'en parlerai pas ici, cette démonstration devant trouver naturellement sa place dans un travail dont la rédaction m'occupe depuis quelque temps.

It is not clear to what planned work Dirichlet refers in this interesting comment. We know that in this period, he facilitated access to many of Gauss's arguments found in the seventh part of the *D.A.* This, the one dealing with cyclotomic equations, naturally was closely related to trigonometric issues, which, in turn would suggest analytic considerations. Whether or not Dirichlet had in mind a larger work in which he would go back to arithmetic and algebraic principles remains unknown. We do know, however, that Kronecker was particularly pleased with Dirichlet's note. In 1883, apropos of Kronecker's own arithmetic approach to the problem of bilinear forms with four variables, he commented on this brief memoir, citing Dirichlet's contribution to the sum of four squares problem as a noteworthy example of "disrobing a number-theoretic problem from an analytic dress."[16]

---

[15] For references to the long history of publications concerning the sum of four squares, see Dickson 1919–23 (2005) 2: Chapter 8.

[16] Kronecker *Werke* 2:428.

1857

The last publication that Dirichlet himself saw into print was a two-page excerpt of the memoir he presented to the Göttingen Royal Society of Science on July 31, 1857. It was the subject of the frustration he felt when he realized that his weakened physical condition no longer allowed him to complete the full proof here indicated. He turned over to Dedekind the task of preparing a completed memoir for publication, thanks to whose endeavor it would be published posthumously.[17]

In his brief abstract, titled "Investigations concerning a problem in hydrodynamics," Dirichlet introduced his topic by stating that the fundamental equations of hydrodynamics in the form one finds in textbooks had been known since Euler and, in modified form, had been known since the appearance of Lagrange's *Mécanique analytique*. In the case where the liquid changes shape while in motion, attempts at basing a solution of these equations for determining the motion had been limited to an approximation of that motion. For that reason, Dirichlet wished to find a rigorous treatment of the problem but, observing that it would involve a system of nonlinear partial differential equations, was prepared to have to restrict himself to treating the problem by imposing severe limiting conditions for the movement of the liquid. He explained that after some initial failed attempts, he was able to treat a case which allowed him to describe the general character of the motion as well as some details. Nevertheless, as he pointed out, this case was not an easy one, as one has to take into account the relative attraction of the elements of the liquid. Complete knowledge of the motion requires nine functions of the time, defined by one that is finite and eight second-order differential equations. Generally, however, the initial shape and condition of motion, which includes eight arbitrary elements, have to be subject to further restrictions before the seven first-order integrals that can be established from the partial differential equations can be reduced to the quadratures that will provide a complete solution of the problem.

Dirichlet stated his limiting conditions and subsequent general result as follows:

> Let a homogeneous incompressible fluid which is subject to a constant pressure or one varying only with time, initially have the shape of an ellipsoid; furthermore, let the initial motion be decomposable into two simpler ones, one a rotation of the fluid like that of a solid about an axis passing through the center of gravity, and a second one which changes the relative position of the elements, where their velocities, decomposed vertically against three definite planes intersecting at the center at right angles, are proportional to the distance of the planes, then the fluid, having the motion resulting from such an initial condition will also, at every later time, have the shape of an ellipsoid, which is concentric with the original one, but whose axes will generally vary over time with respect to direction and size. Of the motion taking place at any time, the same holds that was assumed of the original one, namely that it can be decomposed into two simpler ones, as previously defined, except that the axis of rotation as well as the three mutually perpendicular planes to which the partial motions refer, also generally take on a different position at each moment.

The simplest case, which is the only one he wished to consider in the example he was discussing, is that where the mass initially has the form of an ellipsoid of rotation and

[17] See Chapter 15, **1861**.

there are no initial velocities. The motion then consists of isochronous oscillations, where the fluid, passing through the spherical shape, alternately assumes the form of an elongated or flattened ellipsoid.

1857f

On his last visit to Liouville in Toul, on August 15, 1857, Dirichlet once more showed how to reduce a detailed analysis from the *Disquisitiones arithmeticae* to a short and simple proof. Liouville would publish it in his *Journal* that year as a note by Dirichlet having the title "New demonstration of a proposition pertaining to the theory of quadratic forms."

In articles 164–5 of the *D.A.*, Gauss had stated the following proposition:

> If the form
> $$\text{(F)} \qquad Ax^2 + 2Bxy + Cy^2$$
> contains another form
> $$\text{(F}'\text{)} \qquad A'x'^2 + 2B'x'y' + C'y'^2$$
> both properly and improperly, then one can always find a *forma anceps* which is contained in the form F and contains the form F′.

Dirichlet reformulated the proposition as follows:

> Given an improper substitution $\begin{pmatrix} \alpha & \beta \\ \gamma & \delta \end{pmatrix}$ by which the quadratic form $(a, b, c) = f$ (whose determinant is assumed to be different from zero) is changed into itself, one can always obtain a new form belonging to the same class, and for which twice the middle coefficient is a multiple of the first coefficient.

Dirichlet's translating the statement of the theorem into class-related terms and use of a, by this time, familiar symbolism allowed him to reduce the rather lengthy Gaussian proof to three pages involving simple, algebraic substitutions. This short memoir demonstrates not only the proof of the proposition involved but also his having found the same successful combination of a class-linked approach and a suitable notation that he had employed successfully before.

14.7 Aging

There is no question that Dirichlet had not regained the strength he had lost during his bout with "Roman Fever" more than a decade before. Despite his valiant efforts to maintain his old research routine along with his lectures, the stresses of the intervening years had done nothing to shore up his weakened system. His new students (attending his "old" courses) continued to be impressed by his lecture style and his being able to present ideas, theorems, and proofs without recourse to notes (except for those reminding him where he had left off at the last class meeting). But members of the society saw none of the new research results they had anticipated so eagerly at the time of Dirichlet's arrival, except for that one preliminary two-page announcement of the hydrodynamic memoir. There was no sign of the eulogy for Gauss. Not even review articles, such as those whereby Gauss had kept his promise to let no year go by without a published contribution to the society's *Anzeigen*, ever came from Dirichlet's pen.

Descriptions and drawings of him ever since his return from Italy point to an elderly professor, stately and deliberate in his movements, still displaying his phenomenal memory when lecturing on the topics that were familiar to him, but bearing little other resemblance to the humorous, lanky, quietly engaging young man who had captivated his surrounding social and academic circles upon arrival in Berlin. Rebecca, too, had aged. The strain of the previous decade—the losses of Fanny and Felix, the distressing political events, other deaths, and concerns for their sons and her nephew, the recurring facial neuralgia, along with the changed environment—had taken its toll.

14.8 Travel

As in previous years, Dirichlet used the spring or summer breaks for trips, several times to the clearer climate and views of his beloved mountain tops in Switzerland, often using the solitude of these excursions to attempt some productive work, and reserving any visits to stops to and from his destination.

In 1855, both Dirichlets had taken a short trip to Zurich after settling their affairs in Berlin. In 1856, he spent six weeks after Easter in Paris. While there, he stayed at the Liouvilles, attended a meeting of the Académie des Sciences, was invited by Leverrier to join him in calling on the Minister of Education, and, as Rebecca remarked, "delivered the Empress, made peace, and had a wonderful time." The following year he left for Freiburg and France at the end of July, spending some time in Switzerland before once again visiting the Liouvilles who were back in Toul. Finally, in 1858, after an intermediate visit in Toul, his destination was Montreux. He had planned to use his stay there to work up the long-promised eulogy of Gauss. It did not happen.[18] On the way to his new position in Zurich, Dedekind ran into him accidentally on the street in Basel as Dirichlet was returning home. Dedekind wrote his sister with worry that his admired teacher, though displaying his usual friendliness during their two-minute encounter, had complained about not being well.[19]

On October 17, 1858, Dirichlet was back in Göttingen. It was immediately apparent that he was seriously ill.

[18]Biermann 1971 contains a summary of Dirichlet's notes on Gauss, which give us a glimpse of "what might have been."

[19]Scharlau, ed. 1981:56.

14.9 Illness and Deaths

On October 23, Rebecca informed Paul and Albertine of Dirichlet's return and illness. She reported that after grave, initial concerns he appeared to have improved over the last days, and that Dr. Baum was hopeful he could bring Dirichlet back to full recovery. He had had a widening of the heart muscles that apparently had set in unexpectedly four months before. Because they thought this had happened suddenly, the physicians assumed that it had been caused by a cold or an infection and that it could be healed. Rebecca described the awful state Dirichlet had been in when he arrived, and that everyone, including Baum, "whose heart easily runs away with his head," had expected the worst. She felt, however, that now that there was gradual improvement she could not praise Baum enough; he was in agreement with one of his colleagues and with a physician in Montreux whom Dirichlet had apparently consulted. Meanwhile, Dirichlet appeared pain-free and listened with interest to the accumulated correspondence that she read to him.

Two weeks after his return home it still was not clear just what had caused his problem. On October 30, Rebecca wrote to her nephew that while there could be no talk of his getting up from bed, Dirichlet was free of fever and pain. He was allowed to read and for his "frivolous entertainment" chose Kant, Plato, and Göttingen's weekly newspaper. Rebecca commented that she thought Molière would be more appropriate, because of the physicians. These conceded that they were quite uncertain what had precipitated the episode: a cold, an infection, blood clots, asthma, nerves. None apparently linked his cardiac condition to the "Roman fever" of the previous decade, which, if following the usual course of recurring spiking fevers associated with malaria in addition to the stresses of the period, may well have precipitated a gradual weakening of the heart muscles.

Another week later Rebecca wrote to Albertine that the doctors claimed he was better but that she did not see much improvement: "not to see yet to believe." She noted that this was the beginning of the third week since his return, but that there was no talk of his getting up. He was allowed to see people but not to talk much. He was on a sickness diet consisting of chicken, calf, a light vegetable, and a fruit compote. Rebecca expressed her fear of his still having very short breath resulting in difficulty speaking, but was told this was a secondary concern.

Walter had wanted to come, but his mother kept him in Bretschkehmen, as he and Anna had just had their first child. Neither Dirichlet nor Rebecca would ever see that first grandchild.

During the October rest and careful attention, Rebecca reported that Dirichlet appeared to be improving but, by November 1, still could not get up. By now Göttingen was full of rheumatism and flu; every place, even portions of Italy, was cold and surrounded by snow. Ernst, who had turned eighteen on the ninth of the month, now had a job. They had to give up the dog. Rebecca's other concern had been Mama, who, despite her physical continued well-being, had been indescribably upset; Rebecca wrote that she had never seen someone so distraught. She added, "It must be horrible to be so old. Of that, at least, I believe to be spared."

She was right. On the twenty-ninth of November, Rebecca had a stroke. She remained in a near coma until her death on the morning of December 1. The event shocked not only the family but most of Göttingen. The day after her death, Bertha Wagner, daughter of the physiologist Rudolf Wagner, informed her friend Mathilde Dedekind in Braunschweig of the news, which, she wrote, had affected and upset the whole town. She described Dirichlet's mother sitting at her son's bedside while they had wept together. She noted immediately that, although Dirichlet, despite his great sadness, was composed after Rebecca's death, the physicians feared this event would set him back; and that, although there were as yet no outward signs of a relapse, they had little hope for his own life.[20]

The news spread quickly among Joachim and his fellow musicians outside of Göttingen as well. Immediately after Rebecca's stroke, J. G. Grimm had informed Joachim, noting that the physicians, who now were not leaving the house, had already previously indicated that because of the damage to his heart, a sudden shock could lead to Dirichlet's immediate death. Upon Rebecca's death, Joachim, clearly distressed, passed on the news to Clara Schumann in Vienna: "Who could believe that the lively woman with the intelligent eyes, always striving intellectually, could fall ill and end within two days?!" Schumann replied, relating how much she was saddened by word of Rebecca's death, although, as she acknowledged, they had never been close [the feeling had been mutual]. She wrote that she had slept very little in recent nights, wondering whether Rebecca was now united with her siblings and why heaven took those who loved life while letting live those who feel life's heavy burdens daily. "Oh, had I gone to rest instead of her!"[21]

Dirichlet's progress had come to an end. He and Rebecca had been true to one another. She had combined the quick wit of a Berliner with the rational intelligence of her forbears and an honest understanding of human nature. But now the poet's warning held.[22] Dirichlet's world was no longer "a land of dreams" but one that had "neither joy ... nor hope, nor certitude..." He remained largely bedridden for five more months. There was no struggle nor strife.

Johann Peter Gustav Lejeune Dirichlet died on Thursday, the fifth of May, 1859.

[20] Scharlau, W., ed. 1981:55.

[21] Joachim 1911–13 2:39.

[22] As found in Matthew Arnold's "Dover Beach"; Arnold had spent a brief period in the early 1850s in the environs of Göttingen with his bride.

Chapter 15
Aftermath

Dirichlet's death had an impact on his family, his colleagues and friends, and the future composition of several mathematics departments, notably those in Berlin and Göttingen.

15.1 Family

The family members most closely affected by the two deaths were Dirichlet's mother, aged ninety, and the Dirichlets' fourteen-year-old daughter Flora. According to Sebastian Hensel, both were still living in the Mühlenstraße, the only remaining inhabitants of the once lively home, when he came to Göttingen at Christmas time 1859 to begin packing up the essential belongings of the formerly welcoming household.

Son Walter and nephew Sebastian had for some time built independent existences at a distance from Berlin and Göttingen, in East Prussia, and were busy with their agricultural pursuits and with raising young families. It was the weaker son Ernst who probably had been in greatest need of support and encouragement from his parents.

Mama Dirichlet

Dirichlet's mother lived another ten years. But, according to Sebastian Hensel, the loss of Gustav, the favorite, last surviving, and closest of her children, left her, once unconquerable in mind and spirit, now a weakened shadow of her former self. She is said to have spent her last decade with nieces or grandnieces. Perhaps this refers to Dirichlet's nieces, her granddaughters, among whom there are two likely candidates. One is Mathilde Schoeller, the daughter of Dirichlet's sister Sophie Carstanjen. She lived in Düren in the 1850s and upon her death in 1908 left a substantial philanthropic bequest to the town of Düren and two of its eleemosynary institutions. Dirichlet's mother may also have spent some of her remaining years with the daughter of his

U. C. Merzbach, *Dirichlet*, https://doi.org/10.1007/978-3-030-01073-7_15

sister Caroline Baerns. After the retirement of his father, Dirichlet's parents had frequented Aachen where J. C. A. Baerns had become postmaster. Caroline had died in 1836, a year before Dirichlet's father.

Flora Dirichlet (Baum)

The Dirichlets' daughter, Flora, for several years lived in the household of her uncle Paul Mendelssohn Bartholdy, who had become her legal guardian, as well as that of the minor children of Felix Mendelssohn Bartholdy. In 1870, when Flora was twenty-five, she would marry her childhood friend and neighbor, Wilhelm Georg Baum, son of the surgeon Wilhelm Baum. He followed in his father's footsteps, both by becoming a noted surgeon and by settling in Danzig, where his father had performed notable medical services before moving to Göttingen. Flora and Wilhelm Georg Baum had six children, among whom the best known is the remarkable Marie Baum, family advocate and member of the Weimar Republic's Parliamentary Coalition. In her autobiography, Marie Baum noted that her mother, Flora, never recovered fully from the loss she had suffered as a teenager. Baum also indicated that she, as well as her mother, essentially followed Rebecca Dirichlet's outlook in their social and political concerns.[1]

Ernst Dirichlet

Young Ernst, thanks to his uncle Paul's support, had just begun a career in business when his parents died. He had never been strong physically and his apparently weak immune system seems to have made him susceptible to colds, flus, and various childhood illnesses that his siblings shook off more easily. He survived his mother and father by less than a decade, dying in Berlin in 1868, before his twenty-eighth birthday.

Walter Dirichlet

At the time of his parents' deaths, Walter Dirichlet was still owner of Klein-Bretschkehmen (his property in East Prussia), which he retained for the rest of his life. It was he who arranged for their burial ground in Göttingen's Bartholomäus cemetery.[2]

Beginning in 1877, Walter Dirichlet held various elected political posts. From 1877 to 1886, he was a member of the *Provinziallandtag* of East Prussia, where he represented several small districts. During the same decade, he was a member of the Prussian House of Representatives (the *Abgeordnetenhaus*) and, from 1881 to the time of his death, was a member of the German Reichstag. There, he was first a representative for the German Progress Party, a liberal party formed in 1861, characterized by its opposition to Bismarck; later he served as member of the German Freeminded Party, which had been formed in 1884 through the merger of the Progress Party with

[1] Baum 1950.

[2] For references concerning the upkeep of the cemetery, see Kühn 1999:155–56.

several other liberal groups. These were precursors of the German Democratic Party of the Weimar Republic, which his niece Marie Baum joined after leaving the Social Democrats.

Of Walter's four children, only Georg Lejeune Dirichlet, the oldest, was born while Dirichlet was alive. He studied classical philology, was a secondary school teacher, and became director of the Königsberg Gymnasium. It would have been interesting had he and his grandfather been able to discuss the merits of studying, or conversing in, ancient languages. One of Walter's daughters, Elisabeth, was the mother of the philosopher Leonard Nelson, who became an associate of David Hilbert. Leonard Nelson consolidated a large part of the Dirichlet Nachlass.[3]

Sebastian Hensel

In contrast to the Dirichlets' children, nephew Sebastian Hensel steered a political course between that of his father and the Dirichlets. This may have had economic reasons. He left agriculture. He had sold his property, Gross-Barthen, in 1872 because the climate seemed to be unsuitable for his wife and son, the future mathematician Kurt Hensel. Initially planning to resettle in Thüringen, he made a stopover in Berlin, where he happened to encounter one of Berlin's leading bankers who persuaded him, with reluctant agreement from his uncle Paul, to run Berlin's "Markthallen" (public market halls). About that time, the city took over most municipal enterprises from private industry. Hensel thereupon became director of a large hotel construction corporation, which resulted in the building of the famous *Kaiserhof*, Berlin's largest luxury hotel. In October 1875, the hotel celebrated a well-publicized opening, having been visited by Emperor Wilhelm I the previous day. Before the end of the month, the opulent structure fell victim to a massive fire. Hensel rebuilt, remained its director for five years, and then became head of the "Deutsche Baugesellschaft," another building corporation. In later years, he retained his allegiance to private enterprise, but also returned to writing and earlier artistic endeavors.[4]

Sebastian Hensel's two-volume history of the family Mendelssohn remains the most extensive and personal account of the nineteenth-century offspring of Moses Mendelssohn. It is based largely on first-hand experience and correspondence, but tends to provide an occasionally unduly rosy picture; it omits references to potentially sensitive issues and lacks the accuracy in quoting and dating correspondence found in later, professionally edited, publications of selected family letters. His subsequent autobiographical volume [Hensel, S. 1904] supplements the family history, which had ended with the death of his mother Fanny.

---

[3]Despite Nelson's efforts, the Dirichlet Nachlass was subsequently split and ended up in various locations. This was primarily due to attempts to safeguard it from confiscation by the regime of the Third Reich. For details, see [Schubring 1986]. Current holdings in Berlin, Göttingen, and Kassel are referenced in our bibliography.

[4]For further details, see [Lowenthal-Hensel 1983].

15.2 Associates

Among Dirichlet's surviving friends and acquaintances who mourned him most deeply or were most affected by his life and death, we single out a number of mathematicians and scientists who were particularly influential in passing on his publications, lectures, and methodology to future generations.

Liouville

With the illness and death of Dirichlet, Liouville lost the correspondent with whom he shared not only matters concerning number theory and analysis but also his discouragement brought about by surrounding conflicts among Parisian mathematicians and physical scientists. In his *Journal*, Liouville had already published translations of several of Dirichlet's memoirs. After Dirichlet's death, a volume for 1859 (series 2, volume 4) contained five translations of Dirichlet memoirs previously published in German; in volume 7, these were followed by Dirichlet's proof of Abel's theorem (Dirichlet's convergence test), taken from his correspondence with Liouville. In addition, Liouville dedicated his lectures in 1859/60, and some following ones, to Dirichlet.[5] He would continue to immerse himself in Dirichlet's work, as evidenced by references in his own lectures and publications.

Kummer

Ernst Eduard Kummer, Dirichlet's successor in Berlin and cousin by marriage, was the author of the influential eulogy for Dirichlet offered at the Berlin Akademie. Aside from holding the Berlin professorship for mathematics until his retirement in 1883, by careful choices of recommendations for senior mathematical positions in Germany he succeeded in maintaining a strong Dirichlet tradition until the end of the century.

Over the years, Kummer had repeatedly benefitted from Dirichlet's support. In 1839, Dirichlet had proposed him for membership in the Berlin Akademie; three years later, Dirichlet, along with Jacobi, had recommended him to the professorship in Breslau; and, in 1855, Dirichlet, upon leaving for Göttingen, listed Kummer as his first choice to succeed him in his own position in Berlin.

In his eulogy of Dirichlet, held in a public session of the Akademie on July 5, 1860, Kummer first reminded his listeners that their fatherland within the previous decade had lost the three men who had been responsible for bringing them a new blossoming period of the mathematical sciences. Referring to Gauss, Jacobi, and Dirichlet, he praised these three giants for renewing the reputation their nation had earned, largely through Kepler and Leibniz, for a deep understanding of both the most abstract mathematical truths as well as those realized concretely in nature. He stressed Dirichlet's close relationship to the Berlin Akademie that had lasted even during those last few years when, as successor to Gauss in Göttingen, he had to leave Berlin. This, Kummer noted, gave the Akademie the right still to call Dirichlet its own, combined with the duty to preserve his memory and offer him its last academic

---

[5] Lützen 1990:215–17.

honor by a public memorial tribute. He had agreed to deliver this tribute because of his admiration for Dirichlet, their friendship that had lasted for more than twenty years, and the close relationship of their respective studies. He wished to describe the high scientific importance of Dirichlet's masterpieces as well as to sketch a picture of Dirichlet's life and character, which was "as noble and pure as his writings."

In his biographical narrative, Kummer stressed Dirichlet's early interest and disciplined study of mathematics and history, the important influence of the Foys on his ability to move comfortably in social spheres above that in which he grew up, and to become familiar with the political currents of the time.

In discussing his mathematical work, Kummer managed to convey the salient points of all the major developments in Dirichlet's publications that we have reviewed in the previous chapters. In addition, Kummer stressed the unusual relationship of Dirichlet's and Jacobi's studies, namely, that they worked together closely, often on similar or like problem areas, without methodological overlaps or competitive results. Kummer remarked that this extraordinary phenomenon could not be explained simply by their difference of methodology, but also by their appreciating each other's priorities while consciously trying to avoid any appearance of rivalry.

Kummer was among the first who remarked on the difference between Jacobi and Dirichlet with regard to having founded a "school." Whereas Jacobi's students tended to follow his spirit and orientation, Dirichlet's would follow more individually differentiated directions. Kummer attributed this difference to Dirichlet's preference in his own studies to focus on fundamental difficulties and explore their basis rather than pursuing more widely explored topics and expanding their consequences. Kummer considered as characteristic of Dirichlet's work his complete rigor and demonstration of method and proofs, which, while basic to mathematics, had not been carried out with the same consistency by even the greatest mathematicians of the time. Kummer linked this characteristic to Dirichlet's love for "pure and totally secure" truth, as well as to his work habits, and the care he took in preparing his publications. Dirichlet, according to Kummer, tended not to publish significant results until he had explored their connection to related theorems and found the simplest derivation most appropriate to the innate nature of the subject.

Kummer also remarked on the incomplete last results that Dirichlet had apparently worked out in his head, supposedly ranging from a theory of ternary, second-degree indefinite forms, to asymptotic laws for a variety of number-theoretic functions, leading to the determination of prime number frequency, and to a proof for stability of the system of the world. In particular, he mentioned Dirichlet's having confided to Kronecker his step-by-step approach to the solution of problems in mechanics.[6]

Borchardt

It was Dirichlet who had introduced Borchardt to Jacobi, whose doctoral student Borchardt became. Borchardt had been in regular communication with both Dirichlets since their Italian trip. Largely as result of the urging of Dirichlet, Borchardt would edit the *Journal für reine und angewandte Mathematik* from 1856, shortly

[6]Kronecker clarified this in [Kronecker 1888b]; see Kronecker *Werke* 5:473–76.

after Crelle's death in 1855, until 1880, the year of his own death. He had collaborated with Dirichlet in the managerial task of handling Jacobi's *Nachlass* and family affairs. He published the first fully two-sided edition of the Legendre–Jacobi correspondence, edited the first volume of Jacobi's *Gesammelte Werke*, and published more than two dozen memoirs of his own, among other contributions. Karl Weierstrass, to whom he became personally close, succeeded him in the editorship of Jacobi's *Werke*.

Heine

At the time of Dirichlet's death, having spent the years 1844–1856 as privatdozent and extraordinary professor in Bonn, Heine was one of two ordinary professors of mathematics in Halle, where he remained until his own death in 1881. The other ordinary professor was the Bessel student Otto August Rosenberger, who had been engaged for applied mathematics, built up the observatory, became known as a comet specialist, but after winning the Royal Astronomical Society's Gold Medal in 1837 ceased publishing and concentrated on teaching and various other university- and community-related activities.

Eduard Heine's name nowadays is primarily identified with the so-called Heine–Borel Theorem and the concept of uniform continuity. It is perhaps because of this connection that he has been described as being a member of the "Weierstrass school." This has little justification. As previously noted, Heine had been a student of Dirichlet's, had dedicated his doctoral dissertation to Dirichlet, and had been personally connected through the marriage of his sister to Dirichlet's brother-in-law, Paul Mendelssohn Bartholdy. As for the Heine–Borel theorem, there, too, the original direction is attributable to Dirichlet, as suggested by the details provided in [Dugac 1989].

Until the mid-sixties, Heine continued to study Dirichlet's work and express admiration for him. For example, his respect for Dirichlet's analyses is reflected in the following statement preceding his reproduction of Dirichlet's proof **1837e** concerning spherical functions that, as we noted in Chap. 9, he characterized as providing the mathematical foundation for physical problems:

> If our work were not left with an essential gap were we to neglect to completely firm up this foundation for applications to physical problems, we would have omitted the proof by Dirichlet and only referred to it. Those who know Dirichlet's work know that it is a model of presentation of mathematical topics and could only lose by other than a verbal reproduction.[7]

We must note, however, that two of Heine's publications, in 1869 on trigonometric series, and in 1872 on the elements of function theory, mark the beginning of a period in which Weierstrassian criticisms of Dirichlet and the Dirichlet Principle took center stage in German discussions of the fundamentals of classical analysis. In both of these publications, Heine supported Weierstrass's arguments. It should be mentioned that Georg Cantor, who had been Weierstrass's student, joined Heine in Halle as privatdozent in 1869, afterward rising through the ranks until he assumed

[7] Heine 1861:266.

the other ordinary professorship there in 1879. As previously observed, Heine was just working on his 1869 paper when Cantor arrived and interested Cantor in the subject, which resulted in Cantor's studies of trigonometric series.

Heine's 1869 publication began with the statement that until recent times, it was thought that the integral of a converging series, whose terms between finite limits of integration remain finite, must equal the sum of the integrals of the individual terms. He continued by commenting that only Weierstrass had observed that the proof of this statement requires not merely that the series converge within the limits of integration but that it converge uniformly. Heine concluded that, because of this, Dirichlet, Lipschitz, and Riemann could only establish that in a number of cases a function can be expanded as a trigonometric series with known coefficients but not in how many ways such an expansion can take place. Wishing to reestablish conditions for uniqueness of the expansion, Heine arrived at several results, based in part on conversations with Cantor, who also called his attention to Seidel's memoir of 1847 concerning series that represent discontinuous functions.

Heine proceeded by discussing neither the content of Dirichlet's lectures nor that of his major related publications. Aside from subjecting to particular criticism Riemann's *Habilitationsschrift* on the representation of a function by a trigonometric series, Heine chose to subject to a close scrutiny in this paper, without mentioning Dirichlet's more rigorous 1829 memoir (**1829b**), Dirichlet's *Repertorium* report of 1837. Had he disregarded the purpose of Dove's newly reorganized *Repertorium* to introduce physicists not trained in proof techniques to some of the mathematical concepts and procedures they could apply to solving problems in their areas of interest? Or was he trying to be helpful when he added an appendix to his memoir, less than half the length of his entire paper, in which he sought in a meticulously detailed process to prove the statements of Dirichlet's *Repertorium* report?

Heine's rigorous, source-based, but inflexible, attitude may be a combination of a certain personal characteristic and an occasionally quick reaction to an outer influence such as one by Weierstrass or the intermediary Cantor. The characteristic was described by Rebecca Dirichlet as pedantry.[8]

Christoffel

Elwin Christoffel, who had begun studying with Dirichlet in 1850 and only received his doctorate after Dirichlet had left Berlin, nevertheless considered himself primarily a student of Dirichlet's. Although he is best known for the work that is basic to tensor analysis, the publications in which he showed his affinity to Dirichlet most clearly are his several papers in the late 1860s on potential theory. He spent the years 1862–1869 expanding the mathematical curriculum at the ETH in Zurich and then three

[8] She provided an amusing example of Heine's unexpected (but lasting) reaction to a new encounter, when, at the time of Heine's announced engagement in 1850 to a young woman he had only known for twenty-four hours, she had commented to Sebastian Hensel, in a letter which Sebastian would cite: "What do you say about the engagement per steam of Eduard Heine? I think the calendar listing his due dates had an entry '4 April: Engagement.' Otherwise it is incomprehensible how that pedantic man could have arrived at such a hasty decision. By the way, the girl is pretty." Hensel, S. 1904:114.

years at the Gewerbeakademie in Berlin. He built up the mathematical institute at the university in Straßburg in the remaining years of his career before his death in 1900.

Lipschitz

Rudolf Lipschitz, who had already begun his university studies in Königsberg when only fifteen, attended the university in Berlin in the early 1850s, working primarily with Dirichlet. His August 1853 doctorate was cosigned by Dirichlet along with Ohm. Lipschitz, too, considered himself primarily Dirichlet's student. The clearest illustration of his adherence to Dirichlet's work is his expansion of Dirichlet's convergence proofs in [Lipschitz 1864]. His numerous memoirs show him as a follower of Riemann as well as of Dirichlet, especially in regard to issues of curvature and the related differential equations.[9]

Riemann

Riemann was successor of Dirichlet in his position at the university in Göttingen as well as in his research on complex functions and analytic number theory. In his profound expansion of complex function theory (including his one number-theoretic publication leading to the so-called Riemann hypothesis) he opened wide several doors that Dirichlet and Gauss had unlocked. The first suggestion of Dirichlet's influence in this regard, following Riemann's doctoral dissertation, is found in Riemann's letter to his father concerning Dirichlet's visit to Göttingen in 1852, the time Riemann was working on his *Habilitationsschrift*.[10]

These connections have been overshadowed by a frequently assumed link to Dirichlet through the so-called Dirichlet Principle. As noted in our preceding chapter, Riemann had coined the term "Dirichlet Principle," referring to it in his fundamental memoir on Abelian functions. We shall call attention in the next two chapters to later assumptions, not supported by Dirichlet's publications, concerning the relationship of the Dirichlet Principle to Dirichlet.

Riemann's number-theoretic publication containing the famous "Riemann hypothesis" only appeared in 1860, the year after Dirichlet's death; Riemann himself was dead six years later, having spent much time during those last years in Italy, because of the hope that the milder climate would provide an improvement for a finally fatal lung ailment.

Dedekind

We have previously noted Dedekind's becoming attached to the Dirichlets during their years in Göttingen and last encountering Dirichlet in Basel, as he was leaving for his position at the ETH in Zurich while the ailing Dirichlet was returning home

---

[9]Referring to a listing in the DSB, the authors in Lipschitz's MacTutor biography note as "perhaps the most remarkable fact" about his research that "he worked in areas as diverse as number theory, theory of Bessel functions and of Fourier series, ordinary and partial differential equations, and analytical mechanics and potential theory." We note that these are precisely the areas to which Dirichlet had contributed, making the diversity of Lipschitz's interests less surprising.

[10]See our Chapter 12.

to Göttingen. Dedekind remained in Zurich until 1862, when he returned to his native Braunschweig as professor at the polytechnic school there. He received an emeritus appointment in 1894, although still giving lectures at the school known as the Technische Hochschule since 1877.

While back in Braunschweig, Dedekind became the designated editor of Dirichlet's number theory lectures, discussed in our next chapter. Another publication that he brought to its final form at Dirichlet's request was the hydrodynamic memoir which had troubled Dirichlet so greatly in his last years. In the 1870s and 1880s, Dedekind produced the memoirs introducing the Dedekind cut and his ideal theory, among other contributions to algebraic number theory and the nature of number. With Heinrich Weber, he edited Riemann's works in 1876 and in 1880 also published their joint work on algebraic functions.

Although freely using the new terminology that emerged with the generalized abstract concepts which would dominate algebra and number theory in the following century, Dedekind managed to stay close to Dirichlet's characteristics in his editions and ramifications of Dirichlet's work.

Kronecker

Kronecker, who had lectured at the university in Berlin as a "reading" member of the Akademie, assumed a second chair as ordinary professor there in 1883. He edited the first volume of Dirichlet's collected works and published a number of memoirs dealing with some of Dirichlet's results.[11] Most of these memoirs appeared while Kronecker was occupied with the edition of the collected works and show how carefully he studied the publications that he edited. He completed the first of the two volumes of Dirichlet's memoirs in October 1889; it contains Dirichlet's publications up to the time of the Italian journey as well as his memoir on complex units, which was published in January 1846. Kronecker, who had been ailing for some years, died in 1891. When Lazarus Fuchs inherited the task of publishing the second volume of Dirichlet's works, he had Kronecker's plan for it to guide him and found that fifteen folios had already been printed. Details of the choices made are found in the preface to the second volume, along with the acknowledgments to those who assisted, just as Kronecker had provided similar explanations in the preface to the first volume.

In previous chapters, we have referred to Kronecker's close personal attachment to Dirichlet. It should be noted that the overemphasis by numerous commentators concerning oral statements of Kronecker pertaining to foundations and questions of existence has overshadowed the breadth of his contributions to number theory and other areas of mathematics. He shared Dirichlet's fondness for observing linkages between different branches of mathematics and within number theory.[12] Connected to this and to a variety of misunderstandings about Kronecker's intent are refer-

[11] See, for example, Kronecker 1865, 1885a, 1885b, 1888a, 1888b, 1890.

[12] He expressed this succinctly in his "Antrittsrede" when elected to the Akademie. See Kronecker *Werke*, 5:387–89.

ences to his "Jugendtraum." The clearest extant statement concerning this frequently discussed idea may be found in a letter he wrote to Dedekind in 1880.[13]

Unlike Dirichlet, Kronecker frequently commented on issues of priority, which did not endear him to a number of his contemporaries. Their attitudes may be linked to the persistent acknowledgment and documentation of his own work as well as that of others. Negative statements concerning his personality were reinforced by attitudes of the 1920s and 1930s. Nevertheless, the significance of Kronecker's contributions to mathematics had for its future development and the influence he had on his students and associates has been undeniable.

Bachmann

Paul Bachmann entered the Berlin University in 1855, just as Dirichlet had moved to Göttingen. He then followed to Göttingen where he listened to Wilhelm Weber and to Friedrich Wöhler, as well as to Riemann and Dedekind. It is unclear when he attended lectures by Dirichlet; yet, after returning to Berlin and in 1862 receiving his doctorate on substitution (group) theory under Kummer, in 1864 he credited Dirichlet's lectures as having inspired him with his habilitation topic on complex units [Bachmann 1864]. Whatever may have been the degree of his personal acquaintance with Dirichlet, his subsequent publications show his familiarity with Dirichlet's achievements and his multi-volume, readable work on number theory appearing between 1872 and 1923 did a great deal to make non-specialists aware of Dirichlet's importance in that field and to introduce a new generation of readers to the heritage of Gauss and of Dirichlet.

15.3 Institutions

Dirichlet's chief impact from an institutional viewpoint remained in Göttingen and Berlin. We should note, however, that his methodology and specific contributions resonated in numerous other universities, notably through clusters and lines of succession of some of his former students and their students. Among these, we single out six institutions where mathematical descendants of Dirichlet and Kummer played a significant role during the remainder of the nineteenth century. They are Bonn, Breslau, Halle, Munich, Straßburg, and Zurich (the ETH). Although we here focus on these examples, there were other locations in the nineteenth century where individuals and groups shared Dirichlet's legacy.

Göttingen

Dirichlet's death affected mathematicians, physicists, and astronomers in Göttingen. Initially, the mathematical successorship to Dirichlet in Göttingen seemed ill-fated.

Riemann, the most significant in building on Dirichlet's analytic work but suffering from ill health, survived him by less than a decade, dying in 1866. He spent much of the intervening time in Italy in an attempt to improve a finally fatal lung ailment.

[13] It takes us beyond the bounds of this volume to go into details on this subject, but see that letter and Helmut Hasse's analysis found in Kronecker *Werke*, vol. 5:453 and 510–15, respectively.

Alfred Clebsch died of diphtheria after only four years in Göttingen. A graduate of the Altstädtische Gymnasium and the university in Königsberg, he had received his geometric training through study with Otto Hesse and Friedrich Richelot, but received his doctorate in mathematical physics under Franz Neumann in 1854 with a dissertation on hydrodynamics. After several years teaching in the secondary schools of Berlin, and publishing numerous memoirs on hydrodynamics in Borchardt's *Journal*, he joined the faculty at the university in Giessen, where he collaborated with Paul Gordan on a theory of Abelian functions and established algebraic geometry, heavily influenced by the work of Cayley and Sylvester. After five years in Giessen, in 1868 he filled the vacant professorship in Göttingen, particularly appropriate because of his growing interest in Riemann's function theory. By this time, his work was closer to that of the English algebraic geometers than his original geometric orientation might have suggested.

During his time in Göttingen, Clebsch and Carl Neumann, friends since boyhood, founded the *Mathematische Annalen*. The *Annalen* would show less of the influence of Dirichlet than the *Journal für die reine und angewandte Mathematik* in Berlin. The only area where Clebsch appears to have had overlap with Dirichlet's work was in hydrodynamics, a subject that had interested him even before his university studies. Lazarus Fuchs succeeded Clebsch in Göttingen but followed a call to Heidelberg after only one year, in 1875.

H. A. Schwarz, like Fuchs the holder of a Berlin doctorate under Kummer, his father-in-law, had what one might consider a more normal length of tenure in Göttingen, from 1875 to 1892. Known for his work on conformal mapping, his relationship to Dirichlet is remembered primarily because of his approach to solving the Dirichlet problem by use of his alternating principle.

Heinrich Weber succeeded Schwarz. Although in Göttingen for only three years, from 1892 to 1895, his stay there is noteworthy because during this time he published a paper on Galois theory, which is recalled as the first abstract treatment of the group concept.

A native of Heidelberg, Heinrich Weber had received his doctorate there in February 1862 under the supervision of Otto Hesse. He continued his studies in Königsberg where he wrote his *Habilitationsschrift* on singular solutions of first-order partial differential equations. He served at the ETH in Zurich as full professor from 1869 until 1875, when he succeeded Jacobi's student Richelot in Königsberg. He subsequently held professorships in Berlin and Marburg before coming to Göttingen, after which he settled in Straßburg in 1895.

It was in Königsberg, where he had remained for eight years and began to associate with Hilbert and Minkowski as students and friends, that Weber's lecturing on Dirichlet's number theory appears to have been particularly influential.

By 1905, numerous references to Dirichlet's other work had appeared in Weber's publications such as the Encyclopedia of Elementary Mathematics and his edition of the (Dirichlet–Riemann–Hattendorff) Lectures on Partial Differential Equations, as well as several of his earlier papers. Among his many publications, Weber is remembered especially for his joint work with Dedekind on the theory of algebraic functions, his two editions of Riemann's *Mathematische Werke*, his *Treatise on Algebra*, and his establishing the foundations of abstract field theory.

It was the drive and influence of David Hilbert, who followed Heinrich Weber in Dirichlet's chair, that strengthened Dirichlet's legacy for generations. Born in 1862 near Königsberg, he had received his university education there, except for one semester spent in Heidelberg. His doctoral adviser had been Ferdinand Lindemann, best known for proving the transcendence of π. Lindemann had succeeded Heinrich Weber, whose number theory lectures, strongly influenced by Dirichlet, Hilbert attended. It was at this time that he met Minkowski, with whom he developed a close friendship severed only by Minkowski's early death in 1909. While both were students they also became friends of Adolf Hurwitz, who had joined the faculty as extraordinary professor in 1884. He suggested that Hilbert spend some time with Felix Klein, then still in Leipzig, which became a significant step in Hilbert's future career.

After a visit to Paris, where, with prior guidance from Klein, Hilbert met Poincaré, Jordan, Hermite, and other leading mathematicians, he returned to Königsberg to begin his academic career as privatdozent in 1886 and, rising through the ranks, became an ordinary professor in 1893. After two more years, Klein, who had been in Göttingen since 1886, succeeded in there obtaining a chair for Hilbert, which he retained for the rest of his career.

In the meantime, having begun his studies with invariant theory, Hilbert proved his finite basis theorem in 1886, published the *Zahlbericht* in 1897, and followed this with his axiomatic treatment of geometry two years later. In 1900, he presented his famous lecture on unsolved mathematical problems to the Second International Congress, meeting in Paris that year.

If the academic leadership and members of the *Societät* in Göttingen had been disappointed in Dirichlet's not meeting their expectations, those who remained by the beginning of the twentieth century could thank Hilbert for ensuring that their hopes would be satisfied. While there were a number of men who succeeded in carrying out much of the Dirichlet promise, in the twentieth century it was Hilbert who promoted the careers of those who transformed Dirichlet's contributions to modern heights. He himself devoted several publications to an attempt at clarifying the nature of the maligned "Dirichlet Principle," and directed numerous students to expand its methodology. He brought Minkowski to Göttingen in 1902, and he saw to it that Edmund Landau, who would produce the first major twentieth-century accounts of analytic number theory, succeeded Minkowski there in 1909.

Up to 1883, when he assumed emeritus status, it was the stalwart Moritz Abraham Stern who had continued to lecture on number theory in Göttingen. Stern had been a privatdozent from 1829 until 1848, and then served as extraordinary professor for eleven years. His appointment as ordinary professor thirty years after he began his teaching career in Göttingen finally had been made possible by the creation of a second chair in mathematics upon Dirichlet's death in 1859.[14] Stern also took up the slack during the periods immediately following the deaths of Riemann and Clebsch,

[14] Küssner 1982.

when there was no other ordinary professor except for Ulrich who continued to deal with applications, such as the construction of machinery, until his death in 1879.

In 1886, Stern would be succeeded in the second ordinary professorship by Felix Klein, whose ambiguous impact on Dirichlet's legacy is offset by the fact that it was he who was responsible for Hilbert's coming to Göttingen. It should be recalled that Klein, to whose initiative was largely due the considerable expansion and diversification of mathematical activities in Göttingen, came there at the end of his own most productive research career and retired in 1913. His posthumously (1926) published *Lectures on Nineteenth-Century Mathematics* attempt to give a balanced description of some of Dirichlet's contributions but are apparently based largely on Dirichlet's relatively less productive Göttingen period. This may account for curiosities such as Klein's statements that, in his long teaching career, Dirichlet never served on an educational testing commission and that he never taught any courses except to a select circle of auditors.[15] It is true, of course, that Dirichlet did not participate in the courses and seminars designed specifically as part of teacher training programs. As we noted in previous chapters, however, Dirichlet served on teacher training commissions throughout most of the 1830s, was involved in curriculum development at the Kriegsschule as well as in Göttingen, and taught classes that would show substantial increases in the number of enrollees during his later years in Berlin. The fact that these are not comparable to enrollment figures at the end of the century must be considered in relation to the overall enrollment increase at the university over the intervening period.

The two senior physicists in Göttingen felt the loss of Dirichlet more deeply on a personal level than in its impact on their professional activities. Wilhelm Weber's and Listing's designations, respectively, as professor of experimental and of mathematical physics are misleading, but were chosen to justify the establishment of a second professorship in physics after Weber's return in 1849. In fact, when Dirichlet arrived in 1855, Weber had been a member of the Societät's Mathematical Class, whereas Listing was kept busy trying to obtain funds for his portion of the laboratory equipment in physics. Because of his collaboration with Gauss, Weber came to be identified more exclusively with the experimental side of physics than his broader expertise warranted, but it was he who suggested the designations in use after his return.

Wilhelm Weber and Dirichlet had been close for thirty years; Weber had been instrumental in ensuring Dirichlet's move to Göttingen. They represented a strong alliance in carrying the mantle of Gauss in their respective fields. In addition, it was Weber who collaborated with Paul Mendelssohn Bartholdy in taking care of the administrative and financial needs of the household after Dirichlet's death, ensuring the welfare of Ernst and Flora while they were in Göttingen, and advising on the auction of the estate.

Listing, who lacked Weber's administrative clout, had attached himself to the Dirichlets soon after their arrival; his wife, whose expensive tastes would cause difficulties from time to time, could feel comfortable in their expanded social circle.

[15] Klein 1927 (1956:99).

Nowadays, Listing is remembered for his *Vorstudien zur Topologie* more than for his attempts to build up his portion of the physics enterprise in Göttingen. It was Dirichlet with whom he had been able to discuss and receive encouragement for that pioneering topological research, although in his time he was better known as a physicist.

By 1869, the responsibility for astronomy was separated more clearly than before from that for mathematics. Klinkerfues had become director of the observatory in 1857, in which position he remained until his death in 1884. Ernst Schering, who received his habilitation in 1858, had been a privatdozent for two years, served as extraordinary professor for mathematics in the 1860s, but received his appointment as ordinary professor for astronomy in 1869. He died in 1897. Schering has remained best known for his work in connection with the Gauss *Nachlass* and the related publications of Gauss's works.

Berlin

At the time of Dirichlet's death, the situation in Berlin seemed more stable and promising than that in Göttingen. Having Kummer in place in Dirichlet's chair since 1855 meant that he could arrange the appointments as ordinary (full) professors in 1883 of Kronecker, who had been lecturing as reading member of the Akademie since 1861, and of Weierstrass in 1864. Before leaving, Dirichlet had given a positive evaluation of Weierstrass, who began serving as extraordinary professor the year after Dirichlet's departure.

These three men died within six years of one another. Kronecker, who had been in ill health for a decade, died in 1891; Kummer, who had retired in 1883, last lectured in the summer of 1884, but died in 1893; and Weierstrass followed them in death five years after retiring in 1892.

Lazarus Fuchs succeeded Kummer in 1884 and held the chair until his death in 1902. Fuchs had heard Dirichlet lecture at the university in Berlin, which he had entered in 1854. He had obtained his doctorate in Berlin under Kummer in 1858 and began his academic career as privatdozent in Berlin in 1865, meanwhile having taught in a number of secondary schools. After promotion to the rank of extraordinary professor at the university, while still teaching at some of the secondary schools, he obtained an appointment as ordinary professor in Greifswald in 1869. He left after five years, remaining for a year in Göttingen before settling for a particularly successful stay in Heidelberg until returning to Berlin, where he held Dirichlet's old chair as successor to Kummer. His extensive number of publications would be collected and published in three volumes under the editorship of his son and son-in-law, the mathematicians Richard Fuchs and Ludwig Schlesinger.

As noted previously, it became Fuchs's responsibility to edit the second volume of Dirichlet's collected works, in which he was aided by a number of colleagues. The resulting volume is a mixture of Dirichlet's last memoirs, selected correspondence, and miscellaneous items from the *Nachlass*. It was completed in the summer of 1897.

The algebraist Georg Frobenius, doctoral student of Weierstrass, led the department with a firm hand into the twentieth century, and it would be some of his students, notably Edmund Landau and Robert Remak, who, despite Frobenius's misgivings, pursued the mathematical paths laid out by Dirichlet and Minkowski.

Impact of Mathematical Descendants Prior to 1900

Outside of Göttingen and Berlin, we single out six institutions where mathematical descendants of Dirichlet and Kummer played a significant role during the remainder of the nineteenth century.

Bonn

In Bonn, where the ordinary professorship in mathematics had only recently been separated from physics, it was Rudolf Lipschitz who had begun there as privatdozent in 1857 and, after an interlude in Breslau, would serve as the leading professor for mathematics from 1864 until 1903, the year of his death. Aside from his courses, he expanded mathematical activities in Bonn by financially supporting the foundation of a mathematical seminar and promoting growth of the mathematical library. Lipschitz was instrumental in having Minkowski join the faculty; Minkowski had begun as privatdozent in 1887, and then served Bonn as extraordinary professor from 1892 to 1894.

Breslau

In Breslau, Ferdinand Joachimsthal had succeeded Kummer and held the ordinary professorship at the time of Dirichlet's death. He had been a pupil of Kummer's in the Liegnitz Gymnasium and in 1836 had entered the university in Berlin where he attended lectures by Dirichlet. In 1840, he received his doctorate in Halle, and, after a short time as teacher in several Berlin Gymnasien, he became a privatdozent in Berlin in 1845, where Dirichlet held the ordinary professorship; Joachimsthal assisted him with handling the Jacobi Nachlass. While in Berlin, Joachimsthal had as fellow privatdozenten Eisenstein and Borchardt, among others. After a short stay in Halle as ordinary professor, he had been appointed to the professorship in Breslau as Kummer's successor, on Kummer's recommendation.

Upon his death in 1861, Joachimsthal was succeeded by Heinrich Schröter who had taught at the university since 1855 and would remain until his death in 1892. He, too, had attended lectures by Dirichlet in Berlin after 1850, although in 1854 he received his doctorate in Königsberg under Friedrich Richelot on a geometric topic showing the strong influence of Jacob Steiner. While he was the senior professor in the department, he was joined for short periods by two other followers of Dirichlet: Rudolf Lipschitz and Paul Bachmann. In 1892, Schröter would be succeeded by his own doctoral student Rudolf Sturm, who, in 1911, was to write one of the earliest historical accounts of the mathematics department in Breslau.

Halle

As mentioned, Halle could claim two full professors in mathematics. In addition to Otto Rosenberger initially and later to Georg Cantor, Eduard Heine served as ordinary professor from 1856 until his death in 1881.

Heine was succeeded by Albert Wangerin, his student, who continued to keep Dirichlet's methodology alive by encouraging students to read primary sources. He is still known for editing several volumes of *Ostwald's Klassiker*, including the one containing an excerpt of **1839c**. Wangerin served as ordinary professor in Halle from 1882 until his death in 1933.

Munich

During the second half of the nineteenth century, Munich had two full professors who had attended Dirichlet's lectures in the 1840s. The first was Philipp Seidel; the second was Conrad Gustav Bauer.

Seidel had attended Dirichlet's lectures in Berlin between 1840 and 1842, after which he went to Königsberg. Because of Dirichlet's and Jacobi's absence during the following years, he followed Bessel's advice and went to Munich, where he obtained his doctorate and his habilitation. Always interested in both astronomy and mathematics, he obtained his doctorate in astronomy, while his *Habilitationsschrift* dealt with convergence and divergence of continued fractions. He is remembered primarily for his 1847 memoir on series representing discontinuous functions. After a year as privatdozent, Seidel served as extraordinary professor in Munich until 1855, after which he held the rank of ordinary professor until his retirement in 1891.

As a student, Bauer had attended lectures by Dirichlet in Berlin as well as by Liouville in Paris. He had received his doctorate in Erlangen, where von Staudt had been the ordinary professor, and, not surprisingly, was geometrically oriented. His publications include several analytic studies as well, however, some dealing with spherical and gamma functions. He began his career in Munich as privatdozent in 1857 and was appointed as extraordinary professor in 1865 and ordinary professor four years later. He supervised a substantial number of students, acting as doctoral adviser or coexaminer to as many as thirty-four.

Straßburg

In Straßburg, reorganized as a German university after the Franco-Prussian War, Dirichlet's student E. B. Christoffel held one of its two ordinary professorships from 1872 to 1892, and in 1895 was succeeded by Heinrich Weber who remained active until 1912. As mentioned before, Christoffel is credited with having built up the Mathematical Institute in Straßburg. Weber made it possible for later students schooled in the language of twentieth-century generalized abstract concepts to absorb and expand much of the earlier work in number theory, including that of Dirichlet. His own lectures on number theory held in Königsberg had closely followed Dirichlet's.

Zurich - ETH

In Switzerland, the ETH in Zurich claimed several of Dirichlet's and Kummer's students and associates. Dedekind was active from 1858 to 1862, after which he returned to his native Braunschweig and the newly expanded polytechnic school there. Dedekind was succeeded in the ETH by Christoffel, who, in turn, was followed by H. A. Schwarz. Schwarz was active at the ETH from 1869 to 1875, before returning to Berlin. In addition, Rudolf Clausius, who had studied in Berlin and attended

Dirichlet's lectures between 1840 and 1844, stayed longer in Zurich than these men. He taught at the ETH between 1855 and 1867 and introduced the course on potential theory there during this period.

Ferdinand Rudio, a native of Wiesbaden, in 1880 also had received his doctorate in Berlin, under Kummer. He obtained his habilitation at the ETH the following year and from that time rose through the ranks until in 1889 he assumed a full professorship, which he retained until his retirement in 1928. Among his publications are several memoirs on Eisenstein. Notably, Rudio was a leading organizer of the First International Congress that met in Zurich in 1897.

A native of Switzerland who had had occasion to hear Dirichlet lecture in his younger years was the astronomer Rudolf Wolf. After holding appointments in physics, astronomy, and mathematics in Bern in the 1840s, he came to Zurich in 1855, where he became professor of astronomy at both the University of Zurich and the ETH, was director of Zurich's Observatory, and served as librarian at the ETH. He shared Dirichlet's interest in history, in his case directed primarily to the history of astronomy and mathematics, the disciplines to which he contributed with his own research.

After his retirement from Göttingen in 1884, Moritz Stern, too, moved to Switzerland, where his brother held a professorship of history in Bern. Moritz Stern spent considerable time in Zurich, communicating with his friend Rudio and other mathematicians at the ETH. He died in Zurich in 1894.

International Recognition

By the end of the century, Dirichlet's influence was recognized internationally. We note that when an International Congress of Mathematicians met in conjunction with the Chicago World's Fair in 1893, Dirichlet and a number of his students were represented by a display of portraits and publications. Reporting on the mathematical portion of the German exhibition in Chicago, Walter Dyck singled out a giant bust of Gauss and the images of Jacobi, Dirichlet, and Riemann, as presenting the men whose fundamental works designate the markers (the "Merksteine") for the mathematical work of the century in Germany. When the formally established First International Congress of Mathematicians met in Zurich in 1897, among those present were men who had attended Dirichlet's lectures, as well as their students, and those who had studied their works.

Aside from those who had been active in Germany and Switzerland, by 1900 Dirichlet's legacy was furthered by individuals in France, who had read his publications in earlier issues of Liouville's *Journal*, and, among others, by mathematicians in Scandinavia, Russia, Italy, and the USA. In Dirichlet's lifetime, he had been a foreign member of their respective national (European) academies of science.

Chapter 16
Lectures

A week after Dirichlet's death, the *Vossische Zeitung* in Berlin published a letter Dirichlet had written to his former students in Berlin three years earlier to thank them for the silver flagon that they had sent him for his first Christmas in Göttingen. As the editors of the paper noted, it had not been publicized before. It is perhaps the longest and most revealing letter that Dirichlet ever wrote about the role of the teacher vis-a-vis his students.

> Highly honored Gentlemen!
>
> Permit me, highly honored gentlemen, to express my deepest thanks for the beautiful present and the accompanying so very well-meaning words, whereby you pleased as much as surprised me on Christmas Day. Although I feel strongly how little my scholarly achievements deserve the warm appreciation that you have bestowed on them and that for you, the disciples of scholarship, only the devotion for the teacher could account for it, so this feeling does not diminish the feeling of joy with which the expression of your good will has filled me; this devotion which shows itself in every line of your gracious letter and especially in what you say about my closer personal relationship to you—aside from advancing the scholarly training of the student, for the teacher this is the most beautiful reward for his efforts. I can accept your full appreciation in only one regard, namely as it applies to the conscientious care which I have always devoted to my lectures. Looking back on my activity at the University of Berlin, I can give myself the attestation that during the twenty-seven years of my teaching there I never allowed myself to fail in the highest effort of which I was capable to facilitate [your] entrance to the glorious, unmeasurably growing, world of scholarship to which all my powers are dedicated. Your grateful acknowledgment of my efforts provides me with the highest satisfaction and a no lesser reward awaits me in the near future when I, to whom it was given to guide your first steps, may attribute to myself a part, even though minimal, in the enrichments for which scholarship will thank you.[1]

At the time of his death, none of Dirichlet's lectures had been published. With the exception of portions of Dedekind's notes, none of the notes on which the publications by other editors are based had been reviewed by Dirichlet. In addition, his well-known habit of using no notes to guide himself or his students put a considerable burden

[1]Vossische Zeitung, 12 Mai 1859.

U. C. Merzbach, *Dirichlet*, https://doi.org/10.1007/978-3-030-01073-7_16

on the editors of the courses they had attended. It is important to remember this, since most discussions of Dirichlet's work are based on references to the published lectures, some of which follow his actual presentation more closely than others.

After a cursory overview of the totality of Dirichlet's lectures, we review the published lectures in this chapter.

16.1 Summary of Lectures

The courses Dirichlet taught in the early 1830s reflect the needs of students in astronomy and physics, probably steered to Dirichlet by Encke and Gustav Magnus. In the first place, these included a course each in the theory of series and the theory of partial differential equations. Additionally, there were several courses in probability theory, an introduction to higher analysis à la Euler, and various introductory courses in theory of equations, trigonometry, as well as differential and integral calculus.

Although perhaps best known because of the four editions by Dedekind of the lectures given in Göttingen, Dirichlet's lectures on number theory had a slow start. Announced for the summer 1829, winter 1829/30, and winter 1830/31, these had to be canceled because only two to three students wished to enroll in those years. Finally, they were first given in the summers of 1833 and 1834, followed by regular lectures beginning in the winter term 1837/38. The largest number of students attending (46) was reached in 1853, the next to last summer Dirichlet was in Berlin. Once in Göttingen, he lectured on the theory of numbers in each of the three winter terms that he taught there.

By 1835, definite integrals came to be a regular offer every two to three years. In 1842, the course on definite integrals was expanded to courses on single and multiple integrals, which introduced a discontinuity factor into a variety of applications.

By the late thirties, Dirichlet taught other more specialized courses as well. This reflected the better preparation of the students, which also allowed the inclusion of recent discoveries by himself and others. Courses on the shape and movement of celestial bodies in the winter 1836/37, and on the method of least squares in the summer of 1838, most likely were requested by Encke. Beginning in 1839, there were regular courses dealing with attraction. These were never announced as lectures on potential theory, although the term came to be well-known after Gauss's 1840 memoir on the subject. Instead they would continue to carry the Newtonian label, also used by Lagrange and Gauss, as "Theory of forces which act in inverse proportion to the square of the distance."[2]

[2] A detailed list of the courses announced by Dirichlet in Berlin and Göttingen, with titles and dates, appears in Biermann 1959a:33–39 and 73–74. For the courses taught in Berlin, this list also includes the number of students enrolled in each.

16.2 The Editors

Of the ninety-eight lecture courses announced and given by Dirichlet (eighty-eight in Berlin and ten in Göttingen), editions of only seven were published. Their editors were G. Arendt, Richard Dedekind, Gustav Ferdinand Meyer, and Franz Grube.

Although each of the four editors stressed that his publications were based on his own notes or recollections of the lectures he had attended and that he had attempted to closely reproduce the presentation, there are marked differences in the approach each one took to accomplish this.

Arendt

Arendt was a student of mathematics in Berlin during the early 1850s.[3] In the evaluation of a prize submission, Dirichlet commented that Arendt's work consisted of two parts, both based on Dirichlet's lecture of the winter term 1853/54. The first was the solution of a problem Dirichlet had set as an exercise; the second dealt with items brought up in the lecture. Dirichlet judged that Arendt had shown skill—though somewhat circumstantial—in the first part; the second, according to Dirichlet, was to be regarded as a sign of Arendt's editorial competence. He described Arendt as conscientious, insightful, and worthy of the prize, as he had been on a previous occasion. Dirichlet's evaluation is interesting, as, in Arendt's later editing of Dirichlet's lectures, he appears to have come closest among the four editors in conveying the substance as well as the linguistic expressions of Dirichlet's work. Arendt subsequently became a school teacher in Berlin.

The lectures published by Arendt were those on single and multiple integrals and on applications of the integral calculus, both given in the summer term of 1854 and published in the same volume, Dirichlet–Arendt 1904. A lecture on complex numbers appears in Dirichlet–Arendt 1863.

Arendt's editions are the only publications based on Dirichlet's Berlin lectures. The other published lectures were based on Dirichlet's Göttingen presentations.

Dedekind

Dedekind's editions of the number theory lectures, based largely on the course of the winter term 1856/7 given in Göttingen, are probably the most widely read of Dirichlet's lectures. There were four editions of these: Dirichlet–Dedekind 1863, Dirichlet–Dedekind 1871, Dirichlet–Dedekind 1879, and Dirichlet–Dedekind 1894.

Dirichlet, famous for lecturing only from brief notes and never writing down lectures, had reviewed the notes Dedekind had taken of the number theory courses he had audited in Göttingen; he appeared to be agreeable to Dedekind's using these as a basis for a future publication. As a result, Dedekind published his version of Dirichlet's lectures on number theory beginning in 1863. He added supplements to three further editions. The third and fourth editions would become classics in

[3] Arendt's publications are signed G. Arendt; his first name is given as Gustav by Dirichlet in a document quoted in Biermann 1988:51, and as Georg elsewhere.

the history of modern algebra because of Dedekind's last two supplements (X, first appearing in the third edition, XI in the fourth), to the number theory lectures.

Dedekind was not concerned with a verbatim reproduction of Dirichlet's presentation. In fact, he described his procedure: Already being a privatdozent when he attended Dirichlet's lectures, he felt that he was sufficiently acquainted with the material not to need busying himself with transcribing the lectures while he was listening. Instead he wished to let Dirichlet's penetrating (*eindringliche*) delivery impress itself on him. He stated that he had taken no notes during the lectures either, but simply jotted them down from memory afterward. Dirichlet was interested in seeing these, as he was contemplating a possible publication of his own. They discussed the notes, amplified the outline, and Dirichlet commented that a proper textbook would need the addition of some sections that he had had to omit for lack of time that winter.[4]

Dedekind acknowledged that he had to revise his rough notes thoroughly when undertaking the 1863 publication. He inserted minor additions as well as some entire sections to which he called attention in the editions where they appeared.

In the introduction of 1863 to his first edition of Dirichlet's lectures on number theory, Dedekind had mentioned that he wished to follow this work with a less extensive volume on the lectures concerning the forces acting in inverse proportion to the squares of the distances. When this had not materialized by 1876, Franz Grube inquired whether he still intended to pursue this plan. Dedekind responded that he had no such plans but that Grube should turn to Dirichlet's family for the necessary permissions. This in turn led to correspondence between Walter Dirichlet and Dedekind which included an interesting letter of February 1876 from Dedekind about his relationship to the Dirichlet Nachlass and his editing of Dirichlet's lectures on number theory.[5]

G. F. Meyer

Gustav Ferdinand Meyer had received his doctorate with a dissertation on Bernoulli numbers in 1859. It built largely on work by Stern but also reflected careful study of various works by Grunert and reading of other authors.

Meyer's publication Dirichlet–Meyer 1871 is based primarily on his attendance at Dirichlet's lectures on the definite integral held in the summer of 1858—the last course Dirichlet taught. Meyer's work is more of a presentation of the state of knowledge concerning the subject at the time of his publication's appearance than it is a reproduction of Dirichlet's lecture, however. He freely added details from Dirichlet's memoirs, as well as from commentaries, interpretations, and subsequent additions by other authors and himself. He also included material from notes dealing with Dirichlet's lectures on partial differential equations and potential theory. At the same time, he was the most meticulous of the four editors in coding these various additions to his original notes, so that someone taking the time to disentangle them can

[4] See the prefaces to Dirichlet–Dedekind 1863 and Dirichlet–Dedekind 1871. Dedekind explained his procedure not only in the prefaces to all four editions but also in the *Göttingische Gelehrten Anzeigen* of January 1864 and September 1871.

[5] Scharlau, ed. 1981, esp. pp. 49–53.

get a presumably accurate reproduction of the portions that actually emanated from Dirichlet's lectures. Meyer had been a privatdozent in Göttingen before teaching in Bavaria.

Franz Grube

Franz Grube edited the often cited version of Dirichlet's lectures on potential theory, based on notes taken while attending Dirichlet's course in 1857. It was first published two decades later, in 1876; a second edition followed in 1887. Grube, who stated that he had attended all of Dirichlet's lectures (presumably those in Göttingen between 1855 and 1858), subsequently was a teacher in Silesia. His 1887 edition of the potential theory lectures is a carefully constructed textbook, but the time span between the dates of his publication and the year in which he attended Dirichlet's lectures may be responsible for his introducing material that he studied and that has general historical interest but tends to make it difficult for the reader to discern the nature of Dirichlet's original presentation.

16.3 The Topics

The Introductory Courses

Although no notes have surfaced for the courses of the 1830s, their content can be surmised from their titles and dates taught. An "Introduction to the analysis of the infinite according to Euler" most likely was based on Euler's extensive and influential work by that name (Euler 1748). Dirichlet taught it three times between 1830 and 1836; we recall that it was in 1837 that he produced his L-series and functions, to which, as he frequently noted, he was led by the analogy with the material in the fifteenth chapter of Euler's *Introductio* of 1748.

Dirichlet's courses in that period designated as dealing with series most likely introduced students to Fourier series in addition to the better-established series which had been offered previously. Other members of the department continued to provide traditional introductions to series, probably following Euler and Lacroix (whose work initiated them to Lagrange's contributions), and including Taylor series.

Partial Differential Equations

The most notable course Dirichlet introduced in the early 1830s was the one on partial differential equations. It is stated to have been the first such course announced in German universities in 1830 and was well attended, drawing students from the physical sciences as well as mathematics. He taught it regularly in Berlin and Göttingen until the summer term of 1857.

Ironically, this, the most frequently given and highly attended course of Dirichlet's, is one for which no published edition exists. Riemann, who had attended Dirichlet's course in Berlin in 1847 and may have sat in on Dirichlet's lectures at least one of the three times that Dirichlet offered it in Göttingen, succeeded him in teaching the course there. This began a long tradition of subsequent publications, starting with three editions (1869, 1876, and 1882) prepared by Karl Hattendorff in the nineteenth century, often referred to as Riemann-Hattendorff.

Hattendorff had studied in Göttingen where he earned his doctorate in 1862 and served as privatdozent beginning in 1864. He remained in Göttingen until, in 1870, assuming a professorship at the new Polytechnic School in Aachen. In the preface to his third edition of 1882, Hattendorff noted that he had attended Riemann's course in the winter of 1860/61 and had used his own notes as well as an earlier manuscript of Riemann's. He made a point of commenting not only on Riemann's treating a ring-shaped solid in a moving fluid analogously to Dirichlet's handling of a sphere, but additionally credited Dirichlet as follows:

> Among the mathematicians who have substantially furthered the theory of partial differential equations, Dirichlet assumes an outstanding position. But he not only worked on the improvement of the theory; along with the doctrine of the potential, he was the first to make it the subject of special lectures in German universities. The great merit that he thereby earned for the study of mathematics will certainly be put in the proper light by the circumstance that these lectures did not come to an end with his death but nowadays constitute a regular part of the program in most German universities. Thus, Riemann took over Dirichlet's lectures on the potential and on partial differential equations. Given the accord of the subjects treated, it is natural that there should be much coincidence both with regard to arrangement and execution. But Riemann was not content simply to assume the legacy of his great predecessor. He added an abundance of that which was characteristically his own.[6]

These introductory remarks by Hattendorff were omitted from the subsequent publications of the work, so that most twentieth-century readers were unaware of the debt owed to Dirichlet by Riemann and the later editors for the formulation of these lectures.

Hattendorff's editions of the partial differential course were succeeded by three more editions prepared by Heinrich Weber, starting in 1900 with the fourth edition and ending in 1919 with the sixth. Weber's were the most mathematically oriented, and most clearly showed traces of Dirichlet's influence. They were succeeded by the opposite orientation, reflected in a change of title emphasizing mechanics and physics; this was provided by the two physicists Philipp Frank and Richard Von Mises, starting with a seventh edition in 1927 and ending with the eighth of 1943. Except for the 1943 edition, published in New York, this entire sequence, from Hattendorff to Frank-Von Mises, was offered by the publisher Vieweg in Braunschweig. Beginning in 1924 it was supplanted, and eventually superseded, by the wide-spread Courant-Hilbert volumes on methods of mathematical physics.[7]

Analytical Mechanics

Without having seen any records concerning the content of the course in analytical mechanics, it seems reasonable to assume that it was based on Lagrange's substantive book by the same name. It was taught to thirteen students in the summer of 1834, a time during which Dirichlet first appears to have intensified his direct study of

---

[6]Riemann–Hattendorff 1882:v–vii.

[7]A survey of these publications, emphasizing the Frank-Von Mises editions, is found in Siegmund-Schultze 2007. Also see the bibliography at the end of this volume.

Lagrange's Mechanics. It may have included his critique mentioned in Chap. 13, which would be inserted by J. Bertrand in his carefully produced third and fourth editions of Lagrange's *Mécanique Analytique*. Presumably the course covered many of the topics of interest to physicists to which he himself contributed in the 1840s, and he would have used it at this earlier date to introduce future scientists to the mathematical aspects of these areas while introducing them to properties of functions and applications of integration theory. It is worth recalling in this connection that until the 1830s, courses on mechanics such as that offered in the Kriegsschule did not make use of the differential or integral calculus.

Integration Theory

There are two publications dealing with Dirichlet's lectures on definite integrals. Although only published in 1904, the one by G. Arendt, based on Dirichlet's course of the summer 1854, gives us a presumably reliable idea of the sequence of topics treated by Dirichlet. Arendt also explained carefully in his preface any variations made to Dirichlet's lecture, his orthographic usage, and other editorial details. As mentioned previously, the earlier publication by G. F. Meyer was more elaborate.

According to Arendt, Dirichlet's lectures consisted of two portions, the first one dealing with the theory of definite integrals, the second with applications. We know from the university's register that these were taught separately, the first one listed as private, the second as public. The primary portion was divided into discussion of single and multiple integrals. The study of single intervals was subdivided into five sections. The first dealt with the basic concept and properties of definite integrals. This was followed by a section dealing with an evaluation of an integral when the limits of integration are extended to infinity. The third dealt with Eulerian integrals exclusively. It was followed by one introducing complex imaginary parameters and integrals derived from the gamma function. The final, fifth section singled out several specific integrals, showing how to integrate under the integral sign, how to transform a complicated integral by a suitable substitution, how to use series expansions for integrations, and how to integrate between imaginary limits.

The second, public, part of the course was devoted to multiple integrals. It began with a section on double integrals and their application to determining the content of a curved surface. This was followed by a section on triple integrals and their use in determining the attraction of ellipsoids. Finally, the course concluded with a section showing how a discontinuity factor can be used in dealing with various problems, some discussed previously, some stressing application to potential theory. As time permitted, additional uses of the discontinuity factor, such as in dealing with the gamma function, were discussed.

It is interesting that, in contrast to his memoirs, Dirichlet in his lectures introduced historical remarks only in connection with the various forms of the problem of attraction, noting Lagrange's significant contribution, Legendre's occasionally missing the gist of his extensive integrations, and the weakness in Ivory's Theorem, which had been a topic of conversation years earlier when he first met Gauss face-to-face.

Whereas Arendt provided no cross-references to Dirichlet's own memoirs, G. F. Meyer filled his edition of the lectures published in 1871 with these.

Number Theory

The number theory lectures were divided into five major sections and several supplements. They represent a suitable combination of Legendre's *Théorie des nombres* and Gauss's *D.A.*

Dirichlet's lectures barely touched analytic number theory. The first section of the publications deals with divisibility. The second section is titled "Of the congruence of numbers." The third section is devoted to quadratic residues. The fourth section advances to quadratic forms. The fifth section covers class number determinations for binary quadratic forms with given determinant. Aside from omitting the geometric examples Dirichlet gave frequently, and giving references to only his major memoirs, Dedekind followed Dirichlet's results and arguments closely and smoothly.

By the fourth edition of 1894, Dedekind had added eleven supplements. The first four are based on Dirichlet but deal with topics he did not always cover in the lectures. They are devoted to some theorems from Gauss's cyclotomy, to the limit of an infinite series, to a geometric theorem, and to the number into which the classes of quadratic forms of a given determinant are decomposed. The remaining supplements are largely Dedekind's but still deal with Dirichlet's topics. Even in the famous eleventh supplement of 1894 that would become a cornerstone of twentieth-century algebra as found in the writings of the 1920s by E. Noether, R. Brauer, and their associates, one recognizes the foundation laid by Dirichlet. To cite just one example, Dedekind, while using the new terminology we associate with his ideal theory, in the portion of Eleventh Supplement on units in finite fields, quoted the unit theorem, "Dirichlet's great theorem," of **1846b**.[8] Dedekind remarked that "this theorem, aside from the theory of ideals, forms the most important foundation for the deeper study of the integers in the field Ω and is indispensable for the true determination of the number of ideal classes according to Dirichlet's principles."[9]

Potential Theory

Between 1839 and 1858, Dirichlet gave eight courses on the attraction of forces, or potential theory. As previously noted, these were announced as dealing with forces acting in inverse proportion to the square of the distance, the traditional Newtonian label. The first of the courses dealing with attraction was restricted to the attraction of ellipsoids, the topic Dirichlet once described as most celebrated, and that he had used as an example of his new techniques on a number of occasions in the late 1830s. Beginning with the winter term 1846/47, however, the expanded course was offered every other year while Dirichlet was in Berlin. It was presented twice more in Göttingen: in the summer of 1856 "with applications to electricity and magnetism" and in the winter term 1857/58 "with applications to physical problems."

---

[8] See Sect. 13.2.

[9] Dirichlet–Dedekind 1894:602–3.

Dirichlet's lectures on potential theory soon proliferated. Growing interest in the topic was shown even during his lifetime. Among those who had attended the university at Berlin in the early 1840s and initiated courses on potential theory in other universities is Rudolf Clausius, active at the ETH in Zurich from its founding in 1855 until 1867. Aside from lecturing on the subject, as did several of his contemporaries such as Carl Neumann in Tübingen and in Leipzig, Clausius published a work titled *Die Potentialfunction und das Potential* in 1859, the year of Dirichlet's death. A second edition, designed to make it more appealing to physicists than the more heavily mathematically oriented first edition, followed in 1870.[10]

Although we have no enrollment figures for Göttingen, the number of students in Berlin taking Dirichlet's courses on attraction varied from thirteen in 1839 to three dozen, the peak of thirty-six having been reached in 1852/53.

Grube's two editions dealing with Dirichlet's potential theory, published in 1876 and 1887, were based on a lecture course Dirichlet gave in 1857.[11] Grube's second edition, [Dirichlet–Grube 1887], especially widely circulated, is divided into seven sections. The first four deal with general properties of potentials; the last three are dedicated to applications to electricity and magnetism, which include in section 6 the editor's effort to come to terms with the Dirichlet Principle. The work clearly develops elements of potential theory and certain applications, which may explain its adoption as a textbook thirty years after Dirichlet's course. In the 1887 edition, Grube, however, had used not only Max Bacharach's concise, short history of potential theory, [Bacharach 1883], but in the latter part of the work also threw in extensive hypotheses by Poisson and others, particularly concerning the nature of electricity and magnetism. These have little relationship to content, style, or methodology of Dirichlet.

Grube had read widely in the topic of potential theory; for example, he published a paper on d'Alembert's work in this connection. However, his references suggest a lack of familiarity with certain nineteenth-century developments in the subjects to which the chapters on applications are devoted, or with Dirichlet's response to these as suggested in his publications. Grube, who seems to have studied Poisson extensively while preparing his "Dirichlet" lectures, may have been less familiar with other pertinent nineteenth-century experiments and discoveries.

Of more serious significance because of its impact on later studies pertaining to Dirichlet is Grube's discussion in his sixth section concerning the Dirichlet Principle. Grube emphasized the role of a minimum using it in a general form of the principle without taking into account the conditions that must be imposed to assure the existence of such a minimum. In fact, he states in the fourth paragraph of section six that "it is evident that the integral has a minimum since it cannot become negative." Although, in a note appended to his second edition, Grube mentioned raised

[10]For further interesting details see Jungnickel and McCormmach 1986, esp. pp. 197–98.

[11]Grube stated that his editions were based on notes he took while attending Dirichlet's course in the winter 1856/57; however, the list of courses reproduced in Biermann 1959a and based on the university's official records, as well as the introductory remarks in Dirichlet–Meyer 1871, indicates that, aside from a course in the summer of 1857, the relevant course was given in the winter of 1857/58.

objections and refers to [Bacharach 1883] where these were addressed, he left the emphasis on the minimum while disregarding the difference between the existence of a minimum and a lower bound. Later researchers seem to have been unaware that Grube's edition was based on his general reading in the literature more than a verified set of his two-decade-old notes from Dirichlet's lectures.

There is no known publication by Dirichlet referring to the so-called Dirichlet Principle as it is defined by various authors after Riemann. Elstrodt calls attention to a sheet in the Riemann *Nachlass* according to which Riemann explained, when asked how he arrived at the term "Dirichlet Principle," that Dirichlet mentioned to him in conversation it was something he had used in his lectures.[12]

As previously noted, though mentioning them, Dirichlet frequently paid less attention in his lectures to the limiting conditions imposed for solution of a problem than to the technique of the solution—especially if the problem was one defined and limited by real-world conditions. The, usually public, lecture was geared to a topic in mathematical physics rather than to one in mathematics per se. Yet we have learned from his publications that limiting conditions were a prominent factor in Dirichlet's methodology and thought processes. For that reason, it seems unlikely that Dirichlet would have "forgotten" his earlier reminder to Steiner to distinguish between existence of a minimum value and a lower bound. He simply did not formulate the "Dirichlet Principle" in the manner stated by Grube and others after Riemann's original phrasing, in which Riemann had carefully distinguished between Dirichlet's concern with solutions of Laplace's partial differential equation in three dimensions, and Riemann's own modified two-dimensional applications, in which he adapted a technique from his doctoral dissertation of 1851.[13] Indeed, most of the subsequent criticisms of the Dirichlet Principle were directed at Riemann's productive applications.

It is highly plausible that in the conversation between Dirichlet and Riemann, Dirichlet was referring to his "New Method" of solving multiple integrals with the introduction of a discontinuity factor, a technique he frequently mentioned, and, in more than one published instance, described as using a "principle" that had helped him find otherwise inaccessible solutions to problems.[14]

Probability

Dirichlet's several lectures on probability were not published by those who had attended them. There is, however, a relevant memoir by Hans Fischer that includes the examination of students' notes. Since each of the editors of Dirichlet's lectures only had attended these in the 1850s, it is of special interest to compare their content to that of Dirichlet's earlier courses on the same subjects. Such a comparison must utilize the very disparate students' notes left to us in a number of archives for some of

---

[12] Elstrodt 2007:27 and Elstrodt-Ulrich 1999.

[13] See Chapter 14 for his statement.

[14] See, for example, **1841a**, reproduced in *Werke* 1:393, and a quote by Arendt, according to which Dirichlet concluded his 1854 lecture on integrals by stating that "es liesse sich noch manches über dieses Attractionsproblem und über verwandte nach demselben Prinzip des diskontinuirlichen Faktors zu behandelten Aufgaben beibringen, was uns aber die bis ans Ziel vorgerückte Zeit nicht mehr verstattet." (Dirichlet–Arendt 1904:394).

Dirichlet's courses. [Fischer 1994], dealing with probability, includes the only such examination seen so far.

In addition to discussing Dirichlet's publications relevant to mathematical probability theory, namely **1836**, **1838**, **1839a–c**, and mentioning various notes and letters reproduced in the second volume of Dirichlet's *Werke*, Fischer utilized unpublished lecture notes found in Seidel's *Nachlass*, based on Dirichlet's courses of the 1840s, and other lecture notes, apparently based on Dirichlet's course of 1838, in the *Nachlass* of Borchardt and that of Clausius. Perhaps this interesting article ([Fischer 1994]), which also analyzes Dirichlet's own contributions to the topic against the background of those made by several of his predecessors and contemporaries, will serve as a useful model for similar, future efforts.

Chapter 17
Centennial Legacy and Commentary

This concluding chapter deals with the centennial of Dirichlet's birth in 1905. Consideration is given to some of the mathematicians active at the time who contributed to the two major commemorations that year or were influenced by Dirichlet and perpetuated a part of his legacy. Among these are major figures such as Hilbert and Minkowski who had read and discussed Dirichlet's work attentively, both before and after the centennial. Without straying into details of Dirichlet's post-1905 impact, though noting several significant events of the year 1909, the chapter concludes with some general comments on his methodology and influence.

After 1900, Dirichlet's name appeared in two different contexts. On the one hand, there were those mathematicians whose talks or publications offering new results and mentioning his name were based on having studied his relevant work; on the other, there were those who took concepts attributed to him and recast them into contemporary terms either through generalization, abstraction, or algorithmic extensions. Quite often the latter group, or those merely using his name, chose methodology quite contrary to that he favored.

Dirichlet's name during the latter part of the nineteenth century had frequently appeared in connection with his L-series. Following Dedekind, they would be called "Dirichlet series," but subsequently it became more common to revert to his own use of "L-series" and their variations. After the turn of the century, his name would resurface most often in conjunction with the previously discredited "Dirichlet Principle," despite the tenuous connection Riemann's term had with statements subsequently attributed to Dirichlet. In contrast to the various nineteenth-century critiques, after 1900 studies of the "Dirichlet Principle" would lead not only to its expanded applicability but also to new techniques developed in the course of these efforts.

These trends are both reflected and also foreshadowed in a 1905 centennial volume published in Berlin. It was preceded by a memorial tribute, delivered in Göttingen on the anniversary of Dirichlet's birth, which focused on the significance for the nineteenth century of his life and work.

U. C. Merzbach, *Dirichlet*, https://doi.org/10.1007/978-3-030-01073-7_17

17.1 The Centennial. I: Minkowski's Address

On February 13, 1905, members of the Göttingen Mathematical Society were reminded that this date marked the centennial of Dirichlet's birth. Hermann Minkowski read a tribute to him in honor of the occasion.

What is striking about the memorial tribute in Göttingen is the degree to which Minkowski, who, though born five years after Dirichlet's death, captured the essence of Dirichlet's methodology and outlook more closely than many of his contemporary eulogists and acquaintances had done. Only Kummer, nearly half a century before, had brought together some of the details and issues that Minkowski addressed. They were made particularly vivid by Minkowski's own opinions and sense of style, conveyed more obviously in this tribute than in his other talks or publications.

Minkowski opened his address as follows:

> Thoughts of time and space are the mathematician's constant companions, which he banishes, however, when he enters the realm of pure number. Today we shall approach grandeur deep within the number realm; and the solemn mood of the moment, the influence of the place, are just what guide our steps to it.[1]

Minkowski introduced the subject of his talk by recalling the circumstances surrounding Dirichlet's call to Göttingen. Aside from outlining some of the external details involving Wilhelm Weber and the relevant functionaries, he reminded his audience that von Warnstedt, the "curator" of the university at the time, had stressed the point of view that the university is not only an advanced school for teaching students, or a keeper of previously acquired knowledge, but shall contribute to the progress of science (*Wissenschaft*) as a common property of mankind. He observed that with Dirichlet's appointment, Göttingen's university, which for half a century had enjoyed the fame of possessing the foremost of all living mathematicians, had succeeded in retaining that fame.

Minkowski continued by providing highlights of Dirichlet's life, basing himself largely on the eulogy Kummer had offered to the Berlin Akademie in 1860.

Turning to mathematics, Minkowski noted that since Dirichlet's death nearly half a century of highly intensive development in mathematics had passed. He saw the impulses of Dirichlet's spirit in all its parts and considered Dirichlet's life's work as an uninterrupted testimonial for the brotherhood of mathematical disciplines, and for the unity of the science.

He stated, however, that to see Dirichlet in the proper light one must appreciate the peculiar nature of number theory. He acknowledged that the arithmetic of the day was not exactly the arithmetic of the *D.A.* The inventive, creative imagination of Gauss and Dirichlet, who sought a promised land, had been largely replaced by systematic building on a securely acquired foundation. With an unmistakeable nudge against the prevailing trend for "arithmetization" of all of mathematics, Minkowski noted that there are those who only consider arithmetic a useful constitutional base for the extended realm of mathematics. A few might even see in it no longer anything

---

[1]Minkowski 1905a; see *Ges. Abh.* 2 (1911): 447.

but the police who are authorized to pay attention to all forbidden activities in the widely extended community of quantities and functions. Nevertheless, no one would deny arithmetic's unique attributes: "the simplicity of its foundations, the precision of its concepts, the purity of its truths."

Minkowski suggested taking a moment to awaken the old arithmetic from its sleeping beauty status. First, he observed that those who had seriously cared for arithmetic devoted themselves to it with a certain passion. Modifying an expression of Novalis, he stated that the true arithmetician is an enthusiast: "Without enthusiasm no arithmetic."

He acknowledged that among mathematicians there are those totally lacking a sense for the attraction of the subject of arithmetic. He speculated that this could be either because of the high capacity for abstraction required or because of the impatience with a subject that appears to have no immediate practical usefulness. He expressed the hope that some day pure arithmetic may be praised for relationships to physics and chemistry. Until then, the resigned statements of Gauss and Dirichlet concerning the lack of appreciation for number theory would stand.

Minkowski suggested several points that might bring about a change in attitude. One is an opportunity for experimentation. Here, he compared the work of the number theorist to that of the artist, likening numbers to colors. Another characteristic of experimental science shared by number theory is that obvious connections are frequently recognized before the inner reasons are apparent. As an example of such a connection leading to new progress, he chose Dirichlet's method of computing the class number of quadratic forms with complex coefficients for having opened the gate to the wider realm of number theory; those mathematicians who entered that realm thereby found "new blossoming provinces." Another example from the history of number theory that Minkowski used is one illustrating how often significant progress is tied to phenomena giving the appearance of having little significance; for this, he drew on Dirichlet's work with Gauss sums.

After this general discussion of number theory, Minkowski turned to specific results due to Dirichlet. Referring to the introduction of Hilbert's *Zahlbericht*,[2] he stressed that the focal point of number theory rested on three pillars: the statement of the unique decomposition of prime ideals, the statement of the existence of units, and the statement of the transcendental determination of the class number. He pointed out that the first of these was originally designed by Kummer, after which Dirichlet's two later adherents, Dedekind and Kronecker, independently built it up, the first one displaying his achievement right away for everyone to see, while the other kept it covered for a long time. The other two pillars were erected by Dirichlet alone.

In discussing Dirichlet's arriving at his unit theorem, Minkowski emphasized the difference in approach between Dirichlet and his predecessors. Before Dirichlet, units had been treated only for the simplest case of a real field and dependence on a quadratic equation. Since for this case the existence proof for the unit is joined to a specific procedure for finding the unit, by overestimating the importance of this specific procedure or algorithm much time could have been spent—and wasted—

[2]Hilbert 1897; see Hilbert *Ges. Abh.* 1 (1932).

in attempting to generalize it. Here, Dirichlet saved days, or, as Minkowski put it, "with the entire keenness of his unprejudiced comprehension, he entered, unpeeled the kernel of the problem lying deep within, and laid bare the whole noteworthy organism of units in fullest clarity."[3] Minkowski marveled at the simple form into which Dirichlet knew to pour the proof of his mighty theorem that, in general, a number field contains infinitely many units and that they can all be derived by multiplication and raising to powers.

Minkowski noted that it almost looks as though Dirichlet's most important aid was his everyday box principle (the *Schubfachprinzip*). He tied this to the interesting comment that Dirichlet's boxes are regions of quantities and the items kept within are natural logarithms.

Minkowski repeated the circumstance related by Kummer that Dirichlet found his proof of the unit theorem while listening to the Easter music in the Sistine Chapel. He added cautiously that he did not wish to enter a discussion as to what this may have to do with the elective affinity between mathematics and music.

Turning to Dirichlet's other major contribution to number theory, Minkowski stressed that the formula for the class number grew out of the theory of quadratic forms, quite independently from the theory of algebraic number fields. Yet, he felt that, since that earlier period, the branches of arithmetic had become so closely intertwined that he could discuss this topic while remaining within the later framework of ideas. He therefore began with a review of Kummer's ideal numbers, whereby Kummer had led the way "out of the labyrinth that seemed to envelop number theory," after the recognition that unique decomposition of prime numbers no longer held in expanded number realms such as the quadratic field.

Minkowski next noted Dedekind's formulation of ideals, concluding this reference with one of his only slightly subtle plays on words:

> And so it came about, that mathematicians also have ideals that are presented exclusively through reason; these are not the mathematician's only ideals, but they, too, are such that the one who knows them is enraptured by them.[4]

In briefly discussing the equivalence of ideals, their distribution into classes, and the number of classes, Minkowski told us more about his predilection for Dedekind's ideals than he did about Dirichlet's proof. His only remark directly relevant to Dirichlet's proof is that Dirichlet's method in treating quadratic number fields subsequently remained sufficient for settling more general cases.

In 1905, Minkowski was not alone in favoring Dedekind's ideals. Those who had not been exposed to them in Dedekind's Eleventh Supplement to Dirichlet's Lectures on Number Theory could read about them in Hilbert's *Zahlbericht* (Hilbert 1897) and the second volume of Heinrich Weber's *Algebra* (Weber, H. 1896, or the second edition of 1899).[5]

---

[3]Minkowski 1905a; see *Ges. Abh.* 2 (1911): 454.

[4]Minkowski 1905a; see *Ges. Abh.* 2 (1911): 455.

[5]It was only in the 1930s that Robert Remak, though having followed Minkowski in most of his work, pointed out drawbacks in joining class numbers too closely to ideal theory, as Minkowski had

Returning to Dirichlet, Minkowski called attention to his "macroscopic" view of the lattice-point covering of a surface as opposed to the "microscopic" treatment of a surface, shared by Gauss.[6] Minkowski argued that Dirichlet was led there by use of his series, in particular the need to prove the nonvanishing of a sum that turned out to be a class number of quadratic forms. Referring to the courageous, direct, recent proof by Franz Mertens of the nonvanishing sums,[7] Minkowski compared it to climbing to a mountain peak through briers on the steepest path, instead of gently ascending through a landscape with magnificent views. In concluding the discussion of this part of Dirichlet's work, Minkowski called attention to Dirichlet's constant efforts to make Gauss more accessible, "by turning his rigid proofs into fluid and transparent methods." Not surprisingly, he chose the geometric theory of ternary quadratic forms as the pertinent example of these efforts that had a special impact on future developments.

Minkowski treated the second major direction of Dirichlet's publications, his contributions to mathematical physics, more briefly. He considered these two directions of Dirichlet's research to be "harmoniously united by the band of the integral calculus" and described Dirichlet's use of integrals as often giving his work a characteristic, recognizable stamp.

In discussing this side of Dirichlet's work, Minkowski first dealt with Dirichlet's discoveries pertaining to Fourier series, which, he noted, may be said to have become his most popular contribution, paving the way to the modern treatment of real functions and to be of immeasurable worth because of their impact on the development of set theory.

Minkowski also claimed that the story of trigonometric series, well-known to mathematicians, had been a continuous battle for removal of prejudices. He reminded the listeners that the expansion into trigonometric series of arbitrary functions was accepted only with the most acrimonious differences of opinions. He recalled Cauchy's only partly successful attempt at a convergence proof, likening the failure of "the great analyst" to the lack of success the Montgolfier brothers had when, after their first successful balloon ascent, they could not separate themselves from the heating material they had used, because they were concerned with generating smoke rather than heating the air. By contrast, Dirichlet had recognized the true conditions "for the sought for ascent to the full height of arbitrary functions."

Minkowski commented on Dirichlet's proof concerning the stability of equilibrium as another example of freeing himself from wrong, established methods, by replacing analytic rules for determining the minima of a function with using only the original concept of the minimum. Noting that often mathematical progress is achieved by recognizing that circumstances accidentally connected really have nothing to do with one another, Minkowski suggested that this may be the reason younger mathematicians often are so successful, being less prejudiced:

---

done in the fifth chapter of his *Diophantische Approximationen*; see Merzbach 1992:504. However, note the reference to Kurt Hensel's work below.

[6]Minkowski 1905a; see Minkowski *Ges. Abh.* 2 (1911): 456.

[7]Mertens 1897.

> Every mathematical soldier carries the marshall's baton in his backpack unless, from pure discipline, he swears by all that has gone before.[8]

After mentioning Dirichlet's easily doing away with the notion that rules which apply to finite series can be carried over for infinite ones—showing that series not absolutely convergent can produce any number of sums when its members are interchanged—he turned to Dirichlet's use of a discontinuity factor for determining multiple integrals. It was a technique of which Dirichlet was especially proud. Minkowski repeated the story that Dirichlet would tell people this was really a very simple idea, then add with a smile, "one just has to have it." Minkowski added a little joke of his own, speculating that, considering the hopes Leibniz had had for the use of the binary system, what success he might have achieved in the establishment of the best of all possible worlds with Dirichlet's 0-1 discontinuity factor!

Minkowski concluded his discussion of Dirichlet's contributions to mathematical physics by stressing how much his lectures had contributed to the growth of the subject in Germany. Noting that Dirichlet had paid particular attention to physics during his short period in Göttingen, Minkowski wondered what major achievements he might have left had it not been for his untimely death. Like Kummer, he repeated several statements handed down without documentation that claimed Dirichlet had a rigorous proof for the stability of the planetary system and had a new approach to the solution of differential equations in mechanics.

Minkowski also quoted from one of Wilhelm Weber's reports during the discussion concerning the succession to Gauss. In it, Weber had stressed the importance of having a mathematician working with an astronomer or physicist: Neither one of these would bring the same new essential elements to a joint work as would a mathematician, who, in turn, would be inspired to study new areas of mathematical problems.

Finally, Minkowski turned to Dirichlet's lecture style. He stressed Dirichlet's having a preference for dealing with topics to which he had contributed substantially. As his students had testified, his exposition was so remarkable because it seemed as though he were just in the process of creating the entire structure, and so they were intrigued by following him in this endeavor. Dirichlet developed his subject fully and naturally, without recourse to a deus ex machina, an artifice calculated to surprise by bringing an apparently intractably knotty problem to an unexpected solution.[9]

Minkowski's address is remembered for his description of number theory as much as for his evaluation of Dirichlet's work. Hilbert, in his 1909 eulogy of Minkowski, referred to those remarks concerning number theory as "the best that has ever been said about this wonderful creation of the human spirit." Supporting his enthusiasm,

---

[8]Minkowski 1905a; see *Ges. Abh.* 2 (1911): 459.

[9]Especially in earlier years, Dirichlet's approach had been described as uncertainty rather than as a characteristic device to lay out before his students his procedure of proof so as to allow them to join in that procedure. Even in 1852, Hirst referred to Dirichlet's "hesitancy"; see Gardner and Wilson 1993. Rudolf Wolf in May of 1838 had entered the following vivid description in his diary concerning a classroom lecture by Dirichlet on the method of least squares: "Sein Vortrag ist ein grelles Bild gelehrter Nachlässigkeit"; see Wolf 1993:69.

he quoted from a letter Dedekind had written to Minkowski concerning the address: "I have read your talk with the utmost gratification five times and more frequently, and am especially affected by the vast historical understanding with which your presentation clearly comprehends our science and pursues it in its development."[10]

17.2 The Centennial. II: The Memorial Volume

In Berlin, the editors of the *Journal für die reine und angewandte Mathematik* dedicated volume 129 for 1905 to Dirichlet's memory. Fifteen European mathematicians contributed to that volume during the course of the year.

As the result of invitations sent out by Kurt Hensel, at that time the editor-in-chief of the *Journal*, there were contributors to the memorial volume from France, Switzerland, Austria, and several locations within Germany. A frontispiece was a reproduction of the last sketch of Dirichlet made by Wilhelm Hensel, Rebecca Dirichlet's brother-in-law and Kurt Hensel's grandfather.

The contributors present a curious mix, and in many cases, the volume tells us more about their personalities and own work than it does about Dirichlet's mathematics. Some made no mention of Dirichlet at all. Others only mentioned him in passing. One is led to wonder whether their contributing to the volume, rather than stemming from any particular interest in Dirichlet, had more to do with their wishing to accede to Kurt Hensel's request so as to maintain a friendly relationship with him and with the *Journal*, or to have their names included. A few wrote papers that referenced specific work by Dirichlet and made noteworthy additions to the subjects under consideration. Several of the authors owed their relationship to Dirichlet largely through the study of Riemann's work.

An abbreviated account of the memoirs follows here in the order of their inclusion in the memorial volume, where they appeared in its regular issues throughout the year of 1905. Also included is a brief introduction to those authors not previously discussed in this work.

Richard Dedekind

The volume opened with a memoir by the seventy-three-year-old Dedekind. He was the only contributor who had known Dirichlet personally. Since 1862, he had been living and working in his native Braunschweig.

Dedekind's paper, on "binary trilinear forms and the composition of binary quadratic forms," was based on his prior study of transformations of binary and trilinear forms that he had published in the Eleventh Supplement to Dirichlet's *Vorlesungen über Zahlentheorie*, (Section 182).

Dedekind provided a generalization to a remark Gauss made at the end of Section 235 of the *D.A.* pertaining to the use of a bilinear substitution to transform a single form into the product of two forms. Even a superficial look at Gauss's Section 235

[10]Hilbert 1909; see Minkowski *Ges. Abh.* 1 (1911):xxviii, or Hilbert 1935, 3:361.

shows us that it is a typical example of Gauss's rather extensive computations obscuring his likely awareness of a significant underlying relationship. Sounding very much like Dirichlet, Dedekind now noted that he had been able to considerably simplify and generalize the solution of the problem Gauss had treated. Since his result had been previously unknown, he thought it appropriate to publish his related investigation and to dedicate it to the memory of his "great teacher Dirichlet who himself had made it a point of honor by a series of memoirs to facilitate the understanding of that work of Gauss he had most admired."[11]

Heinrich Weber

Heinrich Weber's memoir was titled "On complex prime numbers in linear forms." In Straßburg since 1895, Weber wished to extend Dirichlet's results of **1841a**, **1841b**, and **1842b** to general class fields.

Weber followed Hilbert's 1894 memoir on the Dirichlet biquadratic number field and the *Zahlbericht* published in 1897. Especially though, Weber built his contribution to the memorial volume on his own memoirs dealing with number groups in algebraic fields that had appeared in the *Mathematische Annalen* in the 1890s. It was in these preceding memoirs that Heinrich Weber had developed his concept of a class field; many of the theorems and proofs required had been set down in greater detail in his three-volume work on algebra, to which he now referred frequently.

The lengthy study provides an uncommon example of translating Dirichlet's extended results on primes in arithmetic progressions to Weber's class field theory. It entails introducing the reader to the entire set of new concepts and definitions of class fields, many of which Weber had set down in his preceding publications. While this accounts for the length of Weber's memoir, it must be noted that it is a rare example of someone providing so close a translation of Dirichlet's familiar number-theoretic concepts to the abstract generalizations that Weber had developed on the path laid out by Dedekind in the linguistic and conceptual development of his field and ideal theory.

David Hilbert

David Hilbert had been in Göttingen for ten years at the time of the centennial. Born in 1862 near Königsberg, he had received his university education there, except for one semester spent in Heidelberg. His doctoral adviser had been Ferdinand Lindemann, best known for proving that π is transcendental. Lindemann had succeeded Heinrich Weber, whose number theory lectures, strongly influenced by Dirichlet, Hilbert attended. It was at this time that Hilbert met Minkowski, with whom he developed a close friendship severed only by Minkowski's early death in 1909. While both were students, they also became friends of Adolf Hurwitz, who had joined the faculty as extraordinary professor in 1884. He suggested that Hilbert spend some time with Felix Klein, then still in Leipzig. This became a significant step in Hilbert's future career. After a visit to Paris, where, with prior guidance from Klein, he met Poincaré, Jordan, Hermite, and other leading mathematicians, he returned to Königsberg to

[11] Dedekind 1905; see Dedekind's *Werke* 2 (1931/1969):307.

begin his academic career as privatdozent in 1886; he rose through the ranks until becoming an ordinary professor in 1893. After two more years, Klein, who had been in Göttingen since 1886, succeeded in obtaining a chair for Hilbert, which he retained until his retirement in 1930.

In the meantime, having begun his studies with invariant theory, Hilbert proved his finite basis theorem in 1886; turning to number theory, he published the *Zahlbericht* in 1897; and following this, with his increased interest in axiomatics, he produced his treatment of geometry two years later. In 1900, he presented his famous lecture on unsolved mathematical problems to the Second International Congress, meeting in Paris that year.

It was at that time Hilbert decided to "resuscitate" Dirichlet's Principle. Hilbert had been dealing with Dirichlet's work for several years. In 1894, he had published a memoir "On Dirichlet's biquadratic number field"; three years later, this would be incorporated in the *Zahlbericht*, his famous memoir on "The theory of algebraic number fields." This was the major exposition which, after a general discussion of number fields, Hilbert divided into four additional parts, discussing the "Galois number field," the quadratic number field (including Dirichlet's biquadratic field), the cyclotomic field, and Kummer's number field.

Hilbert's contribution to the 1905 memorial volume was a short paper on the Dirichlet Principle. It was the reprint of a lecture published in the *Jahresbericht der Deutschen Mathematiker-Vereinigung* five years previously [Hilbert 1900].

Hilbert described the Dirichlet Principle as a method that Dirichlet, prompted by a thought of Gauss, had applied to the solution of the so-called boundary value problem. He started by characterizing it geometrically, for two dimensions, referring at this stage to Riemann but not to Dirichlet. After a brief reference to Weierstrass's critique, he noted that by setting limiting conditions on the boundary values, C. Neumann, H. A. Schwarz, and Poincaré had been able to establish the existence of a minimum. Hilbert observed that only A. Brill and M. Noether had given new hope by their conviction that the Dirichlet Principle, in a certain way formed according to nature, could perhaps be resuscitated in modified form.

Treating the Dirichlet Problem merely as a special problem of the calculus of variations, Hilbert now embarked on the resuscitation by reformulating the Principle in a more general mode:

> Every regular problem of the calculus of variations has a solution as soon as limiting assumptions suitable to the nature of the given boundary conditions are satisfied and, where necessary, the concept of the solution attains a suitable expansion.[12]

He gave two examples showing how this principle can serve as a guide to existence proofs that are both rigorous and simple. The first dealt with drawing the shortest line between two given points on a given surface. The second dealt with finding a potential function $z = f(x, y)$ which assumes given boundary values on a given bounded surface. Hilbert noted that aside from the simplicity and transparency of his procedure he thought that this new procedure has the advantage of using only the

---

[12]Hilbert 1905; see *Ges. Abh.* 3 (1935):11.

minimum property and does not make use of special properties of a geodetic line or the potential function. For that reason, it can be used for more general problems of surface theory and mathematical physics.

Although this short note was only a reprint of his 1901 lecture, it constituted a significant clarification of the discussions concerning the topic. He, too, based his definition of the Principle on a minimum property but then noted that he could also use a direct method. This he set out in a more extensive memorandum of 1904 that utilized the Dirichlet integral and various limiting conditions à la Dirichlet. It would not be referenced in the reprint contained in the centennial volume. However, in that brief note, Hilbert made it clear that he had set out on a course to demonstrate the wide applicability of the Dirichlet Principle once the surrounding conditions were clearly defined. In a footnote, he mentioned two doctoral dissertations (Hedrick and Noble) on the topic he had supervised in 1901, as well as the inclusion of his method in Bolza's *Lectures on the Calculus of Variations*, just published in 1904. He did not mention his having supervised Kellogg's 1902 dissertation, although the present-day reader may still find that Kellogg's *Foundations of Potential Theory* contains the most succinct analysis of approaches to the Dirichlet Principle that takes into account the variety of limiting conditions.[13] Still, by 1905, Hilbert himself had produced a more detailed, generalized, direct approach which involved several concepts due to Dirichlet (Hilbert 1904). Later, more well-founded examples would proliferate, including Richard Courant's 1910 dissertation.[14]

Closely related to his contemplation of the Dirichlet Principle and the calculus of variations is Hilbert's Nineteenth Problem, discussed in his address to the International Congress of 1900, where he raised the question of whether solutions of regular problems of the variational calculus are necessarily analytic. His ruminations on the topic during this time become even more evident in his introduction to the discourse on mathematical problems, where he spoke against the idea that rigor in methods of proof must be the enemy of simplicity. In an interesting implicit closing of the circle with regard to the Dirichlet Principle, he credited its chief critic, the recently deceased Weierstrass, with having shown the way to a new and certain foundation for the calculus of variations; Hilbert wished to follow up his discussion by pointing to the simplifications that no longer require some of the complicated procedures for computing the second variations to establish necessary and sufficient criteria for the existence of a maximum and minimum.[15]

Kurt Hensel

Kurt Hensel, the son of Sebastian Hensel and Julie von Adelson, was born in 1861. He was homeschooled on his father's estate Groß-Barthen until moving to Berlin where he attended a private school and the Friedrich Wilhelm Gymnasium. Except for a short period in Bonn, he attended the university in Berlin where he received his doctorate

[13] Kellogg 1929 (1953): esp. Chapter XI.

[14] According to a conversational remark by the late Mina Rees, Courant never stopped seeking a widely encompassing generalization of the Principle.

[15] Hilbert 1901; see Hilbert *Ges. Abh.* 3 (1935 or 1965): 294; also see his twenty-third problem.

under Kronecker in 1884 and became a privatdozent in 1886. His *Probevortrag* had dealt with "the application of the analysis of the infinite to number theory in its historic development and most important methods and results," ensuring his familiarity with a large portion of Dirichlet's mathematics. In 1901, Hensel became the chief editor of the *Journal für die reine und angewandte Mathematik* and moved to Marburg as ordinary professor, retiring there in 1930.

Hensel included a memoir in the centennial volume titled "On invariants belonging to an algebraic field." He had for half a dozen preceding years published related work, including his book on the theory of algebraic functions [Hensel, K. and Landsberg 1902]. His book on the theory of algebraic numbers, in which he presented his p-adic numbers in full systematic fashion, would only appear in 1908, [Hensel 1908]. It is clear, however, that this was the underlying theory to which he now referred.

He began by stating that Gauss was the first one to provide a rigorous proof that every function $f(x)$ of degree n can be decomposed uniquely into irreducible real factors of the first or second degree. Hensel then noted that Dirichlet, "in one of his shortest and most admirable memoirs," "On the theory of complex units" [**1846b**], had introduced the number h of all the real and pairwise conjugate roots of such a decomposition and thereby expressed the number of fundamental units in the field $K(x)$ defined by the equation $f(x) = 0$.

Hensel wished to point out that Dirichlet's result and the unit theorem he had derived from it can be generalized in such a fashion that it will lead not only to the theory of units but also to the theory of ideal numbers. This, he stressed, is not only noteworthy, but had not been noticed sufficiently. He referred to two of his own memoirs, notably the one [Hensel, K. 1904] titled "New foundations of arithmetic" in which he had demonstrated the theorem he considered equal in importance to Gauss's mentioned above, that every integral function of degree n can be uniquely decomposed both in the fields of rational and irrational real numbers and also the domain of an arbitrary prime number. He showed that the argument for a prime number can be conducted analogously to that for the fields Gauss had treated. As a result, he claimed that the theory of ideal prime factors and the theory of units can be considered as special cases of one and the same general theory and their individual branches derived through specialization from the general statement. He stressed that his new tool cannot only simplify and lead to complete solutions of problems in number theory but can also help tackle problems that could not even be approached through previously existing methods.

He proceeded to discuss one such problem which can be considered a generalization of Dirichlet's treatment of the theory of units. That, he declared, is the reason he dared to publish this short investigation in the volume "dedicated to the memory of this great mathematician."

Hensel and Dmitri Mirimanoff

Hensel followed his paper with excerpts from two letters exchanged with Dmitri Mirimanoff, the Russian mathematician who had resided in Geneva since 1900. Mirimanoff had written to Hensel and mentioned a formula closely related to one that Voronoï had noted at the Third International Congress in Heidelberg and that Hensel had generalized in the preceding memoir. Mirimanoff showed that the law of reciprocity is a result of this formula. Hensel replied, publishing an excerpt of a few lines showing that the formula can also lead to a supplementary theorem.

Henri Poincaré

Poincaré had been a student at the Lycée in Nancy –now named after him– where he stayed until in 1873 he entered the Ecole Polytechnique, advancing from there to the Ecole des Mines, which led to his working as a mining inspector. Although noted as an excellent student in all his subjects since his days at the Lycée, physical awkwardness and weakness, apparently due to a childhood bout with diphtheria, in school at times kept him from doing well in subjects such as geometry and mathematical physics that required drawing ability. He obtained his doctorate in Paris in 1879 with a dissertation on partial differential equations. After an initial teaching position at the University of Caen, he earned a professorship at the Faculté des Sciences in Paris, received the chair for mathematical physics and probability at the Sorbonne, and also taught at the Ecole Polytechnique, holding these positions at the time of the Dirichlet Centennial, in addition to having been in charge of the Paris Congress in 1900 and having received numerous other honors. A considerable number of his publications in a variety of other areas of mathematics and mathematical physics, as well as several of his popular publications, had also appeared by 1905.

Poincaré's paper in the centennial volume was titled "On arithmetic invariants." In his introduction, he called attention to Lejeune-Dirichlet's use of his series in his class number determinations of **1838b** and **1839–40**. Explaining that he himself had used similar series in publications of 1879 and 1881, he asked for consent to refer to those publications in order to present remarks, which perhaps not of great interest in themselves, warranted consideration because of their relationship to Dirichlet's work. He noted that his remarks pertain to Fuchsian functions, abelian functions, elliptic functions, a number of new transcendental functions related to elliptic and Fuchsian functions, as well as Fredholm's function. His stated justification for uniting such a number of diverse functions in one work was the commonality of their arithmetic properties and their connection with the analysis of Lejeune-Dirichlet, but his contemporaries knew also that several of these functions were predominantly associated with his own research.

Poincaré explained the meaning of an arithmetic invariant by contrasting it with an algebraic invariant. Given a linear form $ax + by$, it is not algebraically invariant: There is no uniform function with the two coefficients a and b that remains unchanged when subjected to a linear substitution. There are, however, uniform functions with the two coefficients that do remain unchanged if transformed by a linear substitution with integral coefficients. These are the arithmetic invariants.

As an example, Poincaré called attention to the series

$$\sum \frac{1}{(am+bn)^k} = \phi_k$$

where m and n can take on all integral values, whether positive, negative, or zero, except for both being zero. The series is absolutely convergent if $k > 2$.

In a lengthy, clear exposition of arithmetic invariants, Poincaré first showed how the various functions mentioned in his introduction share the properties of arithmetic invariants. Returning to Dirichlet, he then demonstrated how each of the functions he had discussed also shares the property of arithmetic invariance with Dirichlet's, which remains unchanged.

Felix Klein

Felix Klein, as previously noted, had come to Göttingen in 1886 and, although past his mathematical prime, demonstrated remarkable administrative ability resulting in significant expansion and diversification of mathematical activities in the university.

Klein presented an extensive paper on algebraic solutions of fifth- and sixth-degree equations in the centennial volume but came no closer to Dirichlet than indirectly, by some interesting references to Kronecker's work on the quintic. We do not know whether he realized that Dirichlet disavowed any credit for having influenced Kronecker in his algebraic, as opposed to his number-theoretic and analytic contributions. Klein's memoir was essentially motivated by priority concerns with regard to his own publications but provides some useful details concerning his, Kronecker's, and Gordan's interactions at an earlier time.

Georg Frobenius

Frobenius, best known as the algebraist who made significant contributions to group theory, was a student and protégé of Weierstrass. After the deaths of Kummer, Kronecker, and Weierstrass in the 1890s, he had taken over the leadership of mathematics at the university in Berlin. In sharp contrast to Klein's work in Göttingen, he sought to keep applied mathematics away from the university and relegate it to Berlin's technical schools.

Frobenius offered a paper "On the theory of linear equations" which, consonant with his personality, is dominated by use of the first person singular in outlining some of the history of that topic and concludes with a note giving a "correction" to the *Encyclopédie* concerning the definition of a determinant. Rather than correcting misunderstandings arising from changing meanings and emphases assigned to the term "determinant" in the nineteenth century, Frobenius, too, focused on a previous priority issue. This one involved Kronecker, Weierstrass, and himself. The relevance of his memoir to any specific contribution by Dirichlet is not obvious. We note in passing, however, that three of his students—Landau, Remak, and the younger Carl Ludwig Siegel—would strike out on their own and significantly expand Dirichlet's legacy later in the twentieth century.

Franz Mertens

Although he, too, refrained from mentioning Dirichlet explicitly in his contribution, Franz Mertens in previous publications had shown continual involvement with Dirichlet's work. Mertens, who received his doctorate on potential theory under Kummer and Kronecker, had been teaching at the Jagiellonian University in Cracow until moving to Graz and Vienna.[16] While still in Cracow in the 1870s, he had begun to produce numerous memoirs involving Dirichlet's publications, his series, and his sources.[17] It is of interest to note that a memoir by Mertens of 1874 may be the first to use the expression "analytic number theory" in a title.[18]

Mertens sent in a short memoir from Vienna providing a new proof of a Gaussian theorem. Gauss had proved that every primitive form with positive characters throughout is equivalent to a form created by duplication. Gauss used ternary quadratic forms to prove the theorem. Mertens here provided a proof using only binary forms.

Adolf Hurwitz

Adolf Hurwitz, who had received his doctorate with a thesis on elliptic modular functions under Felix Klein in 1881, had also studied in Berlin in the 1870s. He had been a privatdozent in Göttingen from 1882 to 1884, when he joined the Königsberg faculty as "regular" extraordinary professor. In 1892, he had moved to the ETH in Zurich. He remained there as ordinary professor until his death in 1919.

Hurwitz sent in a paper titled "On a representation of the class number of binary quadratic forms by infinite series." He had done some early work on Chasles and under Klein's influence had become acquainted with Riemann's work. Hurwitz had been the first to spell out the functional equation for the L-function in 1882. It was his close association since the mid-eighties with the younger Hilbert and with Minkowski that further stimulated his interest in this problem area. Despite his not discussing Dirichlet's work explicitly, this put him in clear relationship to their lines of inquiry.

His contribution to the memorial volume deals with "a representation of the class number of binary quadratic forms by infinite series." In it he discussed the issue of binary quadratic forms with negative determinant.

Wilhelm Wirtinger

Wilhelm Wirtinger was an Austrian whose training and career were centered in Vienna, where Franz Mertens had joined him as colleague at the university in 1894. Although a student of Emil Weyr at the University of Vienna, where he received his doctorate in 1887, Wirtinger had spent a postdoctoral term in Göttingen and there became close to Felix Klein who probably influenced his interest in Riemann. He would later become better known than he was in 1905 by having been one of the invited speakers at the 1912 International Congress.

Wirtinger's 1905 paper "On a special Dirichlet series" was prompted by his studies of Riemann's ζ-function. Specifically, he here derived transformation formulas for related functions. The title of his paper was a useful device to remind his readers of the connection between the ζ-function and Dirichlet's L-series and functions; by this

[16]Since the third partition of Poland in 1795, Cracow had been a part of Austria, subsequently becoming an Austrian protectorate, a Free City, and, although remaining a Polish cultural symbol, essentially tied to Austria until after World War I.

[17]Landau 1909, 2:942 lists a dozen relevant publications by Mertens spanning the years 1874 to 1900.

[18]Mertens 1874b.

time, they were likely to have begun spending more time studying Riemann rather than Dirichlet.

Hermann Minkowski

In addition to the memorial address held in Göttingen, Minkowski contributed a major memoir to the Berlin centennial volume. Titled "Discontinuity Domain for Arithmetic Equivalence" he stated at the outset that the work utilized methods that had been formed by Dirichlet. The problem Minkowski wished to address was the following:

First, he defined a system of n linear forms $\xi_1, \xi_2, \ldots, \xi_n$ with n variables $x_1, x_2, \ldots, x_n$, having arbitrary real coefficients $\frac{\partial \xi_h}{\partial x_k} = \alpha_{hk}$ and a nonzero determinant, as arithmetically equivalent to a second such system $\eta_1, \eta_2, \ldots, \eta_n$ if each of the two systems can be transformed into the other by a linear homogeneous substitution having only integral coefficients. In the manifold A of the n^2 real parameters α_{hk}, he wished to construct a domain B in which every complete class of mutually equivalent systems is represented by one point, and if that point lies in the interior of the domain, by only one point.

He noted that this problem can be translated to a corresponding problem concerning positive quadratic forms so that it now consists of seeking a domain B in the manifold A in which every class of positive quadratic forms is represented by one point, and if that point lies in the interior of B, by only one point. That will mean that the domains $A(f)$ form the required domain of discontinuity B in the manifold A for all points f lying in B.

Furthermore, he proposed proving that the domain of discontinuity B for the arithmetic equivalence of positive quadratic forms can be constructed in such a way that in the manifold A it will represent a convex cone bounded by a finite number of planes, with the point of origin $f = 0$ as vertex.

Minkowski now stated that by this theorem he brings the theory of arithmetic equivalence for positive quadratic forms with n variables to the same level at which the theory of ternary forms had arrived in Dirichlet's **1848**.

Minkowski for several years had used the notion of volume as a fundamental object, reminiscent of the kind of "innate" concept to which Dirichlet had so frequently referred. Minkowski here managed to make volume central to his argument.

Minkowski's fifty-four-page-long memoir, a model of rational organization and clarity, is divided into sixteen sections that follow the introductory statement of his topic. Each section carries a heading concerning its content, which is followed by a discussion resulting in the proof of a theorem, usually stated at the end of the section. Many were based on results he had obtained in previous years. He also at several points showed the relationship between his own earlier number-theoretic approach and the later essentially geometric equivalents.

Here are a few sample highlights of the theorems and terminology from this elaborate publication:

Section 1 deals with the character of positive quadratic forms. The resulting theorem states:

> A positive quadratic form f can only assume values for a finite number of integral systems of variables that do not exceed a given boundary L.

The following sections deal with the ordering of an n-dimensional manifold (Section 2), the lowest forms of a class (Section 3), and reduced forms (Section 4). In the discussion of reduced and lower forms, Minkowski referred to Hermite's treatment and definitions.

Section 5 is called "The walls of the reduced space." Note that Minkowski, who used the terminology of walls, edges, rooms, and chambers repeatedly in the following sections, is again enjoying a play with words, as "Raum" has the double meaning of "room" and "space."

Section 7 contains the statement and proof that there are only a finite number of integral substitutions of determinant ± 1 that can transform positive reduced forms to other reduced forms. An alternate formulation states that in the domain of positive forms the reduced space bounds only a finite number of equivalent chambers.

Section 9 treats the problem of the densest packing of spheres. Only the previous year Minkowski had presented to the Göttingen Society a paper on the densest lattice packing of congruent bodies, in which he reached back to the Gauss–Seeber–Dirichlet issue concerning ternary quadratic forms.[19]

Section 15 deals with the maximal density for lattice-packed spheres. Issues of maximal density had been suggested by Hermite, but Minkowski here applied his volume determination. The final Section 16 establishes an asymptotic law for the class number of integral positive forms.

The only explicit references to the name of Dirichlet are Minkowski's introductory, easily corroborated, statement that he followed Dirichlet's methods, and a discussion on the use of Dirichlet series in Section 13. His was the longest memoir in the volume.

Emile Picard

Even without his somewhat general justification (mainly pertaining to Riemann's branch points) for his contribution, Picard's name also would have drawn attention to the *Journal*'s volume of 1905. The precocious author, a protégé (and son-in-law) of Hermite, was nearly fifty years old. He had obtained his "agrégation" the year he turned twenty-one, succeeded Bouquet to the chair of differential geometry in Paris at age thirty (he had been appointed a year before, but was not eligible because of a minimum age requirement), and traded this for the chair in analysis and higher algebra. He had been a member of the Paris Académie since 1889. He had received widespread attention for his three-volume treatise on analysis, which had appeared between 1891 and 1896 and was followed by a two-volume opus, coauthored by Georges Simart, on the theory of algebraic functions of two independent variables. The first volume of this had appeared in 1897, and the contribution to the centennial collection would be reissued in its second volume published in 1907. By 1905, Picard had received major prizes from the Académie and in later years would hold numerous additional honors and influential positions.

[19]Minkowski 1904; see *Ges. Abh.* 2:311–55.

Picard sent in a memoir on "Some questions related to linear connections in the theory of algebraic functions of two variables." As Picard explained in the beginning of his memoir, he had been studying and publishing various results concerning linear connections and two-dimensional connections of algebraic surfaces. The present paper was intended to complete some aspects concerning linear connections. He justified the inclusion of this work in the centennial volume by a certain analogy "with questions to which Riemann had attached the name of Dirichlet."

Ernest Lebon would point out that the main result contained in this memoir was Picard's proof of the theorem that "the adjoints of a surface of order m, which are of an order higher or equal to $m - 2$, on an arbitrary plane give the complete system of adjoints of the same order of the plane section."[20]

Ludwig Schlesinger

Ludwig Schlesinger, born in Hungary, had received his secondary education in Bratislava before matriculating at the university in Berlin. Working primarily with Fuchs as well as Kronecker, he earned his doctorate in 1887 with a dissertation on certain fourth-order linear differential equations. He served in Berlin as privatdozent from 1889 until 1897, although carrying the title of professor from 1894 until leaving for a short stay in Bonn as extraordinary professor. At the time of the centennial, he held a professorship in Cluj (Romania), the former Klausenburg, where he would also serve as dean. In 1911, he followed a call to the university in Giessen, where he taught until his retirement in 1930.

Schlesinger contributed a paper, "On the solutions of certain linear differential equations as functions of singular points," to the centennial volume. This paper was the third of several he had devoted to studying Riemann's determination of a system of functions with given branch points and substitutions, and exploring related analogies between the theory of linear differential equations and the theory of algebraic functions. Here, as elsewhere in his early studies, Schlesinger followed closely his thesis advisor and father-in-law Lazarus Fuchs, editor of the second volume of Dirichlet's *Werke*. Schlesinger explained the relationship of his contribution to Dirichlet by referring to his aim in "fixing functional dependencies conceptually (*begrifflich*)"—a phrase Fuchs had used five years earlier in an address as Rektor at the university in Berlin. Schlesinger stated that this aim "in which, since Riemann, we see the principal feature of modern function theoretic thought, first becomes significant in the methods which Dirichlet created for the theory of the potential."

Ernst Steinitz

Among those mentioning Dirichlet in passing only is Steinitz. Although he would become better known to later mathematicians for his subsequent contributions to abstract field theory, prior to the publication in 1909 of his pioneering paper on that subject he had been working on geometric configurations.

To understand Steinitz's transition from such geometric approaches to his later classic field theory memoir, it is useful to note the following. He began his university

[20]Lebon 1914 (1991:62).

studies in Breslau, but, in the early 1890s, he attended lectures in Berlin, including those of Leopold Kronecker. In 1893, he returned to Breslau, where he studied with Heinrich Schröter, who had attended lectures by Dirichlet in the 1850s, and with Jacob Rosanes, under whom he received his doctorate in 1897. However, Steinitz went back to Berlin as privatdozent at the Technical High School that year. While in Berlin, he became friendly with Kurt Hensel, Landau, and Issai Schur, and before the turn of the century had begun to immerse himself in the study of Dedekind's final Supplement to Dirichlet's Lectures on Number Theory.

In his submission for the Dirichlet memorial volume, titled "On the attraction of hyperboloidal bowls," he listed Dirichlet as one of those dealing with problems of attraction analytically, while he himself wished to extend the work of Chasles in this memoir.

17.3 Voronoĭ

In the published extract of his letter to Kurt Hensel, Dmitry Mirimanoff called attention to the work of Georgĭ Voronoĭ who, although not himself a contributor to the centennial volume, had just published one of his several studies explicitly related to memoirs by Dirichlet.

Voronoĭ was a Ukrainian mathematician who studied at Saint Petersburg University, which he had entered in 1885. There he obtained his doctorate in 1894 with a dissertation on a "generalization of the continued fraction algorithm" in which he already referred to work by Dirichlet. Initially interested in algebra, under the influence of Chebyshev and some of his followers in Saint Petersburg, specifically men like Zolotarev, Korkin, and Markov, Voronoĭ turned to number theory. Beginning no later than 1903, he produced several memoirs based on publications by Dirichlet.

In 1903, Voronoĭ's memoir titled "On a problem of the calculation of asymptotic functions" had appeared in volume 126 of the *Journal für die reine und angewandte Mathematik*, by then published by Kurt Hensel. Dealing with Dirichlet's so-called divisor problem, it was based directly on Dirichlet's two memoirs **1838a** and **1851b**; Voronoĭ not only referred to these two publications by Dirichlet but suggested the wider extent of his familiarity with Dirichlet's work by quoting in extenso from the letter Dirichlet wrote to Kronecker in July 1858 concerning sharpening the error factor in his approximation.[21] Voronoĭ's work resulted in the sharpened error formula for the sum of the divisors:

$$F(n) = n(\log n + 2C - 1) + \theta_n \sqrt[3]{n} \log n,$$

[21] Voronoĭ could have read this in either the second volume of Dirichlet's *Werke*, that had been published in 1897, or in the last section of Paul Bachmann's work on Analytic Number Theory, published in 1893/94.

and Voronoĭ, who throughout his memoir observed that he is following the method of Dirichlet, here opened up a path for the further pursuit of the divisor problem by Erich Hecke, G. H. Hardy, and others. In addition, Boris Delone in 1947 commented that this article of Voronoĭ's "served as one of the starting points for the work of the greatest among the contemporary representatives of the St. Petersburg school, Academician I. M. Vinogradov."[22]

In 1904, Voronoĭ had attended the International Congress of Mathematicians in Heidelberg, where he met Minkowski and presented two papers. By 1907, two years after the centennial, readers were treated to another publication by Voronoĭ titled "On some properties of positive perfect quadratic forms" which would be followed by two very long memoirs on primitive parallelohedra.[23] Through these, his name has been linked to that of Dirichlet by their study of tessellations (tiles) and his coining the term *Dirichlet tessellation*. He died while the last of these publications was in press, preceding Minkowski in death by six weeks.

17.4 1909: Thue and Landau

The year of Minkowski's death marks a convenient end point to the story of Dirichlet's nineteenth-century legacy. Aside from being the year that Poincaré, upon an invitation from Hilbert to speak on an unspecified topic, gave six lectures in Göttingen in which he "intruded" on Hilbert's domain by dealing with Fredholm and integral equations, 1909 brought to the fore two men who were part of a new generation that reestablished Dirichlet's legacy in the twentieth century. They are Axel Thue, primarily associated with Oslo, although having studied in Berlin and Leipzig, and Edmund Landau, a native of Berlin, where he was trained and received his doctorate in 1899.

In 1891–92 at Berlin, Thue had attended lectures by Kronecker and attended Kronecker's mathematical seminar.[24] Among Thue's more than forty publications particularly relevant to Dirichlet are those having to do with diophantine approximations. In his memoir of 1909, Thue sharpened Liouville's approximation theorem of 1844. Liouville's theorem in its most familiar form was only published in 1851:

> If x is an algebraic number of degree n..., then there exists a positive number A such that for all rational numbers $\frac{p}{q}$ with $q > 0$ and $\frac{p}{q} \neq x$ the following inequality holds:[25]
>
> $$\left| x - \frac{p}{q} \right| > \frac{1}{Aq^n}.$$

Thue's 1909 theorem can be stated as follows:

---

[22]Delone 1947 (Delone 2005:168).

[23]Voronoĭ 1908–09 in Hensel's *Journal*.

[24]In Norway, Thue subsequently taught Thoralf Skolem who became Øystein Ore's adviser.

[25]Lützen 1990:524.

If x is an algebraic number of degree n, and if $k > 1 + \frac{n}{2}$, then there exists a constant $c(a, k)$ such that for every rational $\frac{p}{q}$ one has[26]

$$\left| a - \frac{p}{q} \right| > \frac{c(a,k)}{q^k}.$$

Dieudonné, in an extended discussion of Thue's theorem [Dieudonné 1992], noted that Thue's proof has no provision for ensuring the existence of the constant c, a fact that, along with all other ineffective attempts to build on Thue's theorem, "had haunted the theory of numbers" for decades. He elaborated on the relationship of this statement of the theorem to Thue's proof that

If $f(x, y) \in \mathbf{Z}[x, y]$ is a homogeneous form of degree $n > 2$, irreducible on $\mathbf{Q}$, then, for every integer k the equation $f(x, y) = k$ has only a finite number of integral solutions.[27]

The then twenty-six-year-old Landau had come to notice already in 1903 by his simplified proof for the Prime Number Theorem. In 1909, his comprehensive two-volume Handbook of the Distribution of Primes (Landau 1909) appeared. Aside from the earlier volume on analytic number theory by Paul Bachmann, published in 1893/4, Landau's 1909 *Handbuch* is considered the first systematic treatment of analytic number theory. Like his published and unpublished lectures on number theory and on the theory of functions, his many subsequent memoirs were didactic, clear, and important for providing an accessible bridge from Dirichlet's work to the breakthroughs of twentieth-century number theorists.

17.5 Commentary

While bearing in mind certain differences in approach at the beginnings of the nineteenth and the twentieth centuries to the areas treated or affected by Dirichlet, we conclude this volume by noting some characteristics of his work.

Dirichlet's motivation, beginning with his first publications and lectures, included the following:

1. To clarify rigorously, and with suitable examples, existing concepts needed in the classroom as well as in ongoing research.

2. To prove, or to provide more direct proofs for, statements of predecessors previously demonstrated either incompletely or in a cumbersome fashion. This is true particularly with regard to Legendre and Gauss. In addition, theorems or conjectures that Gauss described as particularly difficult served as a frequent challenge to

[26]Dieudonné 1992:224.

[27]The reader interested in pursuing the further history of Thue's Theorem, which takes us well beyond our 1909 cutoff date, may begin by consulting Dieudonné 1992:224–29, which also provides related references to Dirichlet's approximations. Not mentioned there is Carl Ludwig Siegel, who, in 1922, the year of Thue's death, proved a generalized form of Thue's Theorem using Dirichlet's box principle. Perhaps in this connection, Edmund Landau is said to have referred to Thue's 1909 theorem as being the most important discovery in elementary number theory that he knew.

Dirichlet. In numerous cases, Dirichlet's resulting theorems contained the one that presented the challenge as a special case.

3. To demonstrate the unity of various branches of mathematics by the ability to go back and forth between them, often using techniques associated with one of them, rather than founding a separate branch of mathematics. Dirichlet achieved his target of unity by applying successful techniques previously confined to a specific mathematical area to a different branch of mathematics. This accounts for the pride he took in the successful use of a discontinuity factor, a concept that had been effective for the older French analysts in the 1820s; he introduced it for the determination of multiple integrals in a variety of applications. Among numerous other examples of unification, we also note his study of approximations and mean values in analytic geometry, probability, and number theory.

These motivations led to certain aspects of his work:

1. Dirichlet's belief in an inherent unity underlying all branches of mathematics led him to apply a given technique or procedure shown to be successful in one area to another branch of mathematics. It enabled him to expand number theory into two directions barely adumbrated by Gauss. One was the use of analysis in number theory, leading to a formally established analytic number theory. The other was the application of geometric considerations, already present as an earlier heuristic device but becoming particularly apparent in his later work, and coming to the fore in Minkowski's geometry of numbers.

There was another aspect of his seeing mathematics as a unified organism. This is his emphasis on limiting conditions. From his first introduction of such conditions dealing with indeterminate equations, to "Dirichlet conditions" in dealing with the convergence of series, to emphasizing the importance of boundary conditions in partial differential equations, whether in the context of abstract function theory or applications to physical problems — along with the frequent addition of a numerical example — these limiting conditions would ease his readers' and auditors' appreciation for areas such as the partial differential equations, potential theory, and analytic number theory that he and his students were the first to introduce to German-speaking universities in the mid-nineteenth century.

2. Dirichlet sought out more efficient concepts when those that had been useful in establishing an area limited by restrictions such as dimension no longer worked in a broader or different framework. This he frequently mentions in the introductions to his memoirs, where he refers to seeking a more "innate" concept to avoid undue computational or complicated manipulative devices. Yet he demonstrated his command of lengthy algebraic manipulations where necessary, especially in his early publications preceding the use of L-series and L-functions.

3. Dirichlet often started from intuitive ideas to rigorously develop new theorems. This, too, was linked to the quest for suitable "innate" concepts.

4. Dirichlet chose simple examples to illustrate the effectiveness of a concept or of a theorem. These tended to be easy to follow, often consisting only of substituting specific values in a formula.

5. Dirichlet was widely read in the work of his predecessors and did not hesitate to let the readers of many of his memoirs as well as the students in his lectures know of them.

At the time of the centennial, there was a tendency to move away from these characteristics. For example:

1. The trend toward "purification" of specific branches of mathematics ran counter to the idea of "unification." That is why someone like Remak only published some of his most interesting "Minkowskian" investigations (utilizing geometric illustrations) after the death of his purist algebra teacher Frobenius.

2. The search for replacing established concepts by more efficient ones in looking for fundamental elements was frequently stymied by the successes of the axiomatic method at the end of the century first based on traditional concepts.

3. The increased drive toward abstraction tended to run counter to looking for more intuitive elements in building up even a new axiomatic, or eventually abstract, structure.

4. The perceived clarity derived from the increased reliance on abstraction and axiomatisation may have lessened the interest in the kind of "everyday" examples furnished in many of Dirichlet's works. Besides, these examples at times involved unstressed restrictions, suggesting greater generality than intended or demonstrable without specified limitations.

5. Finally, emphasis on historical background appeared to run counter to a purely mathematical approach. It was only in the last decades of the twentieth century that one saw a return to more substantive analyses of historical relationships to new discoveries.

The turning away from elaborations of historical background favored by Dirichlet and several of his predecessors may be largely due to the influence of Weierstrass, whose emphasis on building his systematic lectures from fundamental elements often led to concepts and methods he expounded as being considered original with him rather than as the products of an evolving historical development. As an unfortunate by-product of this approach, by the time historical accounts had been largely expunged from mathematical memoirs emanating from Berlin, specific references to prior results in the literature, such as those made by Dirichlet's adherent Kronecker, were frequently mistaken for an unseemly interest in establishing priorities.

It is important to note, however, that those who preferred more recent methodological trends over those that Dirichlet (and, to a large extent, Minkowski) had favored included men who had carefully studied and admired Dirichlet. Dedekind is the obvious and outstanding example of someone who continued to hold Dirichlet in highest esteem while becoming a leader in the development of modern abstract algebra and in the use of a terminology and of concepts unknown in Dirichlet's time.

After examining his life and work, it becomes clear that Dirichlet's impact affected subsequent teaching as well as research methodology. By 1905, his pedagogical importance was no longer so clearly remembered as it had been by those who had experienced his lectures—never polished, largely informal, frequently described as leading the listener in his train of thought—or had benefited from the extra hours spent with groups that he chose not to call seminars, and from the time taken with individual students, whether at the sickbed of Eisenstein, during visits from Riemann, on walks in Berlin and Göttingen, or in an unoccupied room at either university, or on earlier occasions even in his own home.

His research methodology would remain clearer to those who read his memoirs. His mode of thought was made less transparent by the fact that his posthumously published lectures were more widely known and easier to read than his earlier memoirs. Despite the earnest efforts of his editors to stay close to Dirichlet's presentation, their wording in the published lectures would frequently differ from his, yet be cited as verbatim statements made by Dirichlet. Moreover, the published lectures lacked the explanatory, and frequently historical, explicit introductions to his memoirs that provided guidance not only to the development of mathematical ideas but to the development of his own thought.

There were certain other factors that contributed to Dirichlet's success. Aside from his steady work habits and mathematical aptitude, the following two may be singled out:

He had strong support. For most of his life, this was provided by his mother and his wife. Later, there were men like Alexander von Humboldt and Karl Varnhagen von Ense, as well as Fourier, Jacobi, Wilhelm Weber, Liouville, and younger associates.

In addition, his ease in personal communication was aided by the modesty that Humboldt was fond of touting. Perhaps there is no stronger, early evidence of this characteristic than Dirichlet's signature in a document that had been prepared in 1828, in conjunction with the meeting of the Society of German Scientists and Physicians. The attendees were asked to sign the sheets provided for that purpose "in chemical ink" so that their autographs could be published along with the proceedings of the meeting. The dozens of signatures, familiar and unfamiliar, of varying size and legibility, were characterized by included titles and flourishes. His was a small, legible autograph providing nothing more than his name and location.

17.6 Minkowski: What is a Mathematical School?

Over the years, it has been stressed repeatedly that, in contrast to Weierstrass or Jacobi, Dirichlet did not found a mathematical school. We have previously recalled Dirichlet's satisfaction that, even when his most prominent students did not follow his methodology or subject matter closely, he had planted the seeds for their subsequent success. Kronecker took this attitude further, writing that he did not even like to use the term "student" when describing the relationship between a junior and a senior mathematician. In this relationship, Kronecker saw mathematics as a genuine "Gelehrtenrepublik."[28]

Minkowski enlarged on the theme of a "mathematical school" at the end of his centennial address. His comments provide an interesting antidote to the lack of

[28] Kronecker's letter to Cantor of September 1891. See the extract in Kronecker *Werke* 5:497–98.

attention paid to Dirichlet in some recent histories of nineteenth-century analysis and similar works. Let them serve as a closing tribute:

> Many who were subsequently scattered individually over quite differing paths owed to him the strongest impulses of their scholarly endeavors. The young Eisenstein thought he could not describe with sufficient warmth his enthusiasm for the inspirations provided him by Dirichlet, even had he been granted a heart of bronze and a thousandfold tongue. What mathematician would lack understanding for the fact that the shining path of Riemann, this giant meteor on the mathematical sky, took its start from the starry image of Dirichlet? Even though the sharp sword of the Dirichlet Principle may have been swung first by the youthful arm of William Thomson, the new era in the history of mathematics dates from that other principle of Dirichlet's to vanquish its problems with a minimum of blind calculation, a maximum of perceiving thoughts. Kronecker never forgot to say how much of his mathematical existence he owed to Dirichlet, even though Dirichlet himself claimed that he had only introduced Kronecker to the *lower* regions of *one* of the disciplines on the heights of which the other wandered about as master.
>
> Dedekind established a relationship to Dirichlet only here in Göttingen; we honor him as the only hero left to us of the greatest epoch in arithmetic. Also Lipschitz totally entered the circle of Dirichlet's ideas; in his younger years he made known to Helmholtz, and later to Hertz, his high enthusiasm for Dirichlet.
>
> All these men, of incomparable merits with regard to present-day mathematics, won the best part of their mathematical prowess from Dirichlet, and those of us today who try, more than ever, to recognize and represent the discipline in its simple truth, do we not stand in the school of Dirichlet?[29]

[29]Minkowski 1905a; see Minkowski *Ges. Abh.* 2:460-61.

Bibliography

Introduction

This bibliography consists of four major parts. The first lists nineteenth-century publications by Dirichlet, including translations, the Akademie brief reports, and posthumous publications, in chronological order, as well as the 1969 reissue of his *Werke*. The second lists posthumous publications of Dirichlet's lectures. Since these were based only on the notes taken by their editors when they attended the lectures years before publication and were not reviewed by Dirichlet himself (see Chap. 16), these are given under a separate heading, with each editor's name joined to that of Dirichlet, in order of the editors' surnames. This portion is followed by a third part containing selected secondary sources, sequenced by authors' names. The bibliography concludes with brief references to major sources of manuscript materials, including both his *Nachlass*, students' notes of the lectures, and miscellaneous items.

The "REISSUE" references to Dirichlet's *Werke* pertain to the 1969 Chelsea edition, which is a one-volume reprint of the two volumes (**1889** and **1897a**) of *G. Lejeune Dirichlet's Werke*. The beginning page numbers given for that reprint edition do not correspond to those in its table of contents because those reflect the original page numbering that includes half-title pages.[1] The half-title pages were dropped in the 1969 edition because the information included on them appears in the table of contents.

Whenever a work is mentioned using an English title in the text, this is simply a translation used as a convenience for benefit of the reader. The bibliography lists the actual language used, except for a number of volumes and references to translations ("volume" instead of "band" or "tome," "translation" instead of "Uebersetzung," etc.).

The only actual English phrase quoted in the main text is that cited by Encke in referring to the Shakespearean motto in Chap. 13. I do not know whether Schurz used

[1]In this context, a half-title page of an item is a title page of the item being reprinted. In most instances, the page number that appears in the table of contents is two less than the number of the page on which the article begins.

U. C. Merzbach, *Dirichlet*, https://doi.org/10.1007/978-3-030-01073-7

the original Shakespearean English or the Tieck–Schlegel translation into German when he impressed Kinkel with his declamation of the famous oration from *Julius Caesar*, mentioned in Chap. 12; for that reason, the familiar English verse line is used.

Unless reprinted or reissued in a collection of works by the same author which are listed and were verified, items that have not been consulted are marked with a star (*).

Works by Dirichlet

1826. *Mémoire sur l'impossibilité de quelques équations indéterminées du cinquième degré*. Paris. [Pamphlet] Paris. Read at the Royal Academy of Sciences (Institut de France), 11 July 1825. REISSUE: *Werke* 1:3–20. RELATED ISSUE: **1828c**.

1828a. Recherches sur les diviseurs premiers d'une classe de formules du quatrième degré. *J. reine angew. Math.* 3:35–69. REISSUE: *Werke* 1:65–98.

1828b. *De formis linearibus, in quibus continentur divisores primi quarumdam formularum graduum superiorum commentatio,* quam ad veniam docendi ab amplissimo philosophorum ordine in regia universitate litterarum
Vratislaviensi impetrandam conscripsit Gustavus Lejeune Dirichlet, philosophiae doctor. Vratislavia: Kupfer. Habilitationsschrift. *See Werke* 1:49–62.

1828c. Mémoire sur l'impossibilité de quelques équations indéterminées du cinquième degré. *J. reine angew. Math.* 3:354–75. Expanded version of **1826**. REISSUE: *Werke* 1:23–46.

1828d. Démonstrations nouvelles de quelques théorèmes relatifs aux nombres. *J. reine angew. Math.* 3:390–93. REISSUE: *Werke* 1:101–4.

1828e. Question d'analyse indéterminée. *J. reine angew. Math.* 3:407–8. REISSUE: *Werke* 1:107–8.

1829a. Note sur les intégrales définies. *J. reine angew. Math.* 4:94–98. REISSUE: *Werke* 1:111–16.

1829b. Sur la convergence des séries trigonométriques qui servent à représenter une fonction arbitraire entre des limites données. *J. reine angew. Math.* 4:157–69. REISSUE: *Werke* 1:119–32.

1830. Solution d'une question relative à la théorie mathématique de la chaleur. *J. reine angew. Math.* 5:287–95. REISSUE: *Werke* 1:163–72.

1832a. Démonstration d'une propriété analogue à la loi de réciprocité qui existe entre deux nombres premiers quelconques. *J. reine angew. Math.* 9:379–89. REISSUE: *Werke* 1:175–88.

1832b. Démonstration du théorème de Fermat pour le cas des $14^{\text{ièmes}}$ puissances. *J. reine angew. Math.* 9:390–93. REISSUE: *Werke* 1:191–94.

1835. Untersuchungen über die Theorie der quadratischen Formen. *Math. Abh. Kgl. Preuss. Akad. Wiss. Berlin.* Aus dem Jahre 1833: 101–21. Read at the Akademie, 15 August 1833. REISSUE: *Werke* 1:197–218.

1836a. Einige neue Sätze über unbestimmte Gleichungen. *Abh. Phys.-Math. Klasse Kgl. Preuss. Akad. Wiss. Berlin.* Aus dem Jahre 1834: 649–64. Read at the Akademie, 19 June 1834. REISSUE: *Werke* 1:221–36.

1836b. Ueber die Frage in wie fern die Methode der kleinsten Quadrate bei sehr zahlreichen Beobachtungen unter allen linearen Verbindungen der Bedingungsgleichungen als das vortheilhafteste Mittel zur Bestimmung unbekannter Elemente zu betrachten sei. *Bericht ...Kgl. Preuss. Akad. Wiss. Berlin.* Aus dem Jahre 1836: 67–68. Abstract of a paper read at the Akademie, 28 July 1836. REISSUE: *Werke* 1:281–82.

1836c. Sur les intégrales Eulériennes. *J. reine angew. Math.* 15:258–63. REISSUE: *Werke* 1:273–78.

1837a. Ueber eine neue Anwendung bestimmter Integrale auf die Summation endlicher oder unendlicher Reihen. *Math. Abh. Kgl. Preuss. Akad. Wiss. Berlin.* Aus dem Jahre 1835: 391–407. Read at the Akademie, 25 June 1835. REISSUE: *Werke* 1:239–56. RELATED ISSUE: **1837f**.

1837b. Ueber die Darstellung beliebiger Functionen durch bestimmte Integrale in specieller Anwendung auf die Function P_n, welche bei der Attraction der Sphäroide vorkommt. *Bericht ...Kgl. Preuss. Akad. Wiss. Berlin.* Aus dem Jahre 1837: 79. Title of a paper read at the Akademie, 5 June 1837. REISSUE: *Werke* 2:253. RELATED ISSUE: **1837e**.

1837c. Beweis eines Satzes dass jede unbegrenzte arithmetische Progression, deren erstes Glied und Differenz ganze Zahlen ohne gemeinschaftlichen Factor sind, unendlich viel Primzahlen enthält. *Bericht ...Kgl. Preuss. Akad. Wiss. Berlin.* Aus dem Jahre 1837: 108–10. Abstract of a paper read at the Akademie, 27 July 1837. REISSUE: *Werke* 1:309–12. RELATED ISSUE: **1839a**.

1837d. Ueber die Darstellung ganz willkürlicher Functionen durch Sinus–und Cosinusreihen. *Repertorium der Physik, unter Mitwirkung der Herren Lejeune Dirichlet, Jacobi, Neumann, Riess, Strehlke, herausgegeben von Heinrich Wihelm Dove und Ludwig Moser.* 1:152–74. REISSUES: *Werke* 1:135–60; **1900**.

1837e. Sur les séries dont le terme général depend de deux angles, et qui servent à exprimer des fonctions arbitraires entre des limites données. *J. reine angew. Math.* 17:35–56. REISSUE: *Werke* 1:285–306.

1837f. Sur l'usage des intégrales définies dans la sommation des séries finies ou infinies. *J. reine angew. Math.* 17:57–67. Modified summary of **1837a**. REISSUE: *Werke* 1:259–70.

1837g. Sur la manière de résoudre l'équation $t^2 - pu^2 = 1$ au moyen des fonctions circulaires. *J. reine angew. Math.* 17:286–90. REISSUE: *Werke* 1:345–50.

1838a. Ueber die Bestimmung asymptotischer Gesetze in der Zahlentheorie. *Bericht ...Kgl. Preuss. Akad. Wiss. Berlin.* Aus dem Jahre 1838: 13–15. Abstract of a paper read at the Akademie, 8 February 1838. REISSUE: *Werke* 1:353–56.

1838b. Sur l'usage des séries infinies dans la théorie des nombres. *J. reine angew. Math.* 18:259–74. REISSUE: *Werke* 1:359–74.

1839a. Beweis des Satzes, dass jede unbegrenzte arithmetische Progression, deren erstes Glied und Differenz ganze Zahlen ohne gemeinschaftlichen Factor sind, unendlich viele Primzahlen enthaelt. *Math. Abh. Kgl. Preuss. Akad. Wiss. Berlin.* Aus dem Jahre 1837: 45–71. Expanded version of **1837c**. REISSUE: *Werke* 1:315-42. TRANSL. (Fr.): **1839e**.

1839b. Sur une nouvelle méthode pour la détermination des intégrales multiples. *Comptes rendus hebdomadaires des séances de l'Académie des Sciences* 8:156–60. Read at the French Academy of Sciences, 4 February 1839. REISSUE: *Werke* 1:377–80. TRANSL. OF **1839c**; OTHER ISSUE: **1839d**.

1839c. Ueber eine neue Methode zur Bestimmung vielfacher Integrale. *Bericht ...Kgl. Preuss. Akad. Wiss. Berlin.* Aus dem Jahre 1839: 18–25. Abstract of a paper read at the Akademie, 14 February 1839. REISSUE: *Werke* 1:383–90. TRANSL. (Fr.): **1839b** and **1839d**. RELATED ISSUE: **1841a**.

1839d. Sur une nouvelle méthode pour la détermination des intégrales multiples. *Jl. de math. pures et appliquées.* (1)4:164–68. TRANSL. OF **1839c**; OTHER ISSUE: **1839b**.

1839e. Démonstration de cette proposition: Toute progression arithmétique, dont le premier terme et la raison sont des entiers sans diviseur commun, contient une infinité de nombres premiers. *Jl. de math. pures et appliquées.* (1)4:393–422. TRANSL. (O. Terquem) OF **1839a**.

1839–40. Recherches sur diverses applications de l'analyse infinitésimale à la théorie des nombres. *J. reine angew. Math.* 19 no. 17 (1839): 324–69; 21 no. 1 (1840): 1–12; and 21 no. 9 (1840): 134–55. REISSUE: *Werke* 1:413-96. TRANSL. (Ger.): **1897b**.

1840a. Ueber eine Eigenschaft der quadratischen Formen. *Bericht ...Kgl. Preuss. Akad. Wiss. Berlin.* Aus dem Jahre 1840: 49–52. Abstract of a paper read at the Akademie, 5 March 1840. REISSUES: **1840b**; *Werke* 1:499–502.

1840b. Ueber eine Eigenschaft der quadratischen Formen. *J. reine angew. Math.* 21 no. 7:98–100. Abstract of a paper read at the Akademie, 5 March 1840. REISSUE: *Werke* 1:499–502. OTHER ISSUE: **1840a**.

1840c. Extrait d'une lettre à M. Liouville sur la théorie des nombres. *Comptes rendus hebdomadaires des séances de l'Académie des Sciences* 10:285–88. REISSUES: **1840d**; *Werke* 1:621–23.

1840d. Extrait d'une lettre à M. Liouville sur la théorie des nombres. *Jl. de math. pures et appliquées.* (1)5:72–74. REISSUE: *Werke* 1:621–23. OTHER ISSUE: **1840c**.

1841a. Ueber eine neue Methode zur Bestimmung vielfacher Integrale. *Math. Abh. Kgl. Preuss. Akad. Wiss. Berlin.* Aus dem Jahre 1839: 61–79. Expanded version of **1839c**. REISSUE: *Werke* 1:393–410.

1841b. Untersuchungen über die Theorie der complexen Zahlen. *Bericht ... Kgl. Preuss. Akad. Wiss. Berlin.* Aus dem Jahre 1841: 190–94. Abstract of a paper read at the Akademie, 27 May 1841. REISSUES: **1841d** and *Werke* 1:505–8. RELATED ISSUE: **1843a**.

1841c. Einige Resultate von Untersuchungen ueber eine Classe homogener Functionen des dritten und der höheren Grade. *Bericht ... Kgl. Preuss. Akad. Wiss. Berlin.* Aus dem Jahre 1841: 280–85. Read at the Akademie, 11 October 1841. REISSUE: *Werke* 1:627–32.

1841d. Untersuchungen über die Theorie der complexen Zahlen. *J. reine angew. Math.* 22 no. 15:375–78. Abstract of a paper read at the Akademie, 27 May 1841. REISSUE: *Werke* 1:505–8. OTHER ISSUE: **1841b**. RELATED ISSUE: **1843a**.

1842a. Verallgemeinerung eines Satzes aus der Lehre von den Kettenbrüchen nebst einigen Anwendungen auf die Theorie der Zahlen. *Bericht ... Kgl. Preuss. Akad. Wiss. Berlin.* Aus dem Jahre 1842: 93–95. Abstract of a paper read at the Akademie, 14 April 1842. REISSUE: *Werke* 1:635–38.

1842b. Recherches sur les formes quadratiques á coefficients et á indéterminées complexes. *J. reine angew. Math.* 24:291–371. REISSUE: *Werke* 1:535–618.

1843a. Untersuchungen über die Theorie der complexen Zahlen. *Math. Abh. Kgl. Preuss. Akad. Wiss. Berlin.* Aus dem Jahre 1841: 141–61. Read at the Akademie, 27 May 1841. REISSUE: *Werke* 1:511–32. TRANSL. (Fr.): **1844**.

1843b. Ueber einige Aufgaben, welche die Bestimmung einer unbekannten Function unter dem Integralzeichen erfordern. *Bericht ... Kgl. Preuss. Akad. Wiss. Berlin.* Aus dem Jahre 1843: 152. Title of a paper read at the Akademie, 15 June 1843.

1844. Recherches sur la théorie des nombres complexes. *Jl. de math. pures et appliquées.* (1)9:245–69. TRANSL. (H. Faye) OF **1843a**.

1846a. Ueber die Bedingungen der Stabilitaet des Gleichgewichts. *Bericht ... Kgl. Preuss. Akad. Wiss. Berlin.* Aus dem Jahre 1846: 34–37. Read at the Akademie, 22 January 1846. RELATED ISSUE: **1846e**.

1846b. Ueber die Theorie der complexen Einheiten. *Bericht ... Kgl. Preuss. Akad. Wiss. Berlin.* Aus dem Jahre 1846: 103–7. Read at the Akademie, 30 March 1846. REISSUE: *Werke* 1:641–44.

1846c. Sur un moyen général de vérifier l'expression du potentiel relatif á une masse quelconque, homogène ou hétérogène. *J. reine angew. Math.* 32:80–84. REISSUE: *Werke* 2:11–16.

1846d. Ueber die charakteristischen Eigenschaften des Potentials einer auf einer oder mehreren endlichen Flächen vertheilten Masse. *Bericht ... Kgl. Preuss. Akad. Wiss. Berlin.* Aus dem Jahre 1846: 211–12. Read at the Akademie, 25 June 1846. REISSUE: *Werke* 2:19–20.

1846e. Ueber die Stabilität des Gleichgewichts. *J. reine angew. Math.* 32 no. 8:85–88. Modified version of **1846a**. REISSUE: *Werke* 2:5–8. TRANSL. (Fr.): **1847b**.

1847a. Bemerkungen zu Kummer's Beweis des Fermat'schen Satzes, die Unmöglichkeit von $x^\lambda - y^\lambda = z^\lambda$ für eine unendliche Anzahl von Primzahlen λ betreffend. *Bericht ...Kgl. Preuss. Akad. Wiss. Berlin.* Aus dem Jahre 1847: 139-41. REISSUE: *Werke* 2:254–55.

1847b. Note sur la stabilité de l'équilibre. *Jl. de math. pures et appliquées.* 12: 474–78. TRANSL. (Kopp) OF **1846e**. REISSUE: **1853c**.

1848. Ueber die Reduction der positiven quadratischen Formen mit drei unbestimmten ganzen Zahlen. *Bericht ...Kgl. Preuss. Akad. Wiss. Berlin.* Aus dem Jahre 1848: 285–88. Read at the Akademie, 31 July 1848. REISSUE: *Werke* 2:23–26. RELATED ISSUE: **1850a**.

1849. Ueber die Bestimmung der mittleren Werthe in der Zahlentheorie. *Bericht ...Kgl. Preuss. Akad. Wiss. Berlin.* Aus dem Jahre 1849: 218. Title of a paper read at the Akademie, 9 August 1849. RELATED ISSUE: **1851b**.

1850a. Ueber die Reduction der positiven quadratischen Formen mit drei unbestimmten ganzen Zahlen. *J. reine angew. Math.* 40:209–27. Expanded version of **1848**. REISSUE: *Werke* 2:29–48. TRANSL. (Fr.): **1859a**.

1850b. Ueber die Zerlegbarkeit der Zahlen in drei Quadrate. *J. reine angew. Math.* 40 no. 21:228–32. REISSUE: *Werke* 2:91–96. TRANSL. (Fr.): **1859b**.

1850c. Ueber einen neuen Ausdruck der Massen–Vertheilung auf einer Kugelfläche, wenn das Potential in jedem Punkte der Fläche einen beliebig gegebenen Werth erhalten soll. *Bericht ...Kgl. Preuss. Akad. Wiss. Berlin.* Aus dem Jahre 1850: 464. Title of a presentation to the Akademie, 28 November 1850. RELATED ISSUE **1852a**.

1851a. De formarum binariarum secundi gradus compositione. Commentatio qua ad audiendam orationem pro loco in facultate philosophica rite obtinenda die VI. mens. maii hor. XII. publice habendam invitat auctor P. G. Lejeune Dirichlet, phil. doct. prof. publ. ord. design. Berolini: typis academicis. REISSUES: **1854d** (shortened title); *Werke* 2:107–14. RELATED ISSUE: **1859d**.

1851b. Ueber die Bestimmung der mittleren Werthe in der Zahlentheorie. *Math. Abh. Kgl. Preuss. Akad. Wiss. Berlin.* Aus dem Jahre 1849: 69–83. Presented to the Akademie, 9 August 1849. REISSUE: *Werke* 2:51–66. TRANSL. (Fr.): **1856e**.

1851c. Ueber ein die Theorie der Division betreffendes Problem. *Bericht ...Kgl. Preuss. Akad. Wiss. Berlin.* Aus dem Jahre 1851: 20–25. Read at the Akademie, 20 January 1851. REISSUES: **1854c** (shortened title); *Werke* 2:99–104.

1851d. Nachricht über Jacobi's wissenschaftlichen Nachlass. *J. reine angew. Math.* 42:91–92. REISSUE: *Werke* 2:221–23.

1852a. Ueber einen neuen Ausdruck zur Bestimmung der Dichtigkeit einer unendlich dünnen Kugelschale, wenn der Werth des Potentials derselben in jedem Punkte ihrer Oberfläche gegeben ist. *Math. Abh. Kgl. Preuss. Akad. Wiss. Berlin.* Aus dem Jahre 1850: 99–116. Presented to the

Akademie, 28 November 1850. REISSUE: *Werke* 2:69–88. TRANSL. (Fr.): **1857d**.

1852b. Ueber einige Fälle, in welchen sich die Bewegung eines festen Körpers in einem incompressibeln flüssigen Medium theoretisch bestimmen läßt. *Bericht ...Kgl. Preuss. Akad. Wiss. Berlin*. Aus dem Jahr 1852: 12–17. Presented to the Akademie, 8 January 1852. REISSUE: *Werke* 2:117–20.

1852c. Gedächtnissrede auf das verstorbene Mitglied der Akademie, den Mathematiker Jacobi. *Bericht ...Kgl. Preuss. Akad. Wiss. Berlin*. Aus dem Jahre 1852: 433. Title of a paper read at the Akademie, 1 July 1852. RELATED ISSUE: **1853a**.

1853a. Gedächtnissrede auf Carl Gustav Jacob Jacobi. *Abh. Kgl. Akad. Wiss. Berlin*. Aus dem Jahre 1852: 1–27. Presented to the Akademie, 1 July 1852. REISSUES: **1854e**; **1856d**; *Werke* 2:227–52. TRANSL. (Fr.): **1857e**; TRANSL. (Russ.): **1886**; **1887–94**.

1853b. Ueber eine neue Ableitung von zwei arithmetischen Sätzen aus einer gemeinschaftlichen Quelle. *Bericht ...Kgl. Preuss. Akad. Wiss. Berlin*. Aus dem Jahre 1853: 300. Title of a paper read at the Akademie, 23 May 1853.

1853c. Sur la stabilité de l'équilibre. In Lagange 1853, pp. 399–401. OTHER ISSUE: **1847b**.

1854a. Ueber eine Eigenschaft der Kettenbrüche und deren Gebrauch zur Vereinfachung der Theorie der quadratischen Formen von positiver Determinante. *Bericht ...Kgl. Preuss. Akad. Wiss. Berlin*. Aus dem Jahre 1854: 384. Title of a paper read at the Akademie, 13 July 1854. RELATED ISSUE: **1855a**.

1854b. Ueber den ersten der von Gauss gegebenen Beweise des Reciprocitätsgesetzes in der Theorie der quadratischen Reste. *J. reine angew. Math.* 47:139–50. REISSUE: *Werke* 2:123–38.

1854c. Ueber ein die Division betreffendes Problem. *J. reine angew. Math.* 47:151–54. TRANSL. (Fr.): **1856f**. OTHER ISSUE: **1851c**.

1854d. De formarum binariarum secundi gradus compositione. *J. reine angew. Math.* 47:155–60. TRANSL. (Fr.): **1859d**. OTHER ISSUE (with original title): **1851a**.

1854e. Gedächtnissrede auf Carl Gustav Jacob Jacobi. *Arch. Math. u. Physik.* 22 no. 2:158–82. OTHER ISSUES: **1853a**; **1856d**.

1855a. Vereinfachung der Theorie der binären quadratischen Formen von positiver Determinante. *Math. Abh. Kgl. Preuss. Akad. Wiss. Berlin*. Aus dem Jahre 1854: 99–115. Erratum p. 116. Presented to the Akademie, 13 July 1854. REISSUE: *Werke* 2:141–58. TRANSL. (Fr.): **1857g**.

1855b. Ueber eine Eigenschaft der quadratischen Formen von positiver Determinante. *Bericht ...Kgl. Preuss. Akad. Wiss. Berlin*. Aus dem Jahre 1855: 493–95. REISSUE: *Werke* 2:185–87. RELATED ISSUES: **1856a**; **1857b**.

1856a. Sur une propriété des formes quadratiques à déterminant positif. Extrait des Comptes rendus de l'Académie de Berlin (juillet 1855), et librement traduit par l'auteur. *Jl. de math. pures et appliquées.* (2)1 (Feb): 76–79. REISSUE: *Werke* 2:191–94. RELATED ISSUES: **1855b**; **1857b**.

1856b. Sur un théorème relatif aux séries. *Jl. de math. pures et appliquées.* (2)1 (Feb): 80–81. REISSUE: *Werke* 2:197–200. RELATED ISSUE: **1857c**.

1856c. Sur l'équation $t + u + v + w = 4m$. Extrait dune lettre à M. Liouville, 11 April 1856.. *Jl. de math. pures et appliquées.* (2)1:210–14. REISSUE: *Werke* 2:203–8.

1856d. Gedächtnissrede auf Carl Gustav Jacob Jacobi. *J. reine angew. Math.* 52 no. 13:193–217. OTHER ISSUES: **1853a** q.v.; **1854e**.

1856e. Sur la détermination des valeurs moyennes dans la théorie des nombres. *Jl. de math. pures et appliquées.* (2)1:353–70. TRANSL. (J. Hoüel) OF **1851b**.

1856f. Sur un problème relatif à la division. *Jl. de math. pures et appliquées.* (2)1 (Oct): 371–76. TRANSL. (J. Hoüel) OF **1854c**.

1857a. Untersuchungen über ein Problem der Hydrodynamik. *Nachr. G. A. Univ. und Kgl. Ges. der Wiss. Göttingen.* Jahrgang 1857 (Aug. 10): 205–7. Abstract of a paper read to the Royal Society, 31 July 1857. REISSUE: *Werke* 2:217–18. RELATED ISSUES: **1861a**; **1861b**.

1857b. Ueber eine Eigenschaft der quadratischen Formen von positiver Determinante. *J. reine angew. Math.* 53:127–29. RELATED ISSUES: **1855b** q.v.; **1856a**.

1857c. Sur un théorème relatif aux series. *J. reine angew. Math.* 53:130–32. RELATED ISSUE: **1856b**.

1857d. Sur une nouvelle formule pour la determination de la densite d'une couche spherique infiniment mince, quand la valeur du potentiel de cette couche est donnee en chaque point de la surface. *Jl. de math. pures et appliquées.* (2)2:57–80. TRANSL. (J. Hoüel) OF **1852a**. OTHER RELATED ISSUE: **1850c**.

1857e. Éloge de Charles–Gustave–Jacob Jacobi. *Jl. de math. pures et appliquées.* (2)2:217–43. TRANSL. (J. Hoüel) OF **1853a** q.v.

1857f. Démonstration nouvelle d'une proposition relative à la théorie des formes quadratiques. *Jl. de math. pures et appliquées.* (2)2:273–76. REISSUE: *Werke* 2:211–14.

1857g. Simplification de la théorie des formes binaires du second degré à determinant positif. *Jl. de math. pures et appliquées.* (2)2:353–76. REISSUE: *Werke* 2:161–81. TRANSL. (J. Hoüel) OF **1855a** together with "Addition à ce mémoire; par lateur" (pp. 373–75, dated 14 August 1857) and "Extrait d'une Lettre de M. Dirichlet à M. Liouville" (pp. 375–76, undated).

Posthumous Publications

1859a. Sur la réduction des formes quadratiques positives à trois indéterminées entières. *Jl. de math. pures et appliquées.* (2)4:209–32. TRANSL. (no translator indicated) OF **1850a**.

1859b. Sur la possibilité de la décomposition des nombres en trois carrés. *Jl. de math. pures et appliquées.* (2)4:233–40. TRANSL. (J. Hoüel) OF **1850b**.

1859c. Sur le caractère biquadratique du nombre 2; extrait d'une Lettre de M. Dirichlet á M. Stern. *Jl. de math. pures et appliquées.* (2)4:367–68. TRANSL. (J. Hoüel) OF letter later published as **1860**.

1859d. De la composition des formes binaires du second degre. *Jl. de math. pures et appliquées.* (2)4:389–98. TRANSL. (V.-A. Lebesgue) OF **1854d**. RELATED ISSUE: **1851a**.

1859e. Sur la première démonstration donnée par Gauss de la loi de réciprocité dans la théorie des résidus quadratiques. *Jl. de math. pures et appliquées.* (2)4:401–20. TRANSL. (J. Hoüel) OF **1854b**.

1860. Ueber den biquadratischen Charakter der Zahl "Zwei." Aus einem Briefe Dirichlet's an Herrn Stern zu Göttingen, Göttingen. 21 January 1857. *J. reine angew. Math.* 57:187–88. REISSUE: *Werke* 2:261–62. RELATED ISSUE: **1859c**.

1861a. Untersuchungen über ein Problem der Hydrodynamik. *Abh. Math. Classe Kgl. Ges. Wiss. Göttingen* 8:3–40. Includes a foreword by R. Dedekind. REISSUES: **1861b**; *Werke* 2:265–301. RELATED ISSUE: **1857a**.

1861b. Untersuchungen über ein Problem der Hydrodynamik. *J. reine angew. Math.* 58:181–216. OTHER ISSUE: **1861a**. RELATED ISSUE: **1857a**. *See also* Dedekind 1861.

1862 [1863]. Démonstration d'un théorème d'Abel. *Jl. de math. pures et appliquées.* (2)7: 253–55. Communicated by J. Liouville. REISSUE: *Werke* 2:305–6. [Date in *Werke* is 1863.]

1886; 1887–94 *Karl Gustav Jacob Jakobi. *Fiziko–matem. nauki* 2 (1886); 1887–1894 A:265–70, 349–69.; TRANSL. OF **1853a**.

1889. *G. Lejeune Dirichlet's Werke*. Bd. 1. Herausgegeben auf Veranlassung der Königlich Preussischen Akademie der Wissenschaften. Ed. by L. Kronecker. Berlin: George Reimer. REISSUE with **1897a**: **1969**.

1897a. *G. Lejeune Dirichlet's Werke*. Bd. 2. Herausgegeben auf Veranlassung der Königlich Preussischen Akademie der Wissenschaften. Ed. by L. Kronecker cont. by L. Fuchs. Berlin: George Reimer. REISSUE with **1889**: **1969**.

1897b. Untersuchungen über verschiedene Anwendungen der Infinitesimal-analysis auf die Zahlentheorie. Ed. by R. Haussner. In: *Ostwald's Klassiker der exakten Wissenschaften*, Nr. 91. Leipzig:

Engelmann, 3–110. TRANSL. (R. Haussner) OF **1839–40**. Volume also includes Remarks by R. Hassner, 111–28.

1900. Die Darstellung ganz willkürlicher Funktionen durch Sinus- und Cosinusreihen. Ed. by Heinrich Liebmann. In: *Ostwald's Klassiker der exakten Wissenschaften*, Nr. 116. Leipzig: Engelmann, 3–34. REISSUE OF **1837d**; bound with reissue of Seidel 1847, 35–45, and Remarks 46–58.

1969. *G. Lejeune Dirichlet's Werke.* Ed. by L. Kronecker and L. Fuchs (vol. 2 only). Bronx, NY: Chelsea Publishing Co. 2 volumes bound in 1. REISSUE OF **1889** and **1897a**. Includes photogravure of Dirichlet.

Publications of Dirichlet's Lectures

Dirichlet–Arendt. 1863. *Éléments de la théorie des nombres complexes de la forme $a + b\sqrt{-1}$, d'après un cours de M. Dirichlet. In: *Programme d'invitation a l'examen public du Collége Royale Français, fixé au 30 Septembre 1863*, 1–43. [Prepared by Gustav Arendt.] Berlin: J. F. Starcke.

Dirichlet–Arendt. 1904. *Vorlesungen über die Lehre von den einfachen und mehrfachen bestimmten Integralen*. Ed. by G. Arendt. Braunschweig: Fr. Vieweg & Sohn.

Dirichlet–Dedekind. 1863. *Vorlesungen über Zahlentheorie*. Ed. with additions by R. Dedekind. Braunschweig: Vieweg. TRANSL. (Eng.): Dirichlet–Dedekind 1999.

Dirichlet–Dedekind. 1871. *Vorlesungen über Zahlentheorie*. Second edition. Ed. with additions by R. Dedekind. Braunschweig: Vieweg.

Dirichlet–Dedekind. 1879. *Vorlesungen über Zahlentheorie*. Third revised ed. Braunschweig: Ed. with additions by R. Dedekind. Braunschweig: Vieweg. TRANSL. (It.): Dirichlet–Dedekind 1881.

Dirichlet–Dedekind. 1881. *Lezioni sulla Teoria dei Numeri*. TRANSL. (A. Faifofer) OF Dirichlet–Dedekind 1879. Venice: Emiliana.

Dirichlet–Dedekind. 1894. *Vorlesungen über Zahlentheorie*. Fourth rev. ed. Ed. with additions by R. Dedekind. Braunschweig: Vieweg.

Dirichlet–Dedekind. 1999. *Lectures on Number Theory*. TRANSL. (J. Stillwell) OF Dirichlet–Dedekind 1863. Providence, RI: Amer. Math. Soc.; London: London Math. Soc.

Dirichlet–Grube. 1876. *Vorlesungen über die im umgekehrten Verhältniss des Quadrats der Entfernung wirkenden Kräfte*. Ed. by F. Grube. Leipzig: B. G. Teubner.

Dirichlet–Grube. 1887. *Vorlesungen über die im umgekehrten Verhältniss des Quadrats der Entfernung wirkenden Kräfte*. Ed. by F. Grube. Second edition. Leipzig: B. G. Teubner.

Dirichlet–Meyer. 1871. *Vorlesungen über die Theorie der bestimmten Integrale zwischen reellen Grenzen. Mit vorzüglicher Berücksichtigung der von P. Gustav*

Lejeune–Dirichlet im Sommer 1858 gehaltenen Vorträge über bestimmte Integrale. Ed. by Gustav Ferdinand Meyer. Leipzig: B. G. Teubner.

Secondary Sources

Abel, N. H. 1839. *Oeuvres complètes.* Redigée par B. Holmboe. Christiania: Chr. Gröndahl.

Académie des Sciences. Mémoires, vol. 8 (1829). *Procès-Verbaux* Séance 18 juillet 1825.

Ahrens, W., ed. 1907. *Briefwechsel zwischen C. G. J. Jacobi und M. H. Jacobi.* Leipzig: B. G. Teubner.

Avigad, Jeremy and Rebecca Morris. 2014. The concept of 'character' in Dirichlet's theorem on primes in an arithmetic progression. *Arch. Hist. Exact Scis* 68 no. 3:265–326.

Babbage, Charles. 1833. *On the Economy of Machinery and Manufactures.* 3rd ed. London: C. Knight.

Babbage, Charles. 1864. *Passages from the Life of a Philosopher.* London: Longman, Roberts & Green.

Babbage, Charles. 1989. *Works* 11:148–49. Ed. Martin Campbell-Kelly. New York: New York University Press. [Quoted excerpt from Babbage 1864].

Bacharach, Max. 1883. *Abriss der Geschichte der Potentialtheorie.* Göttingen: Vandenhoeck & Ruprecht's Verlag. REPRINT: nd. Kessinger Legacy.

Bachmann, Paul. *1864. *De unitatum complexarum theoria.* Breslau. *Habilitationsschrift.*

Bachmann, Paul. 1867. Zur Theorie der complexen Zahlen. *J. reine angew. Math.* 67:200–204.

Bachmann, Paul. 1872. *Zahlentheorie. Dritter Teil. Die Lehre von der Kreisteilung und ihre Beziehungen zur Zahlentheorie.* Leipzig: B. G. Teubner. REPRINT: 1927. B. G. Teubner.

Bachmann, Paul. 1892. *Zahlentheorie. Erster Teil. Die Elemente der Zahlentheorie.* Leipzig: Teubner. REPRINT: 1929. B. G. Teubner.

Bachmann, Paul. 1894. *Zahlentheorie. Zweiter Teil. Die analytische Zahlentheorie.* Leipzig: Teubner. REPRINT: 1921. B. G. Teubner.

Bachmann, Paul. 1900. Analytische Zahlentheorie. *Encykl. math. Wiss.* IC3:636–674. Leipzig: B. G. Teubner.

Bachmann, Paul. 1905. *Zahlentheorie. Fünfter Teil. Allgemeine Theorie der Zahlkörper.* Leipzig: Teubner.

Bachmann, Paul. 1911. Ueber Gauß' zahlentheoretische Arbeiten. In: Gauss *Werke* 10.2.

Baeck, Leo. 1958. Moses Mendelssohn. Pp. 15–24 in *Von Moses Mendelssohn zu Franz Rosenzweig.* Stuttgart: W. Kohlhammer.

Baum, Marie. 1950. *Rückblick auf mein Leben.* Heidelberg: F. H. Kerle.

Bertier de Sauvigny, Guillaume. 1967. *The Bourbon Restoration.* Philadelphia: Univ. Pennsylvania Press.

Bessel 1880. *Briefwechsel. See* Gauss 1880.

Biermann, Kurt–R. 1958a. Alexander von Humboldt als Protektor Gotthold Eisensteins und dessen Wahl in die Berliner Akademie der Wissenschaften. *Forschungen und Fortschritte* 32:78–81.

Biermann, Kurt–R. 1958b. Zur Geschichte der Ehrenpromotion Gotthold Eisensteins. *Forschungen und Fortschritte* 32:332–35.

Biermann, Kurt–R. 1959a. Johann Peter Gustav Lejeune Dirichlet: Dokumente für sein Leben und Wirken (Zum 100. Todestag). *Abh. Deutsch. Akad. Wiss. Berlin: Klasse für Mathematik, Physik und Technik.* Jahrgang 1959, Nr. 2. Berlin: Akademie–Verlag.

Biermann, Kurt–R. 1959b. Ueber die Förderung deutscher Mathematiker durch Alexander von Humboldt. Pp. 83–159 in *Alexander von Humboldt. 14. 9. 1769–6. 5. 1859: Gedenkschrift zur 100. Wiederkehr seines Todestages* ed. Alexander von Humboldt-Kommission der Deutsch. Akad. Wiss. Berlin. Berlin: Akademie-Verlag.

Biermann, Kurt–R. 1960. Vorschläge zur Wahl von Mathematikern in die Berliner Akademie. *Abh. Deutsch. Akad.Wiss. Berlin: Klasse für Mathematik, Physik, und Technik.* Nr. 3.

Biermann, Kurt–R. 1964. Gotthold Eisenstein: Die wichtigsten Daten seines Lebens und Wirkens. *J. reine angew. Math.* 214:19–30.

Biermann, Kurt–R. 1971. Zu Dirichlets geplantem Nachruf auf Gauss. *NTM* 8 no. 1:9–12. REISSUE: 1972. *Mitteilungen. Gauss-Gesellschaft* 9:47–50.

Biermann, Kurt–R, ed. 1977. *Briefwechsel zwischen Alexander von Humboldt und Carl Friedrich Gauss.* Berlin: Akademie-Verlag.

Biermann, Kurt–R, ed. 1982. *Briefwechsel zwischen Alexander von Humboldt und Peter Gustav Lejeune Dirichlet.* Beiträge zur A. v. Humboldt-Forschung, 7. Berlin: Akademie-Verlag.

Biermann, Kurt–R. 1988. *Die Mathematik und ihre Dozenten an der Berliner Universität 1810–1933: Stationen auf dem Wege eines mathematischen Zentrums von Weltgeltung*. Berlin: Akademie-Verlag.

Birkhoff, Garrett. 1973. *Source Book in Classical Analysis.* Cambridge, MA: Harvard University Press.

Bjerknes, C. A. 1915. *Hydrodynamische Fernkräfte: Fünf Abhandlungen über die Bewegung kugelförmiger Körper in einer inkompressiblen Flüssigkeit (1863–1890).* Tr. Arthur Korn. Ed. Arthur Korn and V. Bjerknes. Ostwalds Klassiker Nr. 195. Leipzig: Wilhelm Engelmann.

Borchardt, C. W. 1860. Gustav Lejeune-Dirichlet. *J. reine angew. Math.* 57:91–92.

Bottazzini, Umberto. 1986. *The Higher Calculus: a history of real and complex analysis from Euler to Weierstrass.* New York: Springer.

Bruhns, K. 1877, ed. *Briefe zwischen A. v. Humboldt und Gauss.* Leipzig.

Bruner, Jerome S. 1962. The Conditions of Creativity. Pp. 17–30 in *On Knowing: Essays for the Left Hand.* Harvard University Press.

Butzer, P. 1987. Dirichlet and his role in the founding of mathematical physics. *Arch. Int. Hist. Sci.* 37:49–82.

Butzer, P. L., M. Jansen, H. Zilles. 1982. Johann Peter Gustav Lejeune Dirichlet (1805–1859). Genealogie und Werdegang. *Dürener Geschichtsblätter*, Nr. 71: 31–56.

Cauchy, A.-L. 1823. *Résumé des lecons données à l'École royale polytechnique sur le calcul infinitesimal*. Paris: Debure. REISSUE: Oeuvres (2)4:7–261.

Chebyshev, P. L. 1851. Sur la fonction qui détermine la totalité des nombres premiers inférieurs à une limite donnée. *Mém. présentés à l'Acad. Imp. scis. St.-Petersburg par divers savants* VI:141–57. [Preceded by a Russian issue in 1849]. REISSUES: Chebyshev 1852; [Tchebyshev] *Oeuvres* 1:29–48.

Chebyshev, P. L. 1852a. *Jl. de math. pures et appliquées.* (1)17:341–65.

Chebyshev, P. L. 1852b. Mémoire sur les nombres premiers. *Jl. de math. pures et appliquées.* (1)17:366–90. REISSUES: Chebyshev 1854; *Oeuvres* 1:51–70.

Chebyshev, P. L. 1854. Mémoire sur les nombres premiers. *Mém. présentés à l'Acad. Imp. scis. St.-Petersburg par divers savants,* VII:17–33. REISSUE: *Oeuvres* 1:51–70.

Chebyshev, P. L. 1899/1902 Tchebyshev, P. L. *Oeuvres*. Ed. by Markov, A.A. and N. Sonin. 2 vols. REPRINT: New York, Chelsea 1962.

Courant, R. and D. Hilbert. 1931. *Methoden der mathematischen Physik.* Bd. 1. Zweite verbesserte Auflage. Berlin: Julius Springer.

Cournot, A. A. 1913. *Souvenirs*. Paris: Hachette.

Cullen, Michael. 1982. Leipzigerstraße Drei. Eine Baubiographie. *Mendelssohn-Studien* 5:9–77.

Darrigol, O. 2005. *Worlds of Flow: A History of Hydrodynamics from the Bernoullis to Prandtl.* Oxford: Oxford University Press.

Dauben, Joseph Warren. 1979 (1990). *Georg Cantor: His Mathematics and Philosophy of the Infinite.* Chapters 1 and 2. Princeton, NJ: Princeton University Press.

Dauben, Joseph Warren, ed. 1981. *Mathematical Perspectives. Essays on Mathematics and its Historical Development.* New York/London/et al.: Academic Press.

Davenport, Harold. 2000. *Multiplicative Number Theory.* Third edition. rev. by Hugh L. Montgomery. New York/Berlin/Heidelberg: Springer.

Davis, Audrey and Uta Merzbach. 1994. The role of the polarimeter in the diagnosis of diabetes melittus: Biot's bequest to 19th century medical saccharimetry. Pp. 143–52 in *Proceedings of the Eleventh International Scientific Instrument Symposium held at Bologna University, Italy 9–14 September 1991.* Bologna: Grafis Edizioni.

Dedekind, R. 1857a. Abrißeiner Theorie der höheren Congruenzen in bezug auf einen reellen Primzahl-Modulus. *J. reine angew. Math.* 54:1–26. REISSUE: *Ges. math. Werke* 1:40–67 [incl. note by Ore].

Dedekind, R. 1857b. Beweis für die Irreduktibilität der Kreisteilungsgleichungen. *J. reine angew. Math.* 54:27–30. REISSUE: *Ges. math. Werke* 1:68–71.

Dedekind, R. 1861. Zusatz zu der vorstehenden Abhandlung [Dirichlet **1861b**]. *J. reine angew. Math.* 58:217–28.

Dedekind, R. 1876. Bernhard Riemann's Lebenslauf. Pp. 541–58 in Riemann 1892 (1953:541–58).

Dedekind, R. 1905. Ueber binäre trilineare Formen und die Komposition der quadratischen Formen. *J. reine angew. Math.* 129:1–34. REISSUE: *Ges. math. Werke* 2:307–39.

Dedekind, R. 1930–32. *Gesammelte mathematische Werke*. Ed. R. Fricke, E. Noether, Ö. Ore. 3 vols. Braunschweig: Vieweg. REPRINT: 1969. 3 vols in 2. Bronx, NY: Chelsea.
Also see Dirichlet–Dedekind.

Delone, B. N. 2005. *The St. Petersburg School of Number Theory*. Trans. Robert Burns. Providence, RI: Amer. Math. Soc.; London: London Math. Soc.

Demidov, S. S. 1982. The study of partial differential equations of first order in the 18th and the 19th Centuries. *Arch. Hist. Exact Scis.* 26:325–80.

Demidov, S. S. 1983. On the history of the theory of linear differential equations. *Arch. Hist. Exact Scis.* 28:369–87.

Devrient, Eduard. 1964. *Aus seinen Tagebüchern*, Ed. Rolf Kabel. Bd. 1: Berlin–Dresden, 1836–1852. Weimar: Hermann Böhlers Nachfolger.

Devrient, Therese. 1905. *Jugenderinnerungen*. Stuttgart: C. Krabbe.

Dickson, L. E. 1919–23. *History of the Theory of Numbers*. 3 vols.: Vol. 1 (1919); Vol. 2 (1920); Vol. 3 (1923). Washington, DC: Carnegie Institution. REISSUE: 2005. Dover.

Dieudonné, Jean, ed. 1978 (1992) *Abrégé d'histoire des mathématiques. 1700–1800*. Paris: Hermann.

du Bois-Reymond, Paul. 1873. Ueber die Fourierschen Reihen. *Nachr. Kgl. Ges. der Wiss. und G. A. Univ. zu Göttingen*. Jahrgang 1873 (Aug. 6): 571–84.

Dugac, Pierre. 1976. *Richard Dedekind et les fondements des mathématiques*. Préface de Jean Dieudonné. Paris: Vrin.

Dugac, Pierre. 1981. Des fonctions comme expressions analytiques aux fonctions représentables analytiquement. Pp. 13–36 in Dauben 1981.

Duke, William and Yuri Tschinkel, eds. 2007. *Analytic Number Theory. A Tribute to Gauss and Dirichlet*. Providence, RI: Amer. Math. Soc.; Clay Math. Inst.

Eccarius, Wolfgang. 1976. August Leopold Crelle als Herausgeber des Crelleschen Journals. *J. reine angew. Math.* 286:5–25.

Edwards, H. M. 1974. *Riemann's Zeta-Function*. New York: Academic Press.

Edwards, H. M. 1975. The background of Kummer's proof of Fermat's Last Theorem for regular primes. *Archive Hist. Exact Scis.* 14:219–36.

Edwards, H. M. 1977a. *Fermat's Last Theorem. A Genetic Introduction to Algebraic Number Theory*. New York, Heidelberg, Berlin: Springer.

Edwards, H. M. 1977b. Postscript to [Edwards 1975]. *Archive Hist. Exact Scis.* 17:381–94.

Eisenstein, G. 1847. *Mathematische Abhandlungen besonders aus dem Gebiete der Höhern Arithmetik und der elliptischen Functionen*. Mit einer Vorrede von Prof. Dr. Gauss. Berlin: G. Reimer. REPRINT: 1967. Hildesheim: G. Olms.

Eisenstein, G. 1975. *Mathematische Werke*. 2 vols. New York: Chelsea.

Elstrodt, Jürgen. 2007. The life and work of Gustav Lejeune Dirichlet (1805–1859). Pp. 1–37 in Duke and Tschinkel 2007.

Elstrodt, Jürgen and Peter Ulrich. 1999. A Real Sheet of Complex Riemannian Function Theory: A Recently Discovered Sketch in Riemann's Own Hand. *Historia Mathematica* 26:268–88.

Encke, J. F. 1862. Die Erwiderung [to "Antrittsrede" of L. Kronecker.] *Mber. ...Kgl. Preuss. Akad. Wiss. Berlin.* Aus dem Jahre 1861 (July): 640–42. REISSUE: Kronecker *Werke* 5:391–93.

Euler, Leonhard. 1748. *Introductio in analysin infinitorum.* Vol. 1, ch. 15. Lausanne: Bousquet. REISSUE: Opera Omnia 1922: (1)8. Leipzig: B. G. Teubner. TRANSL.: Euler 1988.

Euler, Leonhard. 1849. **Commentationes Arithmeticae collectae*, ed. P. H. and N. Fuss, 2 vols. Petropoli.

Euler, Leonhard. 1954–55, 1960. *Opera omnia.*

Euler, Leonhard. 1988. *Introduction to Analysis of the Infinite.* vol. 1. New York: Springer. TRANSL. OF Euler 1748 by John D. Blanton.

Eytelwein, Johann Albert. 1814. Untersuchungen über die Bewegung des Wassers, wenn auf die Contraction, welche beim Durchgange durch verschiedene Oeffnungen statt findet, und auf den Widerstand, welcher die Bewegung des Wassers längs den Wänden der Behältnisse verzögert, Rücksicht genommen wird. *Abh. Kgl. Preuss. Akad. Wiss. math. Kl.* 1814/15:137–78; also, 1818/19: 9–18.

Eytelwein, Johann Albert. 1825. Recherches sur le mouvement de l'eau .... traduits de l'allemand, par M. LeJeune-Dirichlet, en avril 1823. *Annales des Mines.* (1)11:417–55 (with addenda by Hachette and Bidone, as well as tables).

Feilchenfeldt, Konrad. 1979. Karl August Varnhagen von Ense: Sieben Briefe an Rebecka Dirichlet. *Mendelssohn–Studien* 3:51–79. Berlin: Duncker & Humblot.

Fischer, Gerd, Friedrich Hirzebruch, Winfried Scharlau, and Willi Törnig, eds. 1990. *Ein Jahrhundert Mathematik. 1890–1990.* Dokumente zur Geschichte der Mathematik. Braunschweig/Wiesbaden: Deutsche Mathematiker-Vereinigung. Friedr. Vieweg & Sohn.

Fischer, Hans. 1994. Dirichlet's Contributions to Mathematical Probability Theory. *Historia Mathematica* 21:39–63.

Fontane, Theodor. 1880 (1967). Wanderungen durch die Mark Brandenburg. Spreeland: An der Nuthe. Wilhelm Hensel. *Sämtliche Werke* 2:853–67. München: Carl Hanser.

Fourier, J. B. J. 1808. *See* P[oisson] 1808.

Fourier, J. B. J. 1822. *Théorie analytique de la chaleur.* Paris: Didot. REISSUE: Fourier 1888 (2013): *Oeuvres 1.*

Fourier, J. B. J. 1829a. Analyse des travaux de l'Académie royale des Sciences, pendant l'année 1825 (Partie mathématique). *Mém. Acad. R. Sci. Inst. France*, VIII: i–lxxii.

Fourier, J. B. J. 1829b. Mémoire sur la théorie analytique de la chaleur. *Mém. Acad. R. Sci. Inst. France*, VIII:581–622. REISSUE: Fourier 1890 (2013): *Oeuvres* 2:147–81.

Fourier, J. B. J. 1888/1890. *Oeuvres.* Ed. G. Darboux. 2 v. Paris: Gauthier-Villars et fils. REPRINT: 2013. Cambridge University Press.

Frank, Philipp and Richard v. Mises, ed. 1927. *Die Differential- und Integralgleichungen der Mechanik und Physik, als 7. Auflage von Riemann-Webers Partiellen Differentialgleichungen der mathematischen Physik.* 2 vols. Braunschweig: F. Vieweg.

Frank, Philipp and Richard v. Mises, ed. 1943. *Die Differential- und Integralgleichungen der Mechanik und Physik, als 7. Auflage von Riemann-Webers Partiellen Differentialgleichungen der mathematischen Physik. 2. vermehrte Aufl., zugleich 8. Auflage von Riemann-Webers Partiellen Differentialgleichungen der mathematischen Physik.* 2 vols. New York: Mary S. Rosenberg.

Frobenius, G. 1905. Zur Theorie der linearen Gleichungen *J. reine angew. Math.* 129:175–80.

Gardner, J. Helen and Robin J. Wilson. 1993. Thomas Archer Hirst—Mathematician Xtravagant III. Göttingen and Berlin. *Amer. Math. Monthly* 100:619–25.

Gauss, Carl Friedrich. 1799. *Demonstrationes Nova Theorematis Omnem Functionem Algebraicam Rationalem Integram Unius Variabilis in Factores Reales Primi vel Secundi Gradus Resolvi Posse.* Helmstedt: C. G. Fleckeisen. REISSUE: Gauss 1876 *Werke* 3 (1876):1–30.

Gauss, Carl Friedrich. 1801. *Disquisitiones Arithmeticae.* Leipzig: Fleischer. REISSUE: Gauss *Werke* 1 (1863).

Gauss, Carl Friedrich. 1808. Theorematis Arithmetici Demonstratio Nova. *Commentationes Societatis Regiae Scientiarum Gottingensis* 16. REISSUE: Gauss *Werke* 2 (1863):1–8.

Gauss, Carl Friedrich. 1811. Summatio Quarundam Serierum Singularium. *Commentationes Societatis Regiae Scientiarum Gottingensis Recentiores* 1: Cl. math. 1–40. REISSUE: Gauss *Werke* 2 (1863):9–45.

Gauss, Carl Friedrich. 1820. Theorematis Fundamentalis in Doctrina de Residuis Quadraticis Demonstrationes et Ampliationes Novae. *Commentationes Societatis Regiae Scientiarum Gottingensis Recentiores* 4: Cl. math. 3–20. REISSUE: Gauss *Werke* 2 (1863):47–64.

Gauss, Carl Friedrich. 1828. Theoria Residuorum Biquadraticorum. Commentatio Prima. *Commentationes Societatis Regiae Scientiarum Gottingensis Recentiores* 6:27–56. REISSUE: Gauss *Werke* 2 (1863):65–92.

Gauss, Carl Friedrich. 1831. Review of Seeber's *Untersuchungen ueber die Eigenschaften der positiven ternaeren quadratischen Formen.* Freiburg i. Breisgau, 1831. *Goettingische Gelehrte Anzeigen* Nr. 108 (9 Juli):1065–77. REISSUE: Gauss *Werke* 2 (1863):188–96.

Gauss, Carl Friedrich. 1832a. Principia Generalia Theoriae Figurae Fluidorum in Statu Aequilibrii. *Commentationes Societatis Regiae Scientiarum Gottingensis Recentiores* 7: Cl. math. 39–88. REISSUE: Gauss *Werke* 5 (1867):29–77.

Gauss, Carl Friedrich. 1832b. Theoria Residuorum Biquadraticorum. Commentatio Secunda. *Commentationes Societatis Regiae Scientiarum Gottingensis Recentiores* 7: Cl. math. 89–148. REISSUE: Gauss *Werke* 2 (1863):93–148.

Gauss, Carl Friedrich. 1840. Allgemeine Lehrsaetze in Beziehung auf die im verkehrten Verhaeltnisse des Quadrats der Entfernung wirkenden Anziehungs- und Abstossungs-Kraefte. *Resultate aus den Beobachtungen des magnetischen Vereins im Jahre 1839*: 1–51. REISSUE: Gauss *Werke* 5 (1867):195–242.

Gauss, Carl Friedrich. 1849. Beiträge zur Theorie der algebraischen Gleichungen. *Abh. Math. Classe Kgl. Ges. Wiss. Göttingen* 4:3–34. With illustration on p. 35.

Gauss, Carl Friedrich. 1863–1933. *Carl Friedrich Gauss Werke.* 12 vols. Kgl. Ges. der Wiss. Göttingen.

Gauss, Carl Friedrich. 1880. *Briefwechsel zwischen Gauss und Bessel.* Ed. Kgl. Akad. Wiss. Berlin. Leipzig: Wilhelm Engelmann.

Gesellschaft deutscher Naturforscher und Ärzte. 1829. *Amtlicher Bericht über die Versammlung deutscher Naturforscher und Ärzte zu Berlin im September 1828, erstattet von den damaligen Geschäftsführern A. v. Humboldt und H. Lichtenstein. Nebst einer lithographirten Sammlung eigenhändiger Namenszüge der Theilnehmer.* Berlin: T. Trautwein.

Gilbert, Felix, ed. 1975. *Bankiers, Künstler und Gelehrte. Unveröffentlichte Briefe der Familie Mendelssohn aus dem 19. Jahrhundert.* Schriftenreihe wissenschaftlicher Abhandlungen des Leo Baeck Instituts 31. Tübingen: J. C. B. Mohr.

Goldstein, Catherine, Norbert Schappacher, and Joachim Schwermer, eds. 2010. *The Shaping of Arithmetic after C. F. Gauss's Disquisitiones Arithmeticae.* Berlin/ Heidelberg: Springer.

Grabiner, Judith V. 1981. *The Origins of Cauchy's Rigorous Calculus.* Cambridge, MA: The MIT Press. REISSUE: Dover Publications, 2005.

Grattan-Guinness, I. 1970. *The Development of the Foundations of Mathematical Analysis from Euler to Riemann.* Cambridge, Mass. and London, England: The MIT Press.

Grattan-Guinness, I. 1990. *Convolutions in French Mathematics, 1800–1840.* 3 vols. Basel, Boston, Berlin: Birkhäuser.

Gray, J. J. 1979/80. Olinde Rodrigues' paper of 1840 on transformation groups. *Arch. Hist. Exact Sci.* 21 no. 4:375–85.

Gray, Jeremy. 2013. *Henri Poincaré. A Scientific Biography.* Princeton and Oxford: Princeton University Press.

Hankel, H. 1870. *Untersuchungen über die unendlich oft oszillierenden und unstetigen Funktionen.* Tübingen. REISSUE: 1905. Jourdain, P. E. B., ed. Pp. 44–102 in: *Ostwald's Klassiker der exakten Wissenschaften* no. 153, Leipzig: W. Engelmann.

Harnack, Adolf v. 1900. *Geschichte der Königlich Preussischen Akademie der Wissenschaften zu Berlin.* Vols. 1 and 2. Berlin: Reichsdruckerei.

Heilbronn, H. 1934. On the class-number in imaginary quadratic fields. *Quarterly J. Math.* 5:150–60.

Heine, E. 1861. *Handbuch der Kugelfunctionen.* 2 vols. Berlin: Georg Reimer.

Heine, E. 1870. Ueber trigonometrische Reihen. *J. reine angew. Math.* 71:353-65.

Hensel, Fanny. 2002. *Tagebücher.* Ed. Hans-Günther Klein and Rudolf Elvers. Wiesbaden/ London/ Paris: Breitkopf & Härtel.

Hensel, Kurt. 1904. Neue Grundlagen der Arithmetik. *J. reine angew. Math.* 127:51–84.

Hensel, Kurt. 1905. Ueber die zu einem algebraischen Körper gehörigen Invarianten. *J. reine angew. Math.* 129:68–85.

Hensel, K. and D. Mirimanoff. 1905. *See* Mirimanoff and Hensel 1905.

Hensel, Sebastian. 1904. *Ein Lebensbild aus Deutschlands Lehrjahren.* 2nd ed. With a preface by Paul Hensel. Berlin: B. Behr.

Hensel, Sebastian. 1908. *Die Familie Mendelssohn. 1729–1847. Nach Briefen und Tagebüchern*. Fünfzehnte Auflage. 2 vols. Berlin: B. Behr.
Herivel, John. 1975. *Joseph Fourier. The Man and the Physicist*. Oxford: Clarendon Press.
Hilbert, David. 1894. Ueber den Dirichletschen biquadratischen Zahlkörper. *Math. Ann.* 45:309–40. REISSUE: Hilbert *Ges. Abh.* 1:24–52.
Hilbert, David. 1897. Die Theorie der algebraischen Zahlkörper. *Jber. DMV* 4:175–546. REISSUE: Hilbert *Ges. Abh.* 1:63–363.
Hilbert, David. 1900. Über das Dirichlet'sche Princip. *Jber. DMV* 8:184–88.
Hilbert, David. 1901. *Ueber das Dirichletsche Prinzip. *Festschrift zur Feier des 150 jährigen Bestehens der Kgl. Ges. Wiss. Göttingen*. REISSUES: Hilbert 1904; Hilbert 1935:15–37.
Hilbert, David. 1904. Ueber das Dirichletsche Prinzip. *Math. Ann.* 59:161–86. REISSUE: Hilbert 1935:15–37.
Hilbert, David. 1905. Ueber das Dirichletsche Prinzip. *J. reine angew. Math.* 129: 63–67. REISSUE: Hilbert 1935:10–14.
Hilbert, David. 1909. Hermann Minkowski. *Nachr. Kgl. Ges. der Wiss. zu Göttingen*, Geschäftliche Mitteilungen aus dem Jahre 1909: 72–101. REISSUES: Hilbert 1910; Hilbert 1935:339–64.
Hilbert, David. 1910. *Math. Ann.* 68:445-71. REISSUE of Hilbert 1909.
Hilbert, David. 1932. *Gesammelte Abhandlungen*. Vol. 1. Zahlentheorie. Berlin. Springer. REPRINTS: 1965, 1981. New York: Chelsea.
Hilbert, David. 1935. *Gesammelte Abhandlungen*. vol. 3. Berlin.
Hilbert, David. 1965. *Gesammelte Abhandlungen*. 3 vols. Bronx, NY: Chelsea. REPRINTS of Hilbert 1932, 1933, 1935.
Hurwitz, A. 1882. Einige Eigenschaften der Dirichlet'schen Functionen …die bei der Bestimmung der Classenanzahlen binärer quadratischer Formen auftreten. *Zeitschrift f. Math. u. Physik*. 27:86–101.
Hurwitz, A. 1905. Ueber eine Darstellung der Klassenzahl binärer quadratischen Formen durch unendliche Reihen. *J. reine angew. Math.* 129:187–213.
Jacobi, C. G. J. 1832. Observatio arithmetica de numero classium divisorum quadraticorum formae $yy + Azz$, designante A numerum primum formae $4n + 3$. *J. reine angew. Math.* 9:189–92. REISSUE: *Ges. Werke* 6:240–44.
Jacobi, C. G. J. 1851. Opuscula mathematica 2. Ed. Dirichlet. Berlin: Georg Reimer.
Jacobi, C. G. J. 1881–91. *Gesammelte Werke*. 7 vols. Berlin: Georg Reimer.
Jacobi, C. G. J. 1907. *See* Ahrens, W., ed. 1907.
Jacoud, Gilles. 2010, ed. *Political economy and industrialism: Banks in Saint-Simonian economic thought*. London, New York: Routledge.
Joachim, Joseph. 1911–13. *Briefe von und an Joseph Joachim*. 3 vols. comp. and ed. by Johannes Joachim and Andreas Moser. Vol. 1 (1911) for 1842–1857; vol. 2 (1912) for 1858–1868; vol. 3 (1913) for 1869–1907. Berlin: Julius Bard.
Jourdain, ed. 1905. REISSUE OF Hankel 1870. Pp. 44–102 in: *Ostwald's Klassiker der exakten Wissenschaften* no. 153, Leipzig: W. Engelmann.
Jungnickel, Christa and Russell McCormmach. 1986. *Intellectual Mastery of Nature. Theoretical Physics from Ohm to Einstein*. Vol. 1: *The Torch of Mathematics*

1800–1870. Vol. 2: *The Now Mighty Theoretical Physics 1870–1925*. Chicago and London: The University of Chicago Pres

Kellogg, O(liver) D(imon). 1902. **Zur Theorie der Integralgleichungen und des Dirichlet'schen Prinzips*. Göttingen: Dieterich.

Kellogg, O(liver) D(imon). 1929. *Foundations of Potential Theory*. Berlin: J. Springer. REISSUE: 1953. New York: Dover.

Kessler, Albert. 1968. *Von Karl dem Grossen bis Napoleon Bonaparte. Grundzüge einer Geschichte des Dürener Landes 748–1814*. Beiträge zur Geschichte des Dürener Landes, 11.

Kinkel, Johanna. *1886. *Memoiren*. Ed. by Gottfried Kinkel. [*See* Feilchenfeldt 1979: 56–57 for quoted excerpt.]

Kirchhoff, G. R. 1876. *Vorlesungen über mathematische Physik: Mechanik*. Leipzig: B. G. Teubner.

Klein, Felix. 1905. Ueber die Auflösung der allgemeinen Gleichungen fünften und sechsten Grades. *J. reine angew. Math.* 129:151–74.

Klein, Felix. 1926–27. *Vorlesungen über die Entwicklung der Mathematik im 19. Jahrhundert.*, 2 vols. Vol.1 (1926) ed. by R. Courant and O. Neugebauer; vol. 2 (1927) ed. by R. Courant and St. Cohn-Vossen. REISSUE: 1956.

Königsberger, Leo. 1904. *Carl Gustav Jacob Jacobi.*. Leipzig: B. G. Teubner.

Kronecker, Leopold. 1858. Ueber die elliptischen Functionen, für welche complexe Multiplication stattfindet. *Mber. ...Kgl. Preuss. Akad. Wiss. Berlin.* Aus dem Jahre 1857 (October): 455–60. REISSUE: *Math. Werke* 4:177–83.

Kronecker, Leopold. 1865. Ueber den Gebrauch der Dirichletschen Methoden in der Theorie der quadratischen Formen. *Mber. ...Kgl. Preuss. Akad. Wiss. Berlin.* Aus dem Jahre 1864 (May): 285–303. REISSUE: *Math. Werke* 4:227–44.

Kronecker, Leopold. 1876. Zur Geschichte des Reciprocitätsgesetzes. *Mber. ...Kgl. Preuss. Akad. Wiss. Berlin.* Aus dem Jahre 1875 (April): 267–74. REISSUE: *Math. Werke* 2:3–10.

Kronecker, Leopold. 1882. Grundzüge einer arithmetischen Theorie der algebraischen Grössen. *J. reine angew. Math.* 92:1–122. REISSUE: *Math. Werke* 2: 239–387.

Kronecker, Leopold. 1885a. Ueber das Dirichlet'sche Integral. *Sber. Kgl. Preuss. Akad. Wiss. Berlin*: 641–85. REISSUE: *Math. Werke* 5:237–65.

Kronecker, Leopold. 1885b. Briefwechsel zwischen Gustav Lejeune-Dirichlet und Herrn Leopold Kronecker. Ed. Ernst Schering. *Nachr. Kgl. Ges. der Wiss. und G. A. Univ. Göttingen*, Jahrgang 1885: 361–82. REISSUES: *Math. Werke* 5:407–31. Dirichlet **1897a**:388–411.

Kronecker, Leopold. 1888a. Ueber die arithmetischen Sätze, welche Lejeune Dirichlet in seiner Breslauer Habilitationsschrift entwickelt hat. *Sber. Kgl. Preuss. Akad. Wiss. Berlin vom Jahre 1888*:417–23. REISSUE: *Math. Werke* 3.1:281–92.

Kronecker, Leopold. 1888b. Bemerkungen über Dirichlet's letzte Arbeiten. *Sber. Kgl. Preuss. Akad. Wiss. Berlin* 1888:439–42. REISSUE: *Math. Werke* 5:473–76.

Kronecker, Leopold. 1890. Ueber die Dirichlet'sche Methode der Wertbestimmung der Gausschen Reihen. *Festschr. Hamb. Mitt.* 2:32–36. REISSUE: *Math. Werke* 4:301–8.

Kronecker, Leopold. 1894. *Vorlesungen über die Theorie der einfachen und der vielfachen Integrale.*

Kronecker, Leopold. 1901. *Vorlesungen über Zahlentheorie.*

Kronecker, Leopold. 1895–1930. *Mathematische Werke.* 5 vols. REPRINT: 1968. New York: Chelsea.

Kühn, Helga-Maria. 1999. "In diesem ruhigen Kleinleben geht so schrecklich viel vor." Rebecka Lejeune Dirichlet, geb. Mendelssohn Bartholdy, in Göttingen 1855–1858. *Mendelssohn Studien* 11:145–56. With 7 illustrations (4 preceding page 145, 3 in text).

Küssner, Martha. 1982. Carl Wolfgang Benjamin Goldschmidt und Moritz Abraham Stern. Zwei Gaußschüler jüdischer Herkunft. *Mitt. Gauss-Gesellschaft* 1982:37–62.

Kummer, Ernst Eduard. 1860. Gedächtnisrede auf Gustav Peter Lejeune-Dirichlet. *Abh. Kgl. Preuss. Akad. Wiss. Berlin.* Hist. Einl. 1–36. REISSUE: Dirichlet **1897a**: 311–44.

Kummer, Ernst Eduard. 1975. Collected Papers. Ed. by A. Weil. Berlin/New York: Springer.

Lacroix, S. F. 1810. Traité du calcul différentiel et du calcul intégral. Seconde éd. rev. et augm. Vol. 1: Introduction.

Lagrange, J. L. 1788. **Mechanique Analytique*. Paris: Desaint.

Lagrange, J. L. 1853. *Mécanique Analytique.* $3^{\text{ième}}$ ed. Ed. Bertrand. Paris.

Lagrange, J. L. 1867-69. *Oeuvres.* vols. 2–3. Paris: Gauthier-Villars.

Lampe, Emil. 1906. Dirichlet als Lehrer der Allgemeinen Kriegsschule. *Naturwissenschaftliche Rundschau* 21:482–85.

Landau, Edmund. 1909. *Handbuch der Lehre von der Verteilung der Primzahlen.* 2 vols. Leipzig and Berlin: B. G. Teubner.

Landau, Edmund. 1985. *Collected Works.* Ed. L. Mirsky et al. Essen: Thales.

Langer, William L. 1969. *Political and Social Upheaval: 1832–1852.* New York: Harper & Row (Harper Torchbooks).

Laubenbacher, R. and D. Pengelley. 2010. Voici ce que j'ai trouvé: Sophie Germain's grand plan to prove Fermat's Last Theorem. *Historia Mathematica* 37:641–92.

Lebon, Ernest. 1914. *Émile Picard: Biographie, bibliographie analytique des écrits.* 2nd ed. Paris: Gauthier-Villars. REPRINT: 1991. Cornell University Library.

Legendre, A. M. 1788. Recherches d'Analyse indéterminée. *Hist. Acad. Roy. Scis. Paris, année 1785*: 465–559.

Legendre, A. M. 1797/8. [an 6]. *Essai sur la théorie des nombres.* Paris: Duprat.

Legendre, A. M. 1808. *Essai sur la théorie des nombres.* 2nd ed. Paris: Courcier.

Legendre, A. M. 1816. *Suppléments à l'Essai sur la théorie des nombres.* 2nd ed. Paris: Vve. Courcier.

Legendre, A. M. 1825. *Traité des fonctions elliptiques et des integrales euleriennes.* 3 vols. Paris: Huzard-Courcier.

Legendre, A. M. 1827. Seconde Supplément á l'Essai sur la théorie des nombres. *Académie des sciences. Mémores* 6.

Legendre, A. M. 1830. *Théorie des Nombres.* Troisième édition. 2 vols. Paris: Firmin Didot Frères.

Lenz, Max. 1910–1918. *Geschichte der Kgl. Friedrich-Wilhelms-Universität zu Berlin.* 4 vols. in 5. Halle: Buchhandlung des Waisenhauses.

Lerch, M. 1899. Sur les séries de Dirichlet. *C. R. Paris* 128:1310–11.

Lerch, M. 1902. Sur la formule fondamentale de Dirichlet qui sert à determiner le nombre des classes de formes quadratiques binaires définies. *C. R. Paris* 135:1314–15.

Lipschitz, R. 1864. De explicatione per series trigonometricas instituenda functionum unius variabilis arbitrariarum, et praecipue earum, quae per variabilis spatium finitum valorum maximorum et minimorum numerum habent infinitum, disquisitio. *J. reine angew, Math.* 63:296–308. TRANSL. (Fr.): Lipschitz 1913.

Lipschitz, R. 1913. Recherches sur le développement en séries trigonométriques des fonctions arbitraires d'une variable et principalement de celles qui, dans une intervalle fini, admettent une infinité de maxima et de minima. *Acta Math.* 36:281–295. TRANSL. OF Lipschitz 1864.

Lorey, Wilhelm. 1916. *Das Studium der Mathematik an den deutschen Universitäten seit Anfang des 19. Jahrhunderts.* Abh. math. Unterricht Deutschland, 3(9). Leipzig u. Berlin: B. G. Teubner.

Lowenthal-Hensel, Cécile. 1983. Mutter und Sohn. Fanny und Sebastian Hensel, in *Die Mendelssohns in Berlin: Eine Familie und ihre Stadt*, ed. Rudolf Elvers and Hans-Günter Klein. Berlin: Staatsbibliothek Preußischer Kulturbesitz, pp. 58–73.

Lützen, Jesper. 1990. *Joseph Liouville 1809–1882. Master of Pure and Applied Mathematics.* New York, Berlin, Heidelberg, et al.: Springer.

Mendelssohn, Moses. 1983. *Jerusalem.* Trans. A. Arkush. Intro. and Commentary A. Altmann. Waltham, MA: Brandeis University Press.

Mertens, Franz. 1874a. Ueber einige asymptotische Gesetze der Zahlentheorie. *J. reine angew. Math.* 77:289–338.

Mertens, Franz. 1874b. Ein Beitrag zur analytischen Zahlentheorie. *J. reine angew. Math.* 78:46–62.

Mertens, Franz. 1897. Ueber Multiplication und Nichtverschwinden Dirichletscher Reihen. *J. reine angew. Math.* 117:169–84.

Mertens, Franz. 1905. Ein Beweis des Satzes, daß jede Klasse von ganzzahligen primitiven binären quadratischen Formen des Hauptgeschlechts durch Duplikation entsteht. *J. reine angew. Math.* 129:181–86.

Merzbach, Uta C. 1981. An early version of Gauss's Disquisitiones Arithmeticae. Pp. 167–77 in Dauben, ed. 1981. *Mathematical Perspectives. Essays on Mathematics and Its Historical Development.* New York/London/et al.: Academic Press.

Merzbach, Uta C, comp. 1984. *Carl Friedrich Gauss: A Bibliography*. Wilmington, DE: Scholarly Resources.

Merzbach, Uta C. 1992. Robert Remak and the estimation of units and regulators. Pp. 481–522 in *Amphora: Festschrift für Hans Wussing zu seinem 65. Geburtstag.* Basel: Birkhäuser.

Merzbach, Uta C. 1994. *See* Davis, Audrey and Uta Merzbach.

Michelmann, Emil. 1929. *Agathe von Siebold. Johannes Brahms' Jugendliebe.* 2d ed. Göttingen: Haentzschel & Co., G.m.b.H.

Michelsen, Johann Andreas. 1778. **Sokratische Gespraeche ueber Geometrie.*

Minding, Ferdinand. 1836–38. *Handbuch der Differential-und Integral-Rechnung und ihrer Anwendungen auf Geometrie und Mechanik.* Berlin: F. Dümmler.

Minkowski, Hermann. 1891. Über die positiven quadratischen Formen und über kettenbruchähnliche Algorithmen. *J. reine angew. Math.* 107:278–97. REISSUE: Minkowski 1911, 1:243–60.

Minkowski, Hermann. 1896. Über Eigenschaften von ganzen Zahlen, die durch rämliche Anschauung erschlossen sind. Pp 271–77 in *Mathematical Papers read at the International Mathematical Congress held in connection with the World's Columbian Exposition, Chicago, 1893.* New York: Macmillan for the Amer. Math. Soc. REISSUE: Minkowski 1911, 1:271–77.

Minkowski, Hermann. 1905a. Peter Gustav Lejeune Dirichlet und seine Bedeutung für die heutige Mathematik. *Jber. DMV* 14:149–63. REISSUES: Minkowski 1911, 2:447–61. Also a *Separatum with portrait of Minkowski.

Minkowski, Hermann. 1905b. Diskontinuitätsbereich für arithmetische Aequivalenz. *J. reine angew. Math.* 129:220–74; REISSUE: Minkowski 1911, 2:53–100.

Minkowski, Hermann. 1907. *Diophantische Approximationen: Eine Einführung in die Zahlentheorie.* Leipzig: B. G. Teubner.

Minkowski, Hermann. 1911. *Gesammelte Abhandlungen.* 2 vols. Ed. David Hilbert et al. Leipzig: B. G. Teubner. REPRINT: n.d. Nabu.

Mirimanoff, D. and K. Hensel. 1905. Sur la relation $\left(\frac{D}{p}\right) = (-1)^{(n-k)}$. *J. reine angew. Math.* 129:86–88.

Monna, A. F. 1975. *Dirichlet's Principle: A mathematical comedy of errors and its influence on the development of analysis.* Utrecht: Oosthoek, Scheltema & Holkema.

Müller-Westphal, Lothar. 1989. *Wappen und Genealogien Dürener Familien.* Beiträge zur Geschichte des Dürener Landes, 20.

Musielak, Dora E. 2015. *Prime Mystery. The Life and Mathematics of Sophie Germain.* Bloomington, IN: AuthorHouse.

Nissen, Walter. 1962. *Göttinger Gedenktafeln. Ein biographischer Wegweiser.* Göttingen: Vandenhoeck & Ruprecht.

Ohm, Georg Simon. 1817. *Grundlinien zu einer zweckmäßigen Behandlung der Geometrie als höheren Bildungsmittels an vorbereitenden Lehranstalten.* Erlangen: J. J. Palm und Ernst Enke.

Ohm, Martin. 1828–1830. *Versuch eines vollkommen consequenten Systems der Mathematik.* Berlin: T. H. Riemann.

Patterson, Samuel J. 2010. Gauss sums. Chap. VIII.2 in Goldstein, Schappacher, and Schwermer, 505–28.

Perron, O. 1908. Zur Theorie der Dirichletschen Reihen. *J. reine angew. Math.* 134:95–143.

Picard, Émile. 1905. Sur quelques questions se rattachant à la connexion linéaire dans la théorie des fonctions algébriques de deux variables indépendantes. *J. reine angew. Math.* 129:275–86.

Poincaré, H. 1905. Sur les Invariants Arithmétiques. *J. reine angew. Math.* 129:89–150.

P[oisson, S. D.. 1808.] Mémoire sur la propagation de la chaleur dans les corps solides. *Nouv. Bull. Scis. Soc. philomath. Paris* 1:112–16. REISSUE: Fourier *Oeuvres* 2:215–21.

Poten, Bernhard. 1896. *Geschichte des Militär-Erziehungs- und Bildungswesens in den Landen deutscher Zunge.* Vol. 4. Preussen. Berlin: A. Hofmann & Comp.

Reissner, Hanns Günther. 1965. *Eduard Gans. Ein Leben im Vormärz.* Schriftenreihe wissenschaftlicher Abhandlungen des Leo-Baeck Instituts. Tübingen: Mohr.

Richter, Werner and Martin Engshuber. 2014. *Alexander von Humboldts Messtechnik. Instrumente Methoden Ergebnisse.* Berlin: epubli. (www.epubli.de)

Riemann, Bernhard. 1860. Ueber die Anzahl der Primzahlen unter einer gegebenen Grösse. *Mber. ...Kgl. Preuss. Akad. Wiss. Berlin.* Aus dem Jahre 1859 (November): 671–81. REISSUE: Riemann 1892 (and later reissues):145–53 with notes 154–55. TRANSL. (Eng.): Edwards 1974, 299–305.

Riemann, Bernhard. 1867. Ueber die Darstellbarkeit einer Function durch eine trigonometrische Reihe. *Abh. K. Gesellschaft Wiss. Göttingen Math. Classe* 13:87–132. REISSUE: Riemann 1892 (1953): 227–65 with notes 266–71.

Riemann, Bernhard. 1892. *Gesammelte Mathematische Werke.* Ed. Heinrich Weber in collaboration with Richard Dedekind. REISSUES: Riemann 1902; Riemann 1953.

Riemann, Bernhard. 1902. *Gesammelte Mathematische Werke.* Ed. Heinrich Weber in collaboration with Richard Dedekind. Second edition. REISSUE: Riemann 1953.

Riemann, Bernhard. 1953. *Collected Works of Bernhard Riemann. Gesammelte Mathematische Werke und Wissenschaftlicher Nachlass.* Ed. Heinrich Weber. With a Supplement edited by M. Noether and W. Wirtinger. New introduction by Hans Lewy. New York, NY: Dover.

Riemann–Hattendorff. 1869. Partielle Differentialgleichungen und deren Anwendung auf physikalische Fragen. Ed. Karl Hattendorff. Braunschweig: F. Vieweg. (G. Vieweg zu Wendhausen). REISSUES: 1876 Second ed.; 1882 Third ed. Braunschweig: Friedrich Vieweg u. Sohn. REPRINT: 1938. OTHER EDITIONS: *See* Weber, H. 1900, 1910, 1919 [Fourth–Sixth ed.]; Frank, Philipp 1927, 1943 [Seventh–Eighth ed.].

Rosenhain, George. 1851. Mémoire sur les fonctions de deux variables et à quatre périodes, qui sont les inverses des intégrales ultra-elliptiques de la première classe. *Mem. divers savants Acad. Sci. Inst. France* 11:361–468.

Rowe, David E. 1988. Gauss, Dirichlet, and the Law of Biquadratic Reciprocity. *Mathematical Intelligencer* 10 (2):13–26.

Rudio, Ferdinand. 1895. Eine Autobiographie von Gotthold Eisenstein. *Zeitschrift Math. u. Physik*. 40. Suppl. 143–68.

Sachse, A. 1879. Versuch einer Geschichte der Darstellung willkürlicher Functionen. Göttingen. TRANSL. (Fr.): Sachse 1880.

Sachse, A. 1880. [TRANSL. (Fr.) of Sachse 1879]. *Bull. scis. math. et astron*. 15:43–64, 83–112.

Scharlau, Winfried, ed. 1981. *Richard Dedekind 1831/1981. Eine Würdigung zu seinem 150. Geburtstag*. Braunschweig/Wiesbaden: Friedr. Vieweg & Sohn.

Scharlau, Winfried, ed. 1990. *Mathematische Institute in Deutschland 1800–1945*. Dokumente zur Geschichte der Mathematik. 5. Braunschweig/Wiesbaden: Deutsche Mathematiker-Vereinigung. Friedr. Vieweg & Sohn.

Schnabel, Franz. 1964. *Deutsche Geschichte im neunzehnten Jahrhundert. Die vormärzliche Zeit*. Freiburg i. Br.: Herder.

Schubring, Gert. 1984. Die Promotion von P. G. Lejeune Dirichlet. Biographische Miteilungen zum Werdegang Dirichlets. *NTM* 21 (1):45–65.

Schubring, Gert. 1985. Die Erinnerungen von Karl Emil Gruhl (1833–1917) an sein Studium der Mathematik und Physik in Berlin (1853–1856). *Jahrbuch Überblicke Mathematik*. 18:143–73.

Schubring, Gert. 1986. The three parts of the Dirichlet *Nachlass*. *Historia Mathematica* 13 no. 1 (Feb):52–56.

Schurz, Carl. 1906–12. *Lebenserinnerungen*. Berlin: G. Reimer. [Also see Wersich 1979:50–71 for translated excerpt.]

Seidel, Philipp Ludwig. 1847. Note ueber eine Eigenschaft der Reihen, welche discontinuirliche Functionen darstellen. *Abh. Akad. Wiss. München*. 51:379-394. REISSUE: 1900. Ostwald Klassiker Nr.116 (bound with Dirichlet **1837c**). Leipzig: Engelmann.

Siegmund-Schultze, R. 2007. Philipp Frank, Richard Von Mises and the Frank-Mises.*Physics in Perspective* 9:26–57.

Smith, Henry John Stephen. 1859. Report on the Theory of Numbers. Part I. *Report of the British Association for 1859*:228–67. REISSUE: *Collected Mathematical Papers* 1:38-92.

Smith, Henry John Stephen. 1860. Report on the Theory of Numbers. Part II. *Report of the British Association for 1860*:120–69. REISSUE: *Collected Mathematical Papers* 1:93–162.

Smith, Henry John Stephen. 1894. *Collected Mathematical Papers*. Ed. J. W. L. Glaisher. Oxford: Clarendon Press. REPRINT: n.d. HardPress.

Stark, H. M. 2007. The Gauss Class-Number Problems. Pp. 247–56 in Duke and Tschinkel 2007.

Steinitz, E. 1905. Ueber die Anziehung hyperboloidischer Schalen. *J. reine angew. Math*. 129:295–316.

Sterneck, R. Daublebsky v. 1896. Bemerkung über die von Dirichlet in seiner Breslauer Habilitationsschrift behandelten Functionen. *Monatshefte in Mathematik und Physik*. 7(1):342–46

Taton, René. 1954. Une lettre inédite de Dirichlet. *Revue d'histoire des sciences et de leurs applications* 7(2):172–75.

Thue, Axel. 1909. Ueber Annäherungswerte algebraischer Zahlen. *J. reine angew. Math.* 135:284–305.

Truesdell, Clifford. 1984–60. *See* Euler opera omnia.

Varnhagen von Ense, Karl August. 1861–70. *Tagebücher*. 15 vols. Leipzig: Brockhaus.

Von Mises, Richard. *See* Frank, Philipp.

Voronoĭ, Georgĭ F. 1903. Sur un problème du calcul des fonctions asymptotiques. *J. reine angew. Math.* 126 (4):241–82.

Voronoĭ, Georgĭ F. 1908–09. Recherches sur les parallèloèdres primitifs, *J. reine angew. Math.* 134:198–287 and 136:67–181.

Vossische Zeitung. 1859. 12. Mai. [Letter from Dirichlet of 1856].

Wangerin, A., ed. 1890. *Ueber die Anziehung homogener Ellipsoide. Abhandlungen von Laplace (1782), Ivory (1809), Gauss (1813), Chasles (1838), und Dirichlet (1839).* Leipzig: Engelmann.

Weber, Heinrich. 1897a. Ueber Zahlengruppen in algebraischen Körpern. *Math. Ann.* 48:433–73.

Weber, Heinrich. 1897b. Ueber Zahlengruppen in algebraischen Körpern. [II]. *Math. Ann.* 49:83–100.

Weber, Heinrich. 1898. Ueber Zahlengruppen in algebraischen Körpern. [III]: Anwendung auf die complexe Multiplicationen und Theilung der elliptischen Functionen. *Math. Ann.* 50:1–26.

Weber, Heinrich. 1899. *Lehrbuch der Algebra*, vol. 2. 2nd ed. Braunschweig: Vieweg.

Weber, Heinrich. 1900. *Die partiellen Differential-Gleichungen der mathematischen Physik. Nach Riemanns Vorlesungen in 4. Auflage neu bearbeitet.* Braunschweig: Vieweg.

Weber, Heinrich. 1905. Ueber komplexe Primzahlen in Linearformen. *J. reine angew. Math.* 129:35–62.

Weber, Heinrich. 1910. *Die partiellen Differential-Gleichungen der mathematischen Physik. Nach Riemanns Vorlesungen.* Fifth ed. Braunschweig: Vieweg.

Weber, Heinrich. 1919. *Die partiellen Differential-Gleichungen der mathematischen Physik. Nach Riemanns Vorlesungen.* Sixth ed. Braunschweig: Vieweg.

Weber, Max. 1930. *The Protestant Ethic and the Spirit of Capitalism.* New York: Scribner; London: Allen & Unwin.

Weil, André. 1983. *Number Theory. An Approach through History from Hammurapi to Legendre*. Boston, Basel, Stuttgart: Birkhäuser.

Weisweiler, Eva, ed. 1985. *Fanny Mendelssohn. Ein Portrait in Briefen*. Frankfurt/Main, Berlin, Wien: Ullstein.

Werner, Karl and Konrad Werner. 1976. *Wilhelm Weber*. Leipzig: BSB Teubner.

Wersich, Rüdiger, ed. 1979. *Carl Schurz, Revolutionary and Statesman: His Life in Personal and Official Documents with Illustrations*. München: Heinz Moos.

Whittaker, E. T. and G. N. Watson. 1927. *A Course in Modern Analysis*. 4th ed. Cambridge University Press. REPRINT: 1962.

Wirtinger, W. 1905. Ueber eine besondere Dirichletsche Reihe. *J. reine angew. Math.* 129:214–19.

Wolf, Rudolf. 1993. *Tagebuch*. Schriftenreihe der ETH-Bibliothek. Zürich: ETH-Bibliothek.

MANUSCRIPTS[2]

In addition to the Nachlass locations listed below, additional relevant materials (mostly student notes) have been found in Chicago, Leipzig, and Nürnberg.

Official documentation and correspondence formerly in Merseburg now are most likely available through the Bundesarchiv.

Manuscripts

BERLIN

Berlin-Brandenburgische Akademie der Wissenschaften. Archiv. Dirichlet Nachlass.
Staatsbibliothek. Preußischer Kulturbesitz. Handschriftenabteilung. Dirichlet Nachlass. Also see Nachlass Borchardt for lecture notes referred to in Fischer 1994.
Staatsbibliothek. Preußischer Kulturbesitz. Musikabteilung. Mendelssohn-Archiv. Considerable family correspondence.

BONN

Archiv der Universität.

GÖTTINGEN

Niedersächsische Staats–und Universitätsbibliothek (NSUB). Handschriftenabteilung.
Universitätsarchiv Göttingen. Personalakten.

KASSEL

Dirichlet Nachlass. Correspondence with parents and a variety of documents including certificates and financial records.

MUNICH

Deutsches Museum. Nachlass Clausius.
Universität. Nachlass Seidel.

---

[2]Numerous student notes taken by those attending Dirichlet's lectures are now available; several will be found among the manuscripts listed in this section. These are of uneven quality. Although there may be some interest in the more detailed future comparison of these notes with one another and with his publications, in this volume we focus on the transmission of Dirichlet's ideas through his published writings and those of his editors. Fischer 1994, dealing with Dirichlet's lectures and publications pertaining to probability, is the only study encountered that is based on a comparison that includes the diverse sources of relevant student notes.

Name Index

U. C. Merzbach, *Dirichlet*, https://doi.org/10.1007/978-3-030-01073-7

T

U

V

W

MIX
Papier aus verantwortungsvollen Quellen
Paper from responsible sources
FSC® C105338

If you have any concerns about our products,
you can contact us on
ProductSafety@springernature.com

In case Publisher is established outside the EU,
the EU authorized representative is:
Springer Nature Customer Service Center GmbH
Europaplatz 3, 69115 Heidelberg, Germany

Printed by Libri Plureos GmbH
in Hamburg, Germany